Reliability Management and Engineering

Advanced Research in Reliability and System Assurance Engineering

Series Editor:

Mangey Ram, Professor,

Graphic Era University, Uttarakhand, India

Modeling and Simulation Based Analysis in Reliability Engineering
Edited by Mangey Ram

Reliability Engineering
Theory and Applications

Edited by Ilia Vonta and Mangey Ram

System Reliability Management
Solutions and Technologies

Edited by Adarsh Anand and Mangey Ram

Reliability Engineering
Methods and Applications

Edited by Mangey Ram

Reliability Management and Engineering
Challenges and Future Trends

Edited by Harish Garg and Mangey Ram

For more information about this series, please visit: https://www.crcpress.com/Reliability-Engineering-Theory-and-Applications/Vonta-Ram/p/book/9780815355175

Reliability Management and Engineering

Challenges and Future Trends

Edited by

Harish Garg and Mangey Ram

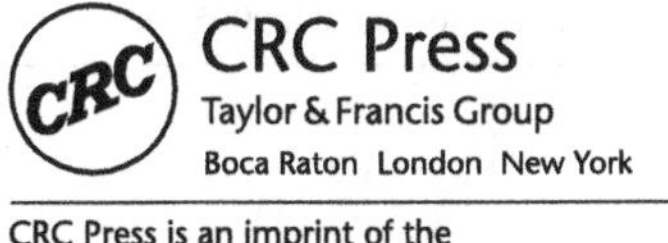

CRC Press is an imprint of the
Taylor & Francis Group, an **informa** business

First edition published 2020
by CRC Press
6000 Broken Sound Parkway NW, Suite 300, Boca Raton, FL 33487-2742

and by CRC Press
2 Park Square, Milton Park, Abingdon, Oxon, OX14 4RN

Library of Congress Cataloging-in-Publication Data

Names: Garg, Harish, editor. | Ram, Mangey, editor.
Title: Reliability management and engineering : challenges and future trends / edited by Harish Garg and Mangey Ram.
Description: First edition. | Boca Raton, FL : CRC Press, 2020. | Series: Advanced research in reliability and system assurance engineering | Includes bibliographical references and index.
Identifiers: LCCN 2020005425 (print) | LCCN 2020005426 (ebook) | ISBN 9780367211530 (hardback) | ISBN 9780429268922 (ebook)
Subjects: LCSH: Reliability (Engineering) | Engineering--Management.
Classification: LCC TA169 .R4425 2020 (print) | LCC TA169 (ebook) | DDC 620/.00452--dc23
LC record available at https://lccn.loc.gov/2020005425
LC ebook record available at https://lccn.loc.gov/2020005426

ISBN: 978-0-367-21153-0 (hbk)
ISBN: 978-0-429-26892-2 (ebk)

Typeset in Times
by Deanta Global Publishing Services, Chennai, India

Contents

Preface

In the present era of industrial growth, maintaining optimal efficiency with minimum hazards is challenging. To address this issue, reliability technology can play an important role. Reliability is measured as the ability of a system to successfully perform its intended function for a specified period under predetermined conditions. This attribute has far-reaching consequences on the durability, availability, and lifecycle cost of a product or system and is of great importance to the end user/engineer. However, failure is an inevitable aspect of products and systems. These failures may be the result of human error, poor maintenance, or inadequate testing and inspection. To improve system reliability and availability, the implementation of appropriate maintenance strategies plays an important role. Consistently good performance of these units can be achieved with highly reliable subunits and perfect maintenance. To this effect, the knowledge of the behavior of a system and their component(s) is required in order to plan and adapt suitable maintenance strategies. Therefore, in recent years, the importance of reliability and maintenance theory has been increasing greatly, especially with the innovations in recent technology for the purpose of making good products with high quality and designing highly reliable systems. Reliability management and engineering is one of the hot topics not only for scientists and researchers but also for engineers and industrial managers. Through this book, titled *Reliability Management and Engineering: Challenges and Future Trends*, engineers will be able gain adequate knowledge on this subject, and this book will aid them in their reliability courses. This book covers the recent developments in reliability and maintenance engineering and is meant for those who to take reliability and maintenance as a subject of study. It presents new theoretical concepts that were not previously published as well as solutions to practical problems and case studies illustrating the applications methodology. This book is written by a number of leading scientists, analysts, mathematicians, statisticians, and engineers who have been working on the front end of reliability science and engineering.

Harish Garg
Patiala, India

Mangey Ram
Dehradun, India

MATLAB® is a registered trademark of The MathWorks, Inc. For product information, please contact:

The MathWorks, Inc.
3 Apple Hill Drive
Natick, MA 01760-2098 USA
Tel: 508 647 7000
Fax: 508-647-7001
E-mail: info@mathworks.com
Web: www.mathworks.com

Acknowledgments

The editors acknowledge CRC Press for this opportunity as well as for providing professional support. Also, we would like to thank all the chapter authors and reviewers for their willingness to contribute to this work.

Editors

Dr. Harish Garg is working as an assistant professor (senior grade) at the School of Mathematics at Thapar Institute of Engineering and Technology (Deemed University), Patiala, India. He earned his Ph.D. degree major in mathematics from the Indian Institute of Technology, Roorkee, Uttarakhand, India. He has authored over 220 plus papers published in refereed international journals including *Information Sciences, IEEE Transactions on Fuzzy Systems, International Journal of Intelligent Systems, Cognitive Computation, Artificial Intelligence Review, Applied Soft Computing (ASOC), Experts Systems with Applications, IEEE Access, Journal of Intelligent and Fuzzy Systems, Expert Systems, Journal of Manufacturing Systems, Applied Mathematics & Computations, ISA Transactions, IEEE/CAA Journal of Automatica Sinica, IEEE Transactions on Emerging Topics in Computational Intelligence, Applied Intelligence, Computer and Industrial Engineering, Soft Computing, Computer and Operations Research, Journal of Experimental & Theoretical Artificial Intelligence, International Journal of Uncertainty, Fuzziness and Knowledge-Based Systems, Journal of Industrial and Management Optimization, International Journal of Uncertainty Quantification*, and many more. He has also authored seven book chapters. His research interests are in the fields of computational intelligence, reliability theory, evolutionary algorithms, multicriteria decision-making, fuzzy decision-making, pythagorean fuzzy sets, computing with words, and soft computing. Dr. Garg is the associate editor of the *Journal of Intelligent and Fuzzy Systems, Technological and Economic Development of Economy, Mathematical Problems in Engineering, Journal of Industrial and Management Optimization, International Journal of Computational Intelligence Systems, Complex and Intelligent Systems, CAAI Transactions on Intelligence Technology*, and so on. He is on the editorial board of several international journals. In 2016–2019, he was the recipient of the outstanding reviewer award for various journals including *ASOC, AMM, EAAI, RESS*, etc. His Google citations are more than 7000. For more details, visit http://sites.google.com/site/harishg58iitr/

Dr. Mangey Ram earned his PhD degree major in mathematics and minor in computer science from G. B. Pant University of Agriculture and Technology, Pantnagar, India. He has been a faculty member for around ten years and has taught several core courses in pure and applied mathematics at the undergraduate, postgraduate, and doctorate levels. He is currently a professor at Graphic Era (Deemed to be University), Dehradun, India. Before joining Graphic Era, he was a deputy manager (probationary officer) with Syndicate Bank for a short period. He is the editor-in-chief of the *International Journal of Mathematical, Engineering and Management Sciences* and is the guest editor and member of the editorial board of various journals. He is a regular reviewer for international journal publishers, including IEEE, Elsevier, Springer, Emerald, John Wiley, and Taylor & Francis. He has published 200 plus research publications in the journals of IEEE, Taylor & Francis, Springer, Elsevier, Emerald, and World Scientific and in many other national and international journals of repute. He has also presented his works at national and international conferences. His fields of research are reliability theory and applied mathematics. Dr. Ram is a senior member of the IEEE, life member of the Operational Research Society of India, Society for Reliability Engineering, Quality and Operations Management in India, Indian Society of Industrial and Applied Mathematics, member of the International Association of Engineers in Hong Kong and the Emerald Literati Network in the UK. He has been a member of the organizing committee of a number of international and national conferences, seminars, and workshops. He was conferred with the Young Scientist Award by the Uttarakhand State Council for Science and Technology, Dehradun, in 2009. He received the Best Faculty Award in 2011, Research Excellence Award in 2015, and recently the Outstanding Researcher Award in 2018 for his significant contribution to academics and research at Graphic Era (Deemed to be University), Dehradun, India.

Contributors

Mohammad H. Ahmadi
Shahrood University of Technology
Shahrood, Iran

Daniel O. Aikhuele
Department of Mechanical and Biomedical Engineering, College of Engineering, Bells
University of Technology
Ota, Nigeria

Rajni Aron
School of Computer Science and Engineering
Lovely Professional University
Punjab, India

Koorosh Aslansefat
Department of Computer Science and Technology
University of Hull
Hull, UK

Anand Bewoor
Mechanical Engineering Department
Cummins College of Engineering for Women
Pune, India

Nabaranjan Bhattacharyee
Department of Mathematics
Sidho-Kanho-Birsha University
Purulia, West Bengal, India

Sachin Chaudhary
Department of Community Medicine
Government Medical College
Kannauj (U.P.), India

Duke E. George
Department of Mechanical and Biomedical Engineering, College of Engineering, Bells
University of Technology
Ota, Nigeria

Youcef Gheraibia
University of York and Assuring Autonomy International Programme
York, UK

Srikant Gupta
Jaipuria Institute of Management
Jaipur (Rajasthan), India

Adlul Islam
Natural Resource Management Division (ICAR)
Krishi Anusandhan Bhawan– II, IARI Campus, Pusa
New Delhi, India

Sahidul Islam
Department of Mathematics
University of Kalyani
Nadia, West Bengal, India

Hanumant Jagtap
Zeal College of Engineering and Research
Narhe, Pune, India

Kanika Jindal
School of Computer Applications
Lovely Professional University
Punjab, India

Sohag Kabir
Department of Computer Science
University of Bradford
Bradford, UK

Manvi Kaushik
Institute of Infrastructure Technology Research and Management (IITRAM)
Ahmedabad, India

Komal
Department of Mathematics, School of Physical Sciences
Doon University
Dehradun, Uttarakhand, India

Jitendra Kumar
Directorate of Economics and Statistics
Planning Department
Delhi, India

Mohit Kumar
Institute of Infrastructure Technology Research and Management (IITRAM)
Ahmedabad, India

Ravinder Kumar
School of Mechanical Engineering
Lovely Professional University
Phagwara, Punjab, India

Deepesh Machiwal
Division of Natural Resources
ICAR-Central Arid Zone Research Institute
Jodhpur, Rajasthan, India

Sanat Kumar Mahato
Department of Mathematics
Sidho-Kanho-Birsha University
Purulia, West Bengal, India

Ram Niwas
Department of Statistics
Goswami Ganesh Dutta Sanatan Dharma College
Chandigarh, India

Yiannis Papadopoulos
Department of Computer Science and Technology
University of Hull
Hull, UK

Rajesh Paramanik
Department of Mathematics
Sidho-Kanho-Birsha University
Purulia, West Bengal, India

Dipen Kumar Rajak
Sandip Institute of Technology and Research Center
Nashik, India

Biswajit Sarkar
Department of Industrial and Management Engineering
Hanyang University
Republic of Korea

Priyanka Sharma
Ground Water Hydrology Division
National Institute of Hydrology (NIH)
Roorkee, Uttarakhand, India

Puja Supakar
Department of Mathematics
Sidho-Kanho-Birsha University
Purulia, West Bengal, India

1 An Integrated Robust Hybrid Fuzzy Reliability Model for Redesigning New Products and Systems

Daniel O. Aikhuele and Duke E. George

CONTENTS

1.1 INTRODUCTION

With the high value placed on quality and reliability by product users/customers and the ever-increasing product information sources available to them (Kostina, 2012), it is necessary nowadays that product development companies are not only able to effectively track, manage, and understand the quality and reliability improvement concerns of their products from their customers, but also anticipate when newly designed and developed products will fail when finally in use and know how to manage the failure in the field (Lu & Anderson-Cook, 2015). This is necessary in order to plan on how to remove them from operations before they put the end user(s)/operator(s) into a life-threatening situation. According to Smith et al. (2012), having a dependable and accurate predictive estimate of when failures may be expected in new complex products and systems can allow for the healthy and efficient management of such products and systems.

The failure or potential failure information of the existing predecessor products, if properly analyzed, could provide the required reliability knowledge and information that can be used for the redesigning of the new product and system. The identification, and the gaining of such reliability knowledge and information for the redesigning of new products and systems, is one of the most critical tasks in today's modern complex product development. Achieving such knowledge and information is capable of improving and enhancing the reliability and quality of the designed and developed product and system (Aikhuele & Turan, 2017c; He et al., 2015). According to Yadav et al. (2003), the understanding of the reliability concerns of existing complex products and systems can lead to improved product design and development, product performance predictability (Zuber & Bajri, 2016), and the opportunities for better product reliability management at the early design stage (Lu & Anderson-Cook, 2015).

The quality, reliability, and safety management of most complex products and systems are extensively managed and handled with the failure mode and effect analysis (FMEA) method (Liu et al., 2015; Liu et al., 2014; Mohammadi & Tavakolan, 2013; Safari et al., 2016). The FMEA method, which was developed by the U.S. aerospace industry, is regarded as a structured and systematic analytical tool that has clear reliability, quality, and safety requirements (Bowles & Pelaez, 1995). Over the years, this quality and reliability tool has been established as a popular analytical tool for reliability engineering management and for the identification, ranking, and evaluation of failures/and potential failures in existing and new products and systems. This has contributed generally to the improvement and enhancement of product reliability and quality determinations. Recently, the method has been extended by integrating it with several other methods, some of which include multi-criteria decision-making (MCDM) methods (Aikhuele et al., 2016; Liu et al., 2012; Liu et al., 2015; Vahdani et al., 2015). This was mainly to address the uncertainty issues in the traditional reliability priority number (RPN) that is normally used in the FMEA method.

In applying these methods (i.e. the FMEA and its extensions), only the root cause of failure and the design risk of product failure are often considered. However, in order to build a complete and a robust design of a reliability knowledge model for the redesigning of new products and systems, the reliability assessment should as well include the actual interactions of the failure modes which resulted in the failure of the system, by quantifying the failure causality relationships (FCRs) within and between the components. That is termed, the internal failure causality relationships

(IFCR) within the components of the product and the external failure causality relationships (EFCR) between components (Ma et al., 2016). The causality relationship or the interaction of failure modes for the two different cases (IFCR and EFCR) are shown in Figure 1.1 for clarity.

In this chapter, however, the study is focused on addressing the root cause of failure issues and the interactions within and between the failure modes of the designed product. In dealing with these issues, a new model which is termed IFWG-TOPSIS (Intuitionistic Fuzzy Weighted Geometric – Technique for Order of Preference by Similarity to Ideal Solution) is proposed. The model which is based on an Intuitionistic Fuzzy Technique for Order of Preference by Similarity to Ideal Solution (IF-TOPSIS) method, Intuitionistic Fuzzy Weighted Geometric operator, and the Dynamic Intuitionistic Fuzzy Einstein Geometric Averaging (DIFWGϵ) operator is used to address the uncertainty and dynamic issues in the evaluation process, while a mathematical model is presented for the evaluation of the interactions within and between the failure modes in the system component(s).

The main advantages and contributions of the proposed new models are listed as follows:

(i) The IFWG-TOPSIS model accounts for the uncertainty and dynamic issues in the decision-making process and is used in the evaluation of the root cause of failure in the designed product.
(ii) The interactions within and between the failure modes are addressed and evaluated adequately using a new mathematical model.
(iii) A complete and a robust design of a reliability knowledge model is achieved, which can be used for the redesigning of new products and systems.

To the best of our knowledge, this is the first study to present both the IFWG-TOPSIS model for addressing dynamic issues in the decision-making process and

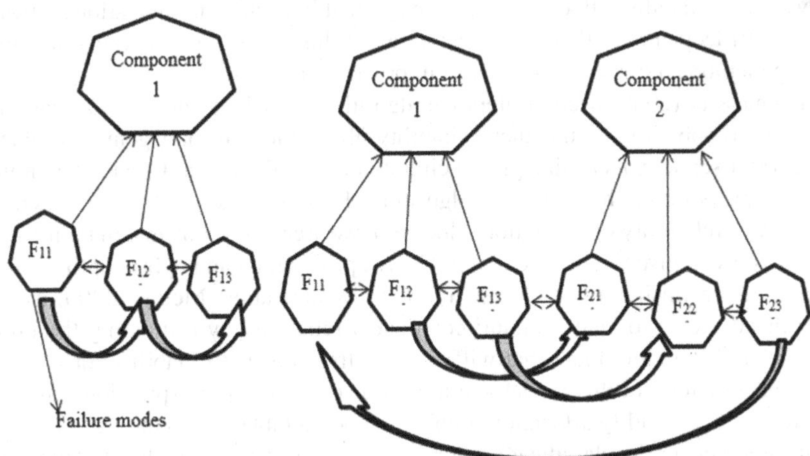

FIGURE 1.1 IFCR and EFCR (interaction within and between the failure modes). Source: (Aikhuele & Turan, 2017c)

a mathematical model for the evaluation of failure modes interaction within and between system components.

1.2 LITERATURE REVIEW

Several studies and methods have been developed and applied in the literature for the reliability management of complex mechanical systems as well as for building reliability knowledge models for the redesigning of complex products and systems from their existing predecessor products and systems. Among them, we can mention the integrated fuzzy logic and expert database method for evaluating the safety and reliability of the hydraulic system by Sharma et al. (2005). Another method uses particle swarm optimization (PSO) and a fuzzy methodology for evaluating the most critical component in the system, which affects the entire system's performance the most, and this achieved by measuring the effects of reliability, availability, and maintainability (RAM) on the failure and repair rate of the system's performance (Garg, 2014).

Zuber and Bajri (2016) applied the artificial neural network (ANN) and the vibration analysis method for faults identification and prediction in an automated roller element bearing system. Shaghaghi and Rezaie (2012) presented a generalized mixture operator for determining and aggregating the risk priorities of failure modes in a Limerick Generating Station (LGS) gas-type circuit breaker product. Kangavari et al. (2015) used the FMEA method for analyzing the risks of some specific systems in the petrochemical industry, starting from the conceptual phase to the system disposal phase. Garg and Sharma (2013) used the PSO method to allocate reliability to a pharmaceutical plant. Furthermore, they considered the reliability of the components as a triangular fuzzy number, which was solved by converting them into a crisp model, where the crisp optimization problem was solved with the PSO method and the result was compared with the genetic algorithm (GA) method (Garg et al., 2014). Lyu et al. (1997) developed a systematic optimization GUI front-end mathematical software for reliability allocation and management in a telecommunication software system. Pardeep et al. (2010) presented an algorithm based on an iterative method, and the method addresses performance improvement issues.

Ha and Kuo (2008) used an iterative algorithm, which is known as a scanning heuristic, to solve the redundancy reliability allocation and management problem in a current semiconductor integrated circuit, and the algorithm depends on an initial starting point of the employed algorithm. Kuo and Wan (2007) presented an overview of reliability optimization allocation research and problems. In their work, recognition was given to the use of ant colony optimization and hybrid optimization methods as promising tools for system reliability allocation. Mettas (2000) used a generalized model to analyze multi-component reliability by estimating the minimum reliability requirement that will produce the expected reliability for a multi-component system. In this model, the nonlinear programming expression was used to address the reliability allocation problem. Other notable contributions to the reliability study cover the classification of reliability allocation methods into "dynamic programming, branch and bound technique, linear programming, Lagrangian multiplier method, heuristic method and optimization method" (Garg & Sharma, 2013;

Govil & Agarwala, 1983; Kuo et al., 2001). Garg (2018) presented a method that addresses the fuzzy system reliability analysis of components with different individual failure probability density functions using the functions of intuitionistic fuzzy numbers, which was calculated with the credibility theory. Similarly, a mathematical model based on the Markov process was developed by Niwas and Garg (2018), where the system reliability, mean time to failure of the system, the system availability, and finally, the expected profit to be derived from the system were used as parameters for analyzing the behavior of an industrial system.

Others include the integration of the FMEA method with some MCDM methods (Adhikary et al., 2014; Chang & Wen, 2010; Liu et al., 2013; Tay et al., 2015), for detecting the failure and for the reliability management of complex mechanical systems. While in other cases, the FMEA method is replaced outrightly with the MCDM methods, some of which include the Analytical Hierarchy Process (AHP) method (Balin et al., 2014), MULTIMOORA method (Zhao et al., 2016), and Grey Theory method (Geum et al., 2011). Although the reliability management methods appear to have met some of the identified shortcomings in the FMEA methods and in reliability management analysis, the literature review indicates that no published study has reported the use of these methods in building appropriate reliability knowledge models for new product and system designs. Also, in the event that dynamic criteria are used in the evaluation process of the reliability of a complex mechanical system, these methods automatically become inefficient. Hence, in the current chapter, some of the above-mentioned problems are addressed. To achieve this, a new model termed IFWG-TOPSIS (Intuitionistic Fuzzy Weighted Geometric – Technique for Order of Preference by Similarity to Ideal Solution) is proposed for addressing the root cause of failure in the products and systems, while a mathematical model is presented for the evaluation of the interactions within and between the failure modes.

1.3 MODEL FORMATION

In this section, the basic definitions for the model (i.e. the concept of intuitionistic fuzzy set (IFS)) (Pérez-Domínguez et al., 2015), the dynamic intuitionistic fuzzy Einstein geometric averaging (DIFWGϵ) operator (Gümüş & Bali, 2017), and the Fuzzy Technique for Order of Preference by Similarity to Ideal Solution (FTOPSIS) model (Aikhuele & Turan, 2017a) are presented. Furthermore, the algorithm of the second model – a mathematical model – which is used in addressing the interactions of the failure modes that result in the actual failure of a system, is introduced. A flow chart of the two proposed models is given in Figure 1.2.

1.3.1 Intuitionistic Fuzzy Set

Definition 1

Let A in $X = \{x\}$ be the nonempty fuzzy set with a closed unit interval $I = [0,1]$. Then the *IFS* is therefore given as:

$$A = \{\langle x, \mu_A(x), v_A(x)\rangle \mid x \in X\} \tag{1.1}$$

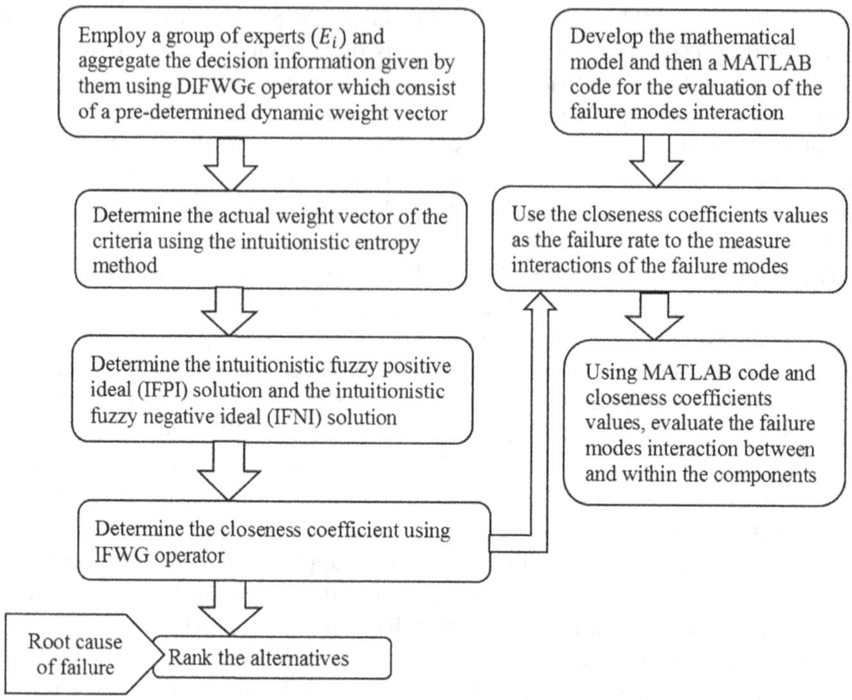

FIGURE 1.2 Flow chart of the proposed models for the root cause of failure and the interactions of the failure modes.

where the mapping $\mu_A : X \to [0,1]$ is the membership function of the fuzzy set A; and $v_A : X \to [0,1]$ is the non-membership function of the fuzzy set, such that $0 \leq \mu_A(x) + v_A(x) \leq 1, \forall x \in X$ for each element $x \in X$ in A.

From the above, it is clear that every fuzzy set A, on the nonempty set $X = \{x\}$ is *IFS* that can be defined as:

$$A = \{\langle x, \mu_A(x), 1 - v_A(x)\rangle | x \in X\} \tag{1.2}$$

The intuitionistic fuzzy hesitation (or the non-determinacy or uncertainty) degree of whether x belongs to A or not is given as:

$$\pi_A(x) = 1 - \mu_A(x) - v_A(x) \tag{1.3}$$

This degree will arise when there is a relative lack of knowledge, personal error, or uncertainty of any form, particularly when $1 - \mu_A(x) - v_A(x) = 0$. For every element $x \in X$ in A, the *IFS A* belongs to the fuzzy set, where $0 \leq \pi_A \leq 1$, hence, the intuitionistic fuzzy number(s) (IFN(s)) is given as $\alpha = (\mu_A, v_A, \pi_A)$ or as $\alpha = (\mu_A, v_A)$.

1.3.2 Aggregation Operators for the Intuitionistic Fuzzy Set

Aggregation operators are the mathematical models used for joining and summarizing the information gathered from a variety of sources and then used for making real-time decisions. Some of the most common and popular aggregation operators that have applications in engineering management include the ordered weighted averaging (OWA) operator (Yager, 1988), the Intuitionistic Fuzzy Weighted Geometric operator (Xu & Yager, 2006), the generalized triangular intuitionistic fuzzy geometric averaging operator (Aikhuele & Odofin, 2017). Uncertain dynamic linguistic weighted harmonic mean (UDLWHM) operator (Park, Kwun, & Koo, 2011) and the dynamic intuitionistic fuzzy Einstein geometric averaging (DIFWGϵ) operator (Gümüş & Bali, 2017). However, in this book chapter, we will be concerned only with the dynamic intuitionistic fuzzy Einstein geometric averaging (DIFWGϵ) operator and the Intuitionistic Fuzzy Weighted Geometric operator for aggregating experts' opinions on the reliability management of a slewing gear system. A detailed definition of the operation is given below.

Definition 2 (Gümüş & Bali, 2017)

Let $\alpha(t_i) = \left(\mu_{\alpha(t_i)}, \nu_{\alpha(t_i)}\right)$ for all $(i = 1,2,3,\ldots,p)$ is a set of Intuitionistic Fuzzy Numbers (IFN) for P different periods $(i = 1,2,3,\ldots,p)$, where $\delta = \left(\delta(t_1),\delta(t_2),\delta(t_3),\ldots,\delta(t_p)\right)^T$ is the weight vector of periods such that $\sum_{i=1}^{n} \delta_{(t_i)} = 1$ and DIFWGϵ: $\Omega^n \rightarrow \Omega$, if

$$\text{DIFWG}\epsilon_{\delta(t)}\left(\left(\delta(t_1),\delta(t_2),\delta(t_3),\ldots,\delta(t_p)\right) = \otimes_{i=1}^{p}\epsilon\cdot\left(\delta_{ij}(t_i)\right)^{\delta(t_i)}\right.$$

$$= \left(\delta_{ij}(t_1)\right)^{\delta(t_1)} \otimes\epsilon\left(\delta_{ij}(t_2)\right)^{\delta(t_2)} \otimes\epsilon\ldots\otimes\epsilon\left(\delta_{ij}(t_p)\right)^{\delta(t_p)}$$

$$= \left(\frac{2\prod_i^p\left(\mu_{ij}(t_i)\right)^{\delta(t_i)}}{\prod_i^p\left(2-\mu_{ij}(t_i)\right)^{\delta(t_i)} + \prod_i^p\left(\mu_{ij}(t_i)\right)^{\delta(t_i)}}, \frac{\prod_i^p\left(1+\nu_{ij}(t_i)\right)^{\delta(t_i)} - \prod_i^p\left(1-\nu_{ij}(t_i)\right)^{\delta(t_i)}}{\prod_i^p\left(1+\nu_{ij}(t_i)\right)^{\delta(t_i)} + \prod_i^p\left(1-\nu_{ij}(t_i)\right)^{\delta(t_i)}}\right) \quad (1.4)$$

Definition 3 (Xu & Yager, 2006)

If $\alpha_i = \left(\mu_{\alpha_i}, \nu_{\alpha_i}\right)$ represents the *IFN* for all $(i = 1,2,3,\ldots,p)$, then the Intuitionistic Fuzzy Weighted Geometric (IFWG) operator of the dimension n is a mapping of IFWG: $\Omega^n \rightarrow \Omega$, such that

$$\text{IFWG}\left(d_1 d_2 d_3,\ldots,d_n\right) = \left(\prod_{i=1}^{n}(\mu_{ij})^{w_i}, 1-\prod_{i=1}^{n}(1-\nu_{ij})^{w_i}\right) \quad (1.5)$$

where $w_i = (w_1, w_2, w_3,\ldots,w_n)^T$ is the weighting vector of $\alpha_i\,(i = 1,2,3,\ldots,n)$ with $w_i \in [0,1]$ and $\sum_{i=1}^{n} w_i = 1$.

1.3.3 TOPSIS and the Proposed Model Algorithm

TOPSIS which was originally proposed by Hwang and Yoon (1981), has remained one of the leading MCDM tools in engineering management. It is based on the concept that the most appropriate decision alternative should have the shortest and farthest distance from the positive and negative ideal solution, respectively. In implementing the TOPSIS model, reliable and practical decision-making information/preference, in the form of an *IFN* when implemented in an intuitionistic environment, is collected from expert(s) and then aggregated before it is used in the evaluation (Aikhuele & Turan, 2017c). From the concept of the *IFS*, the aggregation operators, and the TOPSIS model, the aforementioned proposed model (IFWG-TOPSIS) is described and summarized in the following steps.

Step 1: Employ a group of experts (E_i) with equal weight vectors (i.e. equal work experiences and expertise on the subject matter) to evaluate and give their preference information about the alternatives with respect to the criteria. To ensure accuracy, employ experts from both academia and the industry.

Step 2: Draw up a questionnaire and ask the experts to evaluate the alternatives with respect to the pre-determined risk factors, using the linguistic scale presented in Table 1.1. Similarly, employ the experts to evaluate the failure modes with respect to the risk factors using the same. The collected linguistic decision information/preference is then converted to the *IFN*, respectively, to construct the intuitionistic fuzzy decision matrix $Zn = (x_{ij})_{mxn}$ as shown below:

$$Zn_{mxn}\left(x_{ij}\right) = \begin{bmatrix} \left(\mu_{11}, v_{11}\right) & \left(\mu_{12}, v_{12}\right) & \cdots & \left(\mu_{1n}, v_{1n}\right) \\ \left(\mu_{21}, v_{21}\right) & \left(\mu_{22}, v_{22}\right) & \cdots & \left(\mu_{2n}, v_{2n}\right) \\ \vdots & \vdots & \ddots & \vdots \\ \left(\mu_{m1}, v_{m1}\right) & \left(\mu_{m2,} v_{m2}\right) & \cdots & \left(\mu_{mn,1} v_{mn}\right) \end{bmatrix} \tag{1.6}$$

TABLE 1.1
Linguistic Scale and Its IFN for Data Collection

Linguistic terms	IFN
Moderately (M)	(0.4, 0.1)
Moderately high (MH)	(0.5, 0.2)
High (H)	(0.6, 0.2)
Very high (VH)	(0.7, 0.4)
Hazardous (HZ)	(0.8, 0.2)

Step 3: Using the DIFWGϵ operator presented in Equation 1.4, aggregate the preference information given by the different experts with the pre-determined dynamic weight vector of the criteria to construct the collective intuitionistic fuzzy decision matrix $Zn = (x_{ij})_{mxn}$. The main advantage of the DIFWGϵ operator is that it is able to account and deal with all the dynamic issues in the decision-making process.

Step 4: Calculate the actual weight vector of the criteria from the collective intuitionistic fuzzy decision matrix $Zn = (x_{ij})_{mxn}$, using the intuitionistic entropy method originally presented in the study of Ye 2010).

Step 5: Determine the intuitionistic fuzzy positive ideal (IFPI) solution and the intuitionistic fuzzy negative ideal (IFNI) solution as follows:

$$S^{+} = \left\{ \left\langle \begin{matrix} T_j, \left(\left(\max \mu_{ij}(T_j) \mid j\epsilon Z \right), \left(\min \mu_{ij}(T_j) \mid j\epsilon G \right) \right), \\ \left(\left(\min v_{ij}(T_j) \mid j\epsilon Z \right), \left(\max v_{ij}(T_j) \mid j\epsilon G \right) \right) \end{matrix} \right\rangle \mid i \in m \right\} \tag{1.7}$$

$$S^{-} = \left\{ \left\langle \begin{matrix} T_j, \left(\left(\min \mu_{ij}(T_j) \mid j\epsilon Z \right), \left(\max \mu_{ij}(T_j) \mid j\epsilon G \right) \right), \\ \left(\left(\max v_{ij}(T_j) \mid j\epsilon Z \right), \left(\min v_{ij}(T_j) \mid j\epsilon G \right) \right) \end{matrix} \right\rangle \mid i \in m \right\} \tag{1.8}$$

where Z be a collection of benefit attributes and G the collection of cost attributes when the attributes are categorized into benefit and cost.

Step 6: Calculate the closeness coefficient of the alternatives using the IFWG operator, where the closeness coefficient equation is given as and is calculated using values derived from Equations 1.7 and 1.8, respectively, for the IFPI and IFNI solutions.

$$CC_i = \frac{\text{IFWG}\left(S^{-}\right)}{\text{IFWG}\left(S^{-}\right) + \text{IFWG}\left(S^{+}\right)} \tag{1.9}$$

In using the modified IFWG operator, the $\max \mu_{ij}(T_j)$ in the *IFPI* solution and $\min \mu_{ij}(T_j)$ in the *IFNI* solution are used instead of the weight vector w_i in the IFWG operator.

Step 7: Using the result from the calculation, rank the alternatives in the descending order.

1.3.4 Introduction of the Mathematical Model for Evaluating Failure Modes Interactions

In developing the mathematical model for the evaluation of the failure modes interaction, which may either take the form of the IFCR or the EFCR in the component system (Figure 1.1), the following assumption has been made.

(a) The failure modes in a component of the system can interact with one another, and such interactions are called the IFCR.
(b) The failure modes in a component of the system can interact with the failure modes of other components of the system, and such interactions can be called the EFCR.
(c) The probability that the system with the different interacting failure modes will undergo transitions depends only on the current interactions of the failure modes in the components and not on the previous interactions of the failure modes in the system.
(d) The transition probabilities are constant over time for the system.
(e) The components of the system are assumed to have an exponential lifetime distribution.

As an example, let's consider a simple mechanical system with several components which have failure modes that interact within and between each other. If the failure model has a constant failure rate (i.e. lifetimes are exponential), then their interactions can be modeled as follows:

Mathematical model: Let the failure modes of the components of the system be represented in the set $D = \{SG_1, SG_2, SG_3, SG_4\}$. If $P_i(t)$ is the probability that the failure modes SG_i, will interact with each other at a time (t), then the failure rate from the interaction can be represented as $\lambda_i, (i = 1,2,3,4)$.

Case 1: When the failure mode SG_1, interacts with the failure modes within the same component or between the failure modes of other components at a time $(t, t+\Delta t)$, then the probability that the interaction of the failure modes with failure mode SG_1, will cause the entire system to fail is given as:

$$P_1(t+\Delta t) = P_1(t)(1-\lambda_1\Delta t-\lambda_2\Delta t-\lambda_3\Delta t-\lambda_4\Delta t)$$

$$\frac{P_1(t+\Delta t)-P_1(t)}{\Delta t} = -(\lambda_1+\lambda_2+\lambda_3+\lambda_4)P_1(t)$$

when $\Delta t \to 0$

$$\frac{d}{dt}P_1(t) = \acute{P}_1(t) = -(\lambda_1+\lambda_2+\lambda_3+\lambda_4)P_1(t) \tag{1.10}$$

Case 2: If the failure mode SG_1, interacts with the failure mode SG_2, within the same component or between other components at time $(t, t+\Delta t)$, then the probability that the interaction of the failure mode SG_1, with failure mode SG_2, will cause the entire system to fail is given as:

$$P_2(t+\Delta t) = P_2(t)(1-\lambda_2\Delta t)+P_1(t)\lambda_1\Delta t$$

$$\frac{P_2(t+\Delta t)-P_2(t)}{\Delta t} = -\lambda_2P_2(t)+\lambda_1P_1(t)$$

when $\Delta t \to 0$

$$\acute{P}_2(t) = -\lambda_2 P_2(t) + \lambda_1 P_1(t) \tag{1.11}$$

Case 3: When the failure mode SG_1, interacts with the failure mode SG_3, within the same component or between other components at time $(t, t+\Delta t)$, then the probability that the interaction of the failure mode SG_1, with failure mode SG_3, will cause the entire system to fail is given as:

$$\acute{P}_3(t) = -\lambda_3 P_3(t) + \lambda_1 P_1(t) \tag{1.12}$$

Similarly, other interactions of the failure modes can be modeled using the same format. If we assumed that the components of the system have an exponential lifetime distribution, then the above equations can be rewritten as:

$$\begin{cases} \acute{P}_1(t) = e^{-(\lambda_1+\lambda_2+\lambda_3+\lambda_4)t} \\ \acute{P}_2(t) = e^{-(\lambda_2)t}\left(1-e^{-(\lambda_1)t}\right) \\ \acute{P}_3(t) = e^{-(\lambda_3)t}\left(1-e^{-(\lambda_1)t}\right) \end{cases} \tag{1.13}$$

If the reliability of a system with several failure modes and components is defined as the probability of the system working in a perfect condition without failing, then the reliability of the system can be expressed as:

$$R_{sys}(t) = \sum_{i \in D} P_i(t) \tag{1.14}$$

Using the reliability equation above, the IFCR of the system which consists of the interactions of the failure modes SG_i, at a time (t) within the component of a system can be quantified as follows:

$$R_{sys}(t) = \sum_{i \in D} P_i(t) = P_1(t)$$

$$= e^{-(\lambda_1+\lambda_2+\lambda_3+\lambda_4)t}$$

$$= e^{-(\lambda_1)t} \cdot e^{-(\lambda_2)t} \cdot e^{-(\lambda_3)t} \cdot e^{-(\lambda_4)t} \tag{1.15}$$

Similarly, the EFCR of the system which consists of the interactions of the failure modes SG_i, at a time (t) between the failure modes of several other components can be quantified as follows:

$$R_{sys}(t) = \sum_{i \in D} P_i(t) = P_1(t) + P_2(t) + P_3(t) + P_4(t) + ..$$

$$= e^{-(\lambda_1+\lambda_2+\lambda_3+\lambda_4)t} + e^{-(\lambda_2)t}\left(1-e^{-(\lambda_1)t}\right)+e^{-(\lambda_3)t}\left(1-e^{-(\lambda_1)t}\right)+\cdots$$

$$= e^{-(\lambda_1)t}\cdot e^{-(\lambda_2)t}\cdot e^{-(\lambda_3)t}\cdot e^{-(\lambda_4)t} + e^{-(\lambda_2)t}\cdot\left(1-e^{-(\lambda_1)t}\right)+e^{-(\lambda_3)t}\cdot\left(1-e^{-(\lambda_1)t}\right)+\cdots$$

$$= e^{-(\lambda_1)t}\cdot e^{-(\lambda_2)t}\cdot e^{-(\lambda_3)t}\cdot e^{-(\lambda_4)t} + e^{-(\lambda_2)t}\cdot\left(1-e^{-(\lambda_1)t}\right)+e^{-(\lambda_3)t}\cdot\left(1-e^{-(\lambda_1)t}\right)+\cdots \quad (1.16)$$

To provide a better interpretation of the interactions within and between the failure modes in the system component(s), a MATLAB code (see Appendix) has been developed based on the proposed mathematical model.

1.4 NUMERICAL ILLUSTRATION

In this section, the proposed IFWG-TOPSIS model algorithm is applied for building a reliability knowledge model for the redesigning of a complex system (a slewing gear system used in a marine crane vessel), by determining the root cause of failure in a predecessor slewing gear system design. This is important since failure (reliability) information/knowledge is scarce at the early phases of the product design (Aikhuele & Turan, 2017b; Sanchez & Pan, 2011). This is mainly to allow the designer to learn and allocate adequate reliability values to the different components of the system, by transferring the reliability knowledge gained from the evaluation into the new design through a more robust reliability allocation process. Finally, the interactions of the failure modes within and between the components are determined using the mathematical model.

1.4.1 Implementing the IFWG-TOPSIS Model Algorithm for Building Reliability Knowledge

In implementing the IFWG-TOPSIS model algorithm, the four most important failure modes of the slewing gear system (SG_1, SG_2, SG_3, and SG_4) are evaluated with respect to some dynamic reliability criteria (DC_1, DC_2, DC_3, DC_4, and DC_5). If the following pre-determined dynamic weight vectors $\delta = (0.17, 0.33, 0.52, 0.30, 0.52)^T$ are assigned to the dynamic reliability criteria, respectively, then the algorithm of the IFWG-TOPSIS model can be applied for determining the root cause of failure in the slewing gear system.

Using the algorithm presented in Section 1.3, the linguistic scale in Table 1.1 is used for collecting the preference rating and the opinions of the group of experts. The preference rating and the opinions which are used to construct the individual intuitionistic fuzzy decision matrix $Z = (r_{ij})_{mxn}$ of the experts were obtained from five experts (mainly academic professors from two universities in Nigeria). Since all the experts are of equal rank, they have been assigned equal weight vectors.

The linguistic results derived from the evaluation are shown and presented in Table 1.2. The results which are then converted to the IFN are aggregated using the DIFWGϵ operator (Equation 2) along with the pre-determined dynamic weight

TABLE 1.2
Linguistic Results Derived from Evaluation by Experts

	E1	E2	E3	E4	E5	E1	E2	E3	E4	E5	E1	E2	E3	E4	E5
			DC_1					DC_2					DC_3		
SG_1	VH	MH	MH	H	MH	HZ	MH	M	H	VH	MH	H	VH	M	H
SG_2	HZ	H	H	VH	H	M	H	MH	VH	M	H	VH	HZ	MH	VH
SG_3	HZ	H	M	M	VH	VH	H	VH	MH	HZ	H	VH	HZ	MH	HZ
SG_4	H	VH	MH	MH	HZ	HZ	VH	HZ	H	M	VH	HZ	M	H	VH
			DC_4					DC_5							
SG_1	MH	H	H	M	H	M	H	VH	M	VH					
SG_2	H	VH	VH	MH	VH	MH	VH	HZ	MH	HZ					
SG_3	HZ	H	M	VH	H	HZ	H	M	VH	H					
SG_4	M	VH	MH	HZ	VH	MH	VH	MH	HZ	VH					

vector of the dynamic reliability criteria to construct the collective intuitionistic fuzzy decision matrix. The use of the DIFWGϵ operator in this chapter is mainly to address and account for the uncertainty and dynamic related issues in the criteria and in the entire evaluation process, which to the best of our knowledge is novel in reliability study and research.

Moving on, with the collective intuitionistic fuzzy decision matrix, the actual weight vector (av_i) of the dynamic reliability criteria is calculated using the intuitionistic entropy method, and thereafter it is used to construct the weighted normalization matrix. The results of the evaluations are shown in Table 1.3.

With the construction of the weighted normalization matrix, the intuitionistic positive and negative ideal solutions are determined and then calculated with the IFWG operator. Finally, the relative closeness coefficient of the alternatives, which is based on the intuitionistic positive and negative ideal solutions (IFWG operator), is then determined. The results of the computations are given in Table 1.4, which shows the component with the failure mode (SG_I) as the most affected.

TABLE 1.3
Results from the Aggregation and the Weighted Normalization Matrix

	DC_1	DC_2	DC_3	DC_4	DC_5
Aggregated expert's preference ratings and opinions and the actual weight vector of the criteria					
SG_1	(0.2747, 0.3094)	(0.4047, 0.3890)	(0.2610, 0.3680)	(0.2804, 0.3680)	(0.4591, 0.4526)
SG_2	(0.3495, 0.4131)	(0.3513, 0.4131)	(0.4592, 0.5086)	(0.4211, 0.5887)	(0.4838, 0.4185)
SG_3	(0.4294, 0.5036)	(0.4464, 0.4819)	(0.4661, 0.4041)	(0.3155, 0.3680)	(0.3101, 0.3680)
SG_4	(0.4770, 0.4526)	(0.4705, 0.4185)	(0.4967, 0.4526)	(0.4134, 0.5086)	(0.3938, 0.5086)
av_i	0.105579	0.074609	0.18307	0.44284	0.193903
Weighted normalization matrix of the failure modes					
SG_1	(0.0290, 0.0327)	(0.0302, 0.0290)	(0.0478, 0.0674)	(0.1242, 0.1629)	(0.0890, 0.0878)
SG_2	(0.0369, 0.0436)	(0.0262, 0.0308)	(0.0841, 0.0931)	(0.1865, 0.2607)	(0.0938, 0.0812)
SG_3	(0.0453, 0.0532)	(0.0333, 0.0360)	(0.0853, 0.0740)	(0.1397, 0.1629)	(0.0601, 0.0713)
SG_4	(0.0504, 0.0478)	(0.0351, 0.0312)	(0.0909, 0.0828)	(0.1831, 0.2252)	(0.0764, 0.0986)

TABLE 1.4
Result of the Root Cause of Failure Assessment

	IFWG(S^+)	IFWG(S^-)	CC_i	Ranking
SG_1	0.513084	0.405806	0.507964	1
SG_2	0.550999	0.436698	0.49289	3
SG_3	0.518107	0.408637	0.500124	2
SG_4	0.544327	0.430658	0.492524	4

1.4.2 A Mathematical Model for Evaluating Failure Modes Interactions

In applying the mathematical model for evaluating the interactions within and between the failure modes in the component(s) of the slewing gear system, the following values $SG_1 = 0.507964$, $SG_2 = 0.49289$, $SG_3 = 0.500124$, and $SG_4 = 0.492524$ from Table 1.4 are assumed to be the failure rates of the system.

With these values, the probabilities that the system will fail when they interact within the context of the IFCR and EFCR are determined by using the developed MATLAB algorithm (see the pseudo MATLAB code). The result of the computation has been presented in Table 1.5, which is further shown in Figure 1.3.

Figure 1.3 shows a graph of the reliability of the slewing gear system for a marine crane vessel, when the failure modes interact within a component in the system (IFCR) and when the failure modes interact between several other failure modes of other components in the system (EFCR). From the graph, it is not hard to see that the reliability of the system is higher when the failure modes interact only within a component in the system (IFCR), unlike when it interacts within and between several other failure modes of other components of the system (EFCR).

This shows that the tendency for a system to fail is higher when the failure modes of a component interact with and between the failure modes of several other components in a system (EFCR).

TABLE 1.5
Result of the Computation with the Mathematical Model

EFCR	**1**	**0.601942**	**0.253889**	**0.09803**	**0.036771**	**0.013653**	**0.00505**	**0.001866**	**0.000689**	**0.000254**
IFCR	1	0.715678	0.362388	0.158592	0.064494	0.025196	0.009621	0.003625	0.001355	0.000504

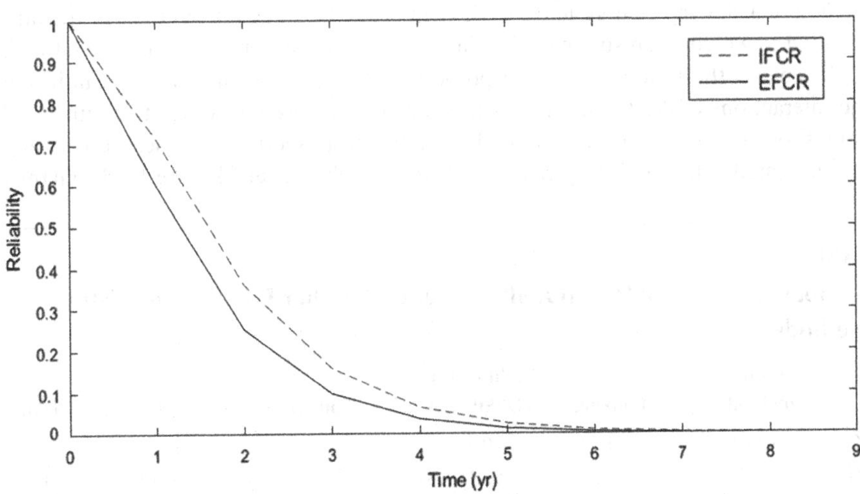

FIGURE 1.3 The failure modes interact within and between other component(s) of a system.

1.4.3 Discussion of the Results

Failure in the slewing gear system, if not managed properly, can lead to a catastrophic damage and total breakdown of the entire system including the marine crane vessel to which it is attached. From the results presented in Table 1.4, it is not hard to see that the component which housed the failure mode (SG_1) is the most affected and can be described as the potential root cause of the failure to the entire slewing gear system based on the expert's opinion presented with linguistic variables in Table 1.2.

It is of interest to note here that the result in Table 1.4 is based on the dynamic criteria adopted, the dynamic weight vector of the criteria, the actual weight vectors determined, and the opinion and weight vectors of the experts. However, to test the feasibility and rationality of the result, it is compared with other results obtained when solved with the established existing methods including the traditional TOPSIS model (Chu, 2002), and the value and ambiguity based ranking method originally proposed in the study of Li et al. (2010). From the comparison results presented in Table 1.6, we can see that the results are consistent with those of the proposed model.

With this information about the root cause of failure of the slewing gear system, the product designer can easily put up a plan to allocate more reliability values to the area covered with the failure modes. However, to ensure a robust reliability plan, the designer needs to know the exact states of the failure modes, how the failure mode interacts with the other failure modes within the component and between several other failure modes of other components, which otherwise is believed to be the main reason for the failure of the entire system. And this has been achieved using the mathematical model developed and presented in Sections 1.3 and 1.4.2, respectively, where the interaction issues of the failure modes in the system components have been addressed.

The simulated results have demonstrated how a failure occurs and the need to measure the physics of a failure in a system. In this study, however, we have been able to quantify the interactions of the failure modes that cause total system failure. With the present results, a more robust design of a reliability knowledge model for the redesigning of a slewing gear system can be built since the actual interactions of the failure modes, which can result in total failure of the entire system, have been quantified.

To ensure the rationality of the proposed mathematical model for quantifying the interactions of the failure modes in a mechanical system, some other simulated values for the failure rates are applied using the proposed mathematical model, and the simulated values and the graphs are shown in Table 1.7 and Figure 1.4. From this,

TABLE 1.6

Comparison of the IFWG-TOPSIS Model with Other Established Existing Methods

	Proposed method CC_i	Ranking	Traditional fuzzy TOPSIS method	Ranking	(Li et al., 2010)	Ranking
SG_1	0.507964	1	0.5558	1	0.265993	1
SG_2	0.49289	3	0.415998	4	0.213212	3
SG_3	0.500124	2	0.52444	2	0.221841	2
SG_4	0.492524	4	0.418283	3	0.205919	4

TABLE 1.7
Results of the Different Computations with the Mathematical Model

EFCR (a)	1	0.42751	0.11496	0.02863	0.00701	0.00171	0.00042	0.00010	0.00002	0.00001
IFCR (a)	1	0.56300	0.19484	0.05814	0.01650	0.00462	0.00130	0.00037	0.00010	0.00003
The following are the simulated failure rate $SG_1 = 0.6544, SG_2 = 0.7663, SG_3 = 0.8625$, and $SG_4 = 0.5432$										
EFCR (b)	1	0.95027	0.84396	0.72134	0.60225	0.49541	0.40367	0.32694	0.26383	0.21247
IFCR (b)	1	0.96988	0.89667	0.80116	0.69792	0.59632	0.50186	0.41733	0.34372	0.28089
The following are the simulated failure rates: $SG_1 = 0.09544, SG_2 = 0.06563, SG_3 = 0.04625$, and $SG_4 = 0.05332$										

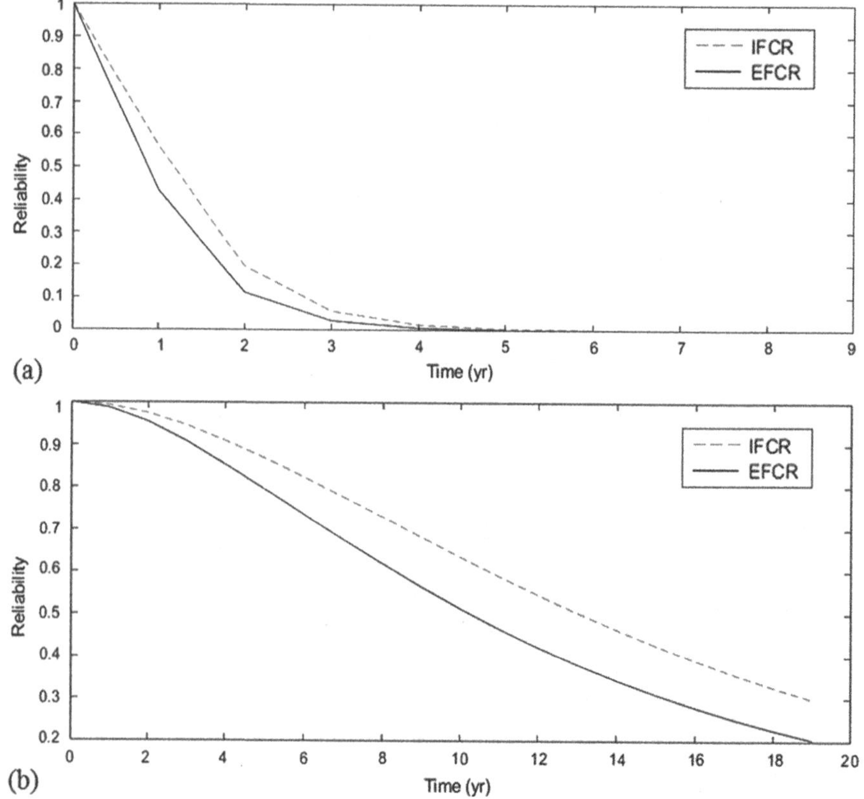

FIGURE 1.4 (a & b) Simulated failure modes interact within and between the component(s) of the system.

it is not hard to see that the reliability of the slewing gear system is higher when the failure modes interact only within a component in the system (IFCR) unlike when it interacts between several components of the system (EFCR).

1.5 CONCLUSIONS

The FMEA method is a popular tool that has been extensively used for the reliability and quality engineering management of complex products and systems. The method, which was developed by the U.S. aerospace industry, is a structured and systematic tool that has clear reliability and safety requirements for the identification, ranking, and evaluation of a failure or potential failure in new and existing products and systems. In applying the FMEA method, only the root cause of failure and the design risk of product failure are often considered. However, in order to build a complete and a robust design of a reliability knowledge model for the redesigning of new products and systems, the reliability assessment should as well include the measurement of the actual interactions of the failure modes which results in the failure of a system, by quantifying the FCRs within (IFCR) and between (EFCR) the system component(s).

To achieve this goal, a new model termed IFWG-TOPSIS (Intuitionistic Fuzzy Weighted Geometric – Technique for Order of Preference by Similarity to Ideal Solution) has been proposed for determining the root cause of failure and for building reliability knowledge model for the to-be-designed product and system. To ensure the feasibility and rationality of the models, the proposed algorithm of the IFWG-TOPSIS model is applied for building reliability knowledge model for the redesigning of a complex system (slewing gear system used in marine crane vessel) by determining the root cause of failure in a predecessor slewing gear system design. This is important since failure (reliability) information/knowledge is scarce at the early phases of the design. The mathematical model is used for evaluating the interactions within and between the failure modes in system component(s).

The main advantages and contributions of the proposed new models in this study are three-fold:

(i) The IFWG-TOPSIS model accounts for the uncertainty and dynamic issues in the decision-making process and is used in the evaluation of the root cause of failure in the designed product.
(ii) The interactions within and between the failure modes are addressed and evaluated adequately using the new mathematical model.
(iii) A complete and robust design of the reliability knowledge model is achieved, which can be used for the redesigning of new products and systems.

In the future, the proposed models will be integrated with some other existing intelligent-based algorithms for solving product reliability problems.

APPENDIX: MATLAB® CODE

```
a1=0.29544;% a1=Component 1 failure mode's rate of occurrence
a2=0.3563;% a2=Component 1 failure mode's rate of occurrence
a3=0.6324;% a3=Component 2 failure mode's rate of occurrence
a4=0.3432;% a4=Component 2 failure mode's rate of occurrence
t=0;% time counter
i=0;
M1=abs((1/a1)+(1/a2))% MTTF of Component 1
M2=abs((1/a3)+(1/a4))% MTTF of Component 2

RP1=zeros(1,50);%Variable for internal interaction of
Component 1
RS1=zeros(1,50);%Variable for external interaction of
Component 1
RP2=zeros(1,50);%Variable for internal interaction of
Component 2
RS2=zeros(1,50);%Variable for external interaction of
Component 2
time=zeros(1,10);%Variable for time

%%WHILE LOOP FOR COMPUTING THE INTERNAL AND EXTERNAL
INTERACTIONS OF THE FAILURE MODES OF COMPONENT 1
```

```
while t<10
    R1=exp(-a1*t)
    R2=exp(-a2*t)
    RS1(t+1)=R1*R2;
    RP1(t+1)=1-((1-R1)*(1-R2));
    time(t+1)=t;
    t=t+1;
end

%%WHILE LOOP FOR COMPUTING THE INTERNAL AND EXTERNAL
INTERACTIONS OF THE FAILURE MODES OF COMPONENT 2
while I <10
    R4=exp(-a4*i)
    R3=exp(-a3*i)
    RS2(i+1)=R4*R3;
    RP2(i+1)=1-((1-R3)*(1-R4));
    time(i+1)=i;
    i=i+1;
end

Rsystem=zeros(1,length(time));     %Variable for IFCR of the
system
RPsystemS=zeros(1,length(time));  %Variable for EFCR of the
system
for i=1:length(time)
    Rsystem(i)=RP2(i)*RP1(i);
    RPsystemS(i)=1-((1-RS1(i))*(1-RS2(i)));
    plot(time,Rsystem,'--r');hold on;
    plot(time,RPsystemS,'k');
    hold off;
end
```

NOMENCLATURE

D = $\{SG_1, SG_2, SG_3, SG_4\}$ = Set of failure modes of components of the system
$P_i(t)$ = The probability that the failure modes SG_i, will interact with each other at a time (t),
λ_i, (i = 1, 2, 3, 4) = Failure rate from the interactions
e = Exponential lifetime distribution
$R_{sys}(t)$ = Reliability of the system
SG_i = Failure modes of the slewing gear system
DC_i = Dynamic reliability criteria
FCRs = Failure causality relationships within and between the components.
IFCR = Internal failure causality relationships within the components
EFCR = External failure causality relationships
IFS = Intuitionistic fuzzy set
IFN = Intuitionistic fuzzy number
δ = Pre-determined dynamic weight vector
CC_i = Closeness coefficient of the alternatives

$Z = (x_{ij})_{mxn}$ = Intuitionistic fuzzy decision matrix
S^+ = Intuitionistic fuzzy positive ideal (IFPI) solution
S^- = Intuitionistic fuzzy negative ideal (IFNI) solution
μ_A = The membership function,
V_A = The membership function,
DIFWGϵ = Dynamic intuitionistic fuzzy Einstein geometric averaging operator
IFWG = Intuitionistic Fuzzy Weighted Geometric operator

REFERENCES

Adhikary, D. D., G. K. Bose, D. Bose, and S. Mitra. 2014. "Multi Criteria FMECA for Coal-Fired Thermal Power Plants Using COPRAS-G." *International Journal of Quality and Reliability Management* 31(5): 601–14. doi:10.1108/IJQRM-04-2013-0068.

Aikhuele, Daniel O., and Faiz B. M. Turan. 2016. "Intuitionistic Fuzzy-Based Model for Failure Detection." *Springer Plus* 5(1): 1–15. doi:10.1186/s40064-016-3446-0.

Aikhuele, Daniel O., and Faiz B. M. Turan. 2017a. "Extended TOPSIS Model for Solving Multi-Attribute Decision Making Problems in Engineering." *Decision Science Letters* 6: 365–76. doi:10.5267/j.dsl.2017.2.002.

Aikhuele, Daniel O., and Faiz B. M. Turan. 2017b. "Need for Reliability Assessment of Parent Product Before Redesigning a New Product." *Current Science* 112(1): 10–11. doi:10.1007/s00170-011-3234-5.D.

Aikhuele, Daniel O., and Faiz B. M. Turan. 2017c. "A Modified Exponential Score Function for Troubleshooting an Improved Locally Made Offshore Patrol Boat Engine." *Journal of Marine Engineering and Technology* 17(1): 52–58. doi:10.1080/20464177.2017.1286841.

Aikhuele, Daniel O., and Sarah Odofin. 2017. "A Generalized Triangular Intuitionistic Fuzzy Geometric Averaging Operator for Decision-Making in Engineering and Management." *Information* 8(3): 1–17. doi:10.3390/info8030078.

Aikhuele, Daniel O., Faiz M. Turan, S. M. Odofin, and Richard H. Ansah. 2016. "Interval-Valued Intuitionistic Fuzzy TOPSIS-Based Model for Troubleshooting Marine Diesel Engine Auxiliary System." *International Journal of Maritime Engineering-Part A* 159: 1–8. doi:10.3940/rina.ijme.2016.a1.402.

Balin, Abit, Hakan Demirel, and Alarçin Fuat. 2014. "A Hierarchical Structure for Ship Diesel Engine Trouble-Shooting Problem Using Fuzzy Ahp and Fuzzy Vikor Hybrid Methods." *Brodogradnja* 66(1): 54–65.

Bowles, John B., and C. Enrique Pelaez. 1995. "Fuzzy Logic Prioritization of Failures in a System Failure Mode, Effects and Criticality Analysis." *Reliability Engineering and System Safety* 50(2): 203–13. doi:10.1016/0951-8320(95)00068-D.

Chang, Kuei Hu, and Ta Chun Wen. 2010. "A Novel Efficient Approach for DFMEA Combining 2-Tuple and the OWA Operator." *Expert Systems with Applications* 37(3): 2362–70. doi:10.1016/j.eswa.2009.07.026.

Chu, T. C. 2002. "Selecting Plant Location via a Fuzzy TOPSIS Approach." *International Journal of Advanced Manufacturing Technology* 20(11): 859–64. doi:10.1007/s001700200227.

Garg, Harish. 2014. "Reliability, Availability and Maintainability Analysis of Industrial Systems Using PSO and Fuzzy Methodology." *MAPAN Journal of Metrology Society of India, Springer* 29(2): 115–29.

Garg, Harish. 2018. "A Novel Approach for Analyzing the Reliability of Series-Parallel System Using Credibility Theory and Different Types of Intuitionistic Fuzzy Numbers." *Journal of the Brazilian Society of Mechanical Sciences and Engineering* 38(3): 1021–35.

Garg, Harish, Monica Rani, S. P. Sharma, and Yashi Vishwakarma. 2014. "Bi-Objective Optimization of the Reliability-Redundancy Allocation Problem for Series-Parallel System." *Journal of Manufacturing Systems* 33(3): 335–47.

Garg, Harish, and S. P. Sharma. 2013. "Reliability – Redundancy Allocation Problem of Pharmaceutical Plant." *Journal of Engineering Science and Technology* 8(2): 190–98.

Garg, Harish. 2018. "A Novel Approach for Analyzing the Reliability of Series-Parallel System Using Credibility Theory and Different Types of Intuitionistic Fuzzy Numbers." *Journal of the Brazilian Society of Mechanical Sciences and Engineering* 38(3): 1021–35.

Geum, Youngjung, Yangrae Cho, and Yongtae Park. 2011. "A Systematic Approach for Diagnosing Service Failure: Service-Specific FMEA and Grey Relational Analysis Approach." *Mathematical and Computer Modelling* 54(11–12): 3126–42. doi:10.1016/j.mcm.2011.07.042.

Govil, K. K., and R. A. Agarwala. 1983. "Lagrange Multiplier Method for Optimal Reliability Allocation in a Series System." *Reliability Engineering* 6(3): 181–90.

Gümüş, S., and O. Bali. 2017. "Dynamic Aggregation Operators Based on Intuitionistic Fuzzy Tools and Einstein Operations." *Fuzzy Information and Engineering* 9(1): 45–65. doi:10.1016/j.fiae.2017.03.003.

Ha, Chunghun, and Way Kuo. 2008. "Initial Allocation Compensation Algorithm for Redundancy Allocation: The Scanning Heuristic." *IIE Transactions* 40(7): 678–89.

He, Yi-Hai, Lin-Bo Wang, Zhen-Zhen He, and Min Xie. 2015. "A Fuzzy TOPSIS and Rough Set Based Approach for Mechanism Analysis of Product Infant Failure." *Engineering Applications of Artificial Intelligence* 47: 1–13. doi:10.1016/j.engappai.2015.06.002.

Hwang, C. L., and K. Yoon. 1981. *Multiple Attribute Decision Making Methods and Applications*. Berlin: Springer.

Kangavari, Mehdi, Sajad Salimi, Rohallah Nourian, Leila Omidi, and Alireza Askarian. 2015. "An Application of Failure Mode and Effect Analysis (FMEA) to Assess Risks in Petrochemical Industry in Iran." *Iranian Journal of Health, Safety and Environment* 2(2): 257–63.

Kostina, Marina. 2012. "Reliability Management of Manufacturing Processes in Machinery Enterprises." *Theses of Tallinn University of Technology. ISSN 1406-4766 71*. http://linda.linneanet.fi/F/?func=direct&doc_number=006392022&local_base=fin01.

Kuo, W., V. R. Prasad, F. A. Tillman, and C. Hwang. 2001. *Optimal Reliability Design-Fundamentals and Application*. Cambridge, UK: Cambridge University Press.

Kuo, W., and R. Wan. 2007. "Recent Advances in Optimal Reliability Allocation." *IEEE Transactions on Systems, Man, and Cybernetics-Part A: Systems and Humans* 37(2): 143–56.

Li, Deng Feng, Jiang Xia Nan, and Mao Jun Zhang. 2010. "A Ranking Method of Triangular Intuitionistic Fuzzy Numbers and Application to Decision Making." *International Journal of Computational Intelligence Systems* 3(5): 522–30. doi:10.1080/18756891.2010.9727719.

Liu, Hu-Chen, Long Liu, and Qing-Lian Lin. 2013. "Fuzzy Failure Mode and Effects Analysis Using Fuzzy Evidential Reasoning and Belief Rule-Based Methodology." *IEEE Transactions on Reliability* 62(1): 23–36. doi:10.1109/TR.2013.2241251.

Liu, Hu-Chen, Long Liu, Nan Liu, and Ling-Xiang Mao. 2012. "Risk Evaluation in Failure Mode and Effects Analysis with Extended VIKOR Method Under Fuzzy Environment." *Expert Systems with Applications* 39(17): 12926–34. doi:10.1016/j.eswa.2012.05.031.

Liu, Hu-Chen, Jian-Xin You, Xue-Feng Ding, and Qiang Su. 2015. "Improving Risk Evaluation in FMEA with a Hybrid Multiple Criteria Decision Making Method." *International Journal of Quality and Reliability Management* 32(7): 763–82. doi:10.1108/09564230910978511.

Liu, Hu-Chen, Jian-Xin You, Qing-Lian Lin, and Hui Li. 2014. "Risk Assessment in System FMEA Combining Fuzzy Weighted Average with Fuzzy Decision-Making Trial and Evaluation Laboratory." *International Journal of Computer Integrated Manufacturing* 28(7): 701–14. doi:10.1080/0951192X.2014.900865.

Lu, Lu, and Christine M. Anderson-Cook. 2015. "Improving Reliability Understanding Through Estimation and Prediction with Usage Information." *Quality Engineering* 27(3): 304–16. doi:10.1080/08982112.2014.990033.

Lyu, Michael R., Sampath Rangarajan, and Aad P. A. Van Moorsel. 1997. "Optimization of Reliability Allocation and Testing Schedule for Software Systems." In *ISSRE '97 Proceedings of the Eighth International Symposium on Software Reliability Engineering*, 336.

Ma, Hongzhan, Xuening Chu, Deyi Xue, and Dongping Chen. 2016. "Identification of To-Be-Improved Components for Redesign of Complex Products and Systems Based on Fuzzy QFD and FMEA." *Journal of Intelligent Manufacturing* 30: 623–39. doi:10.1007/s10845-016-1269-z.

Mettas, Adamantios. 2000. "Reliability Allocation and Optimization for Complex Systems." In *Proceedings of Annual Reliability and Maintainability Symposium*, Los Angeles, California, USA, January 24–27, 2000, 1–7.

Mohammadi, A., and M. Tavakolan. 2013. "Construction Project Risk Assessment Using Combined Fuzzy and FMEA." In *IFSA World Congress and NAFIPS Annual Meeting (IFSA/NAFIPS), 2013 Joint*, 232–37. doi:10.1109/IFSA-NAFIPS.2013.6608405.

Niwas, Ram, and Harish Garg. 2018. "An Approach for Analyzing the Reliability and Profit of an Industrial System Based on the Cost Free Warranty Policy." *Journal of the Brazilian Society of Mechanical Sciences and Engineering* 40(5): 1–9. doi:10.1007/s40430-018-1167-8.

Pardeep, Kumar, D. K. Chaturvedi, and G. L. Pahuja. 2010. "An Efficient Heuristic Algorithm for Determining Optimal." *Reliability: Theory and Applications* 1(18): 15–28.

Park, Jin Han, Young Chel Kwun, and Ja Hong Koo. 2011. "Dynamic Uncertain Linguistic Weighted Harmonic Mean Operators Applied to Decision Making." In *Proceedings 2011 International Conference on System Science and Engineering, ICSSE 2011*, 101–6. doi:10.1109/ICSSE.2011.5961882.

Pérez-Domínguez, Luis, Alejandro Alvarado-Iniesta, Iván Rodríguez-Borbón, and Osslan Vergara-Villegas. 2015. "Intuitionistic Fuzzy MOORA for Supplier Selection." *Dyna* 82(191): 34–41. doi:10.15446/dyna.v82n191.51143.

Safari, Hossein, Zahra Faraji, and Setareh Majidian. 2016. "Identifying and Evaluating Enterprise Architecture Risks Using FMEA and Fuzzy VIKOR." *Journal of Intelligent Manufacturing* 27(2): 475–86. doi:10.1007/s10845-014-0880-0.

Sanchez, Luis Mejia, and Rong Pan. 2011. "An Enhanced Parenting Process: Predicting Reliability in Product's Design Phase." *Quality Engineering* 23(4): 378–87. doi:10.1080/08982112.2011.603110.

Shaghaghi, Mahdi, and Kamran Rezaie. 2012. "Failure Mode and Effects Analysis Using Generalized Mixture Operators." *Journal of Optimization in Industrial Engineering* 11: 1–10.

Sharma, Rajiv Kumar, Dinesh Kumar, and Pradeep Kumar. 2005. "Systematic Failure Mode Effect Analysis (FMEA) Using Fuzzy Linguistic Modelling." *International Journal of Quality and Reliability Management* 22(9): 986–1004. doi:10.1108/02656710510625248.

Smith, Shana, Gregory Smith, and Ying Ting Shen. 2012. "Redesign for Product Innovation." *Design Studies* 33(2): 160–84. doi:10.1016/j.destud.2011.08.003.

Tay, Kai Meng, Chian Haur Jong, and Chee Peng Lim. 2015. "A Clustering-Based Failure Mode and Effect Analysis Model and Its Application to the Edible Bird Nest Industry." *Neural Computing and Applications* 26(3): 551–60. doi:10.1007/s00521-014-1647-4.

Vahdani, Behnam, M. Salimi, and M. Charkhchian. 2015. "A New FMEA Method by Integrating Fuzzy Belief Structure and TOPSIS to Improve Risk Evaluation Process." *International Journal of Advanced Manufacturing Technology* 77(1–4): 357–68. doi:10.1007/s00170-014-6466-3.

Xu, Zeshui, and Ronald R. Yager. 2006. "Some Geometric Aggregation Operators Based on Intuitionistic Fuzzy Sets." *International Journal of General Systems* 35(4): 417–33. doi:10.1080/03081070600574353.

Yadav, Om Prakash, Nanua Singh, Ratna Babu Chinnam, and Parveen S. Goel. 2003. "A Fuzzy Logic Based Approach to Reliability Improvement Estimation during Product Development." *Reliability Engineering and System Safety* 80(1): 63–74. doi:10.1016/S0951-8320(02)00268-5.

Yager, R. R. 1988. "On Ordered Weighted Averaging Aggregation Operators in Multi Criteria Decision Making." *IEEE Transactions on Systems, Man, and Cybernetics: Systems* 18(1): 183–90. doi:10.1109/21.87068.

Ye, Jun. 2010. "Two Effective Measures of Intuitionistic Fuzzy Entropy." *Computing (Vienna/New York)* 87(1–2): 55–62. doi:10.1007/s00607-009-0075-2.

Zhao, Hao, Jian-Xin You, and Hu-Chen Liu. 2016. "Failure Mode and Effect Analysis Using MULTIMOORA Method with Continuous Weighted Entropy Under Interval-Valued Intuitionistic Fuzzy Environment." *Soft Computing* 21: 5355–67. doi:10.1007/s00500-016-2118-x.

Zuber, Ninoslav, and Rusmir Bajri. 2016. "Application of Artificial Neural Networks and Principal Component Analysis on Vibration Signals for Automated Fault Classification of Roller Element Bearings." *Eksploatacja i Niezawodnosc – Maintenance and Reliability* 18(2): 299–306.

2 Reliability and Cost-Benefit Analysis of a Repairable System under a Cost-Free Warranty Policy with the Repairman Taking Multiple Vacations

Ram Niwas

CONTENTS

2.1 INTRODUCTION

The reliability of a system is very important for adequate performance over its expected life period. In fact, uninterrupted service and failure-free operation is an indispensable requirement of large complex systems. The system may fail due to poor designing, incorrect manufacturing, inappropriate testing, and improper installation and maintenance of its components, in addition to other unforeseen reasons. Customers need affirmation that the product they are buying will perform adequately, and the warranty guarantees this affirmation. An effective way to ensure the reliability of a sold product (or system) is by providing a warranty on the system for a certain period of operation. Product warranty plays an increasingly important role in consumer and commercial transactions as discussed by Murthy and Djamaludin (2002). According to Jahromi and Vahdani (2009), when an item fails during the warranty period, there are two variables that influence the manufacturer's decision on whether to repair it using minimal repair or to replace it with a new one free of charge to the customer. The repairman can also contribute to the economic benefit of the system directly or indirectly. Therefore, to improve the profitability of the system, the repairman might take a sequence of vacations during the system's idle time. Repairman's vacations are beneficial for the system when the repairman utilizes this idle time for different purposes (Levy & Yechiali, 1975). When there is no failed unit in the system, the repairman leaves for a vacation or may take other assigned jobs, which can have a considerable influence on the performance of the system (Niwas & Kadyan, 2018). Although the vacation model originally arises from the queueing theory, it can successfully be applied in manufacturing and communication network systems. On the basis of the vacation queueing theory, some researchers have applied the vacation model to repairable systems for analyzing their reliability and performance.

By using various approaches such as the Markov process, supplementary variable technique, and so on, a variety of methods have been developed for warranty as well as vacation analysis, which are summarized in the following sections.

2.2 BACKGROUND AND LITERATURE REVIEW

In this section, a brief literature review regarding warranty, reliability, repairable systems, multiple vacations, and maintainability is given. The gaps found in the literature review are also addressed.

2.2.1 The Concept of Warranty

A warranty can be defined as an assurance from a manufacturer/system provider to a system user that the system sold will perform satisfactorily up to a certain length of time, which is the warranty period. In other words, warranty is a contract offered by a manufacturer/system provider to system users/buyers to replace or repair a faulty item, or to partially or fully reimburse the user, in the event of a failure up to a certain length of time. Warranties are widespread and serve many purposes such as protection for manufacturers and users, assurance of the quality and reliability of the system, and as a marketing strategy.

2.2.1.1 Role of Warranty

(i) **Consumer/Customer**: When a consumer purchases a product, the warranty document acts as a source of information about the product characteristics, and it also acts as an indicator of the quality and reliability of the product in the context of complex or innovative products when the customer is unable to evaluate the product performance due to lack of knowledge, expertise, or experience. It protects the interests of the customer by acting as an insurance against early failures of an item due to poor designing, manufacturing defects, or quality assurance problems during the warranty period. If the warranty is optional (such as an extended warranty), the consumer has to decide whether the warranty is worth the additional cost. For commercial products, the cost of repairs not covered by a warranty policy can significantly impact profits due to downtimes.

(ii) **Manufacture/Dealer**: One of the main functions of a warranty policy from the system provider's point of view is protection. The terms of the warranty specify the use and the conditions of the use for which the system is intended. They also provide for limited coverage or no coverage at all in the case of misuse of the system/product. The second important purpose of a warranty is promotional. As a user often infers a more reliable system when a long warranty is offered, this has been used as an effective advertising tool.

2.2.1.2 Warranty Cost

For products sold with a warranty, the manufacturer incurs additional costs resulting from the servicing of claims covered under the warranty. Warranty claims occur due to item failures which are influenced by several factors/reasons such as the engineering decision during designing and manufacturing, which determines the inherent reliability of the system, the usage intensity and environment, and the maintenance effort expended by users. Despite the fact that warranties are so commonly used, the accurate pricing of warranties in many situations remains an unsolved problem. It may be surprising that the entitlement of warranty claims may cost companies large amounts of money. The data relevant to the warranty costs in a particular industry are usually highly confidential since they are commercially sensitive.

2.2.2 Warranty Cost Analysis

Blischke (1990) was the first to review papers on warranties, and it dealt with mathematical models for warranty cost analysis. Murthy (1991) studied a usage dependent model for warranty costing. Chun (1992) studied an optimal number of periodic preventive maintenance operations under a warranty. Murthy et al. (1995) studied two-dimensional failure-free warranty policies and developed two-dimensional point process models. Sahin and Polatoglu (1996) investigated maintenance strategies following the expiration of a warranty.

Monga and Zuo (1998) studied an optimal system design considering maintenance and warranty. Kim and Rao (2000) discussed the expected warranty cost of two-attribute free-replacement warranties based on a bivariate exponential distribution. Ja et al. (2001) developed a renewable minimal repair warranty policy with time-dependent costs. Bai and Pham (2004) studied repair-limit risk-free warranty policies with imperfect repair. Chukova and Johnston (2006) developed a two-dimensional warranty repair strategy based on minimal and complete repairs. Rahman (2007) discussed the modeling analysis of reliability and costs for lifetime warranty and service contract policies. Huang and Yen (2009) studied a two-dimensional warranty policy that takes both time and usage into account would be more realistic, and the authors concluded that its flexibility would allow manufacturers to increase profits by attracting customers with different usage modes . Vahdani et al. (2011) developed a replacement-repair model to study a renewing free replacement warranty (RFRW) for a class of multi-state deteriorating repairable products. Varnosafaderani and Chukova (2012) discussed an intensity reduction approach to model imperfect repairs with a specified warranty servicing strategy and identified the optimal strategy from the manufacturer's point of view.Kadyan and Ramniwas (2013) analyzed the reliability and profit of a single unit repairable system with a warranty for repair using a supplementary variable technique. Niwas et al. (2015 & 2016) discussed reliability models of a single-unit system with the concept of preventive maintenance, inspection beyond warranty, and degradation. The dependence of the accessible reliability of products on the expenses level of the manufacturer offering the warranty was determined by Podolyakina (2017). A new warranty policy wherein the buyer invests in the preventive maintenance (PM) cost within the product's life cycle to reduce the losses from production downtime was presented by Mo et al. (2017). Preventive maintenance (Garg et al., 2013; Garg, 2014), which is implemented for both manufacturers and users, is often used to increase product reliability and extend product life because too many claims during the warranty term result in high warranty costs for manufacturers and poor availability for users as reported by Huang et al. (2017). Warranty as a marketing strategy for remanufactured products was discussed by Alqahtani and Gupta (2017). An approach for analyzing the behavior of an industrial system under the cost-free warranty policy was presented by Niwas and Garg (2018). Under this policy, a mathematical model of the system was developed based on the Markov process, and various parameters such as reliability, mean time to system failure, availability, and expected profit of the system were derived.

2.2.3 Vacations Model Analysis

Vacation is a phenomenon by which a component of a system cannot be immediately repaired after it fails (Takagi, 1991).When there is no failed unit in the system, the repairman leaves for a vacation or may utilize his idle time for different purposes. If the machine fails while the repairman is on vacation, the system is in the state of waiting for repair. When the repairman comes back from his vacation, he starts repair immediately if the system has failed and is waiting for repair; otherwise, he will wait idly in the system for the occurrence of a failure/fail arrival, upon which he starts repairing it immediately. However, when the failed machine has been repaired, the repairman will go for a vacation again. In reliability theory, the vacation concept was studied by Su and Shi (1995), who discussed an n-unit series system with multiple vacations. Chao and Zhao (1998) analyzed multi-server queues with station and server vacations. Ke (2003) introduced the concept of vacation policy into the modeling analysis of a queue system in which two standard vacation policies were defined: single and multiple vacations. Tian and Zhang (2006) discussed vacation queueing models theory and applications. Ke and Wang (2007) developed vacation policies for machine repair problems with two types of spares. Jia and Wu (2009) developed a replacement policy for a repairable system with its repairman taking multiple vacations. Hu et al. (2010) analyzed a series-parallel repairable system with three units and a vacation. A machine repair problem with homogeneous machines and available standbys, in which multiple technicians are responsible for supervising these machines, was investigated by Ke et al. (2011). An optimal replacement policy of a deteriorating system with its repairman taking multiple vacations was evaluated by Yuan and Xu (2011). Wu et al. (2013) analyzed the reliability of a K-out-of-n:G repairable system with multiple vacations. Yuan and Cui (2013) analyzed the reliability of a consecutive k-out-of-n:F system with repairmen taking multiple vacations. Sun and Ma (2014) analyzed a deteriorating system with a single vacation based on the assumption that the system cannot be repaired as good as new after the failure. A warm maintenance repairable system with continuous switch and multiple vacations for the repairman was discussed by Zhang and Wu (2015). Qiao and Guo (2015) developed a repairable system with a repairman vacation and a warning device which can send an alarm when the system is not in good condition. Dongliang et al. (2017) analyzed the reliability of a series system containing n identical components with a repairman taking a single vacation. He et al. (2018) studied a repairable k-out-of-n:G system with failure dependencies, N-policy, and the repairman's multiple vacations. Yang and Tsao (2019) analyzed the reliability and availability of standby systems with working vacations and the retrial of failed components by using the matrix-analytic method.

2.2.4 Overcoming the Limitations of the Literature

From the above discussion, we can find out that no study has been undertaken to consider the reliability and profit analysis of the system based on the cost-free warranty policy where the repairman takes multiple vacations. In this chapter, we propose a repairable system with a repairman who can take multiple vacations within and beyond the cost-free warranty period by using the Markov process (Jain & Bura, 2011; Bura, 2018; Bura & Gupta, 2019). In it, if the system fails and the repairman is on vacation, it will wait for the inspection of the failed unit until the repairman is available. When the repairman comes back from his vacation, he starts inspection immediately if the system waits for inspection to check whether the system is within warranty or not. On the other hand, if the repairman comes back from his vacation and system is working well, then he will wait idly for fail arrival. The failure, inspection, and repair rates of the system follow a negative exponential distribution. Various reliability measures which depict the performance of the system have been derived. Finally, the effect of various parameters on the system reliability and the expected profit has been analyzed through an illustrative example.

The remainder of this chapter is described as follows. Section 2.2 discusses the background and reviews the literature on warranty cost analysis and vacation model analysis. Section 2.3 gives a description of the system including the assumptions of the model, state-specifications, and notations related to the proposed system model. Section 2.4 presents the model analysis in which different performance measures of the system are computed, such as the reliability of the system, mean time to system failure (MTSF), availability, busyness, vacation, and idle times of the repairman within and beyond the warranty, and profit function. The numerical results based on system reliability and profit function in the form of tables and graphs and their interpretations are shown in Section 2.5. Section 2.6 presents the concluding remarks. Finally, future research directions are presented in Section 2.7.

2.3 DESCRIPTION OF THE SYSTEM

2.3.1 Assumptions

(1) The system consists of a single machine and a single repairman; the machine is new and starts to work at the initial time $t = 0$, and the repairman takes his first vacation after the system has started.
(2) If the machine fails during the repairman's vacation, the system is in the state of waiting for repair or inspection.
(3) The repairman goes for vacation after repairing the failed machine.
(4) All the repairs are cost-free to the users during the warranty, provided failures are not due to the negligence of users.
(5) During the warranty, the repairman inspects the failed unit to check whether the system is under a warranty policy or not.
(6) The machine works as good as new after the repair.
(7) The distribution of failure, inspection, repair time of a machine, and arrival time of the repairman after his vacations are taken as the negative exponential.

2.3.2 State-Specifications

Let $\{N(t),\, t \geq 0\}$ be a stochastic process with state space $E = \{S_0, S_1, ..., S_8\}$, and the specifications of the states are defined as follows:

S_0/S_6: The system is working within/beyond the warranty and the repairman is on vacation;
S_1/S_7: The system is working within/beyond the warranty and the repairman waits idly in the system;
S_2/S_8: The system fails within/beyond the warranty and the repairman is on vacation;
S_3: The failed machine is under inspection during the warranty to check whether the system is under the warranty policy or not;
S_4/S_5: The failed machine is under repair within/beyond the warranty.

2.3.3 Notations

$\lambda\,/\,h\,/\,\mu$: Constant failure/inspection/repair rate of the machine.
α: Constant arrival rate of the repairman after his vacation.
p/q: Probability that the warranty is completed/not completed.
$p_0(t)\,/\,p_6(t)$: Probability density that at time t, the system is in good/working state within/beyond the warranty and the repairman is on his vacation.
$p_1(t)\,/\,p_7(t)$: Probability density that at time t, the system is in good/working state within/beyond the warranty and the repairman waits idly in the system.
$p_2(t)\,/\,p_8(t)$: Probability density that at time t, the system is in failed state within/beyond the warranty and the repairman is on his vacation.
$p_3(t)$: Probability density that at time t, the system is under inspection during the warranty.
$p_4(t)\,/\,p_5(t)$: Probability density that at time t, the system is under repair within/beyond the warranty.
$p(s)$: Laplace transform of function $p(t)$.

2.4 SYSTEM ANALYSIS

The system consists of a single unit in which there is a single repairman who can take multiple vacations. Consider that initially the system is in good working condition during the warranty and repairman is on vacation. If the machine fails within the warranty while the repairman is on vacation, the system is in the state of waiting for inspection. When the repairman comes back from his vacation, there are two cases: one is that the system is in the state of waiting for the inspection, and the repairman starts inspection immediately to check whether the system is under the warranty or not; the other is that the machine is running, and the repairman waits idly in the system for occurrence of a failure/fail arrival, upon which he starts repairing it immediately. If the system fails during the warranty period, then all the repairs are

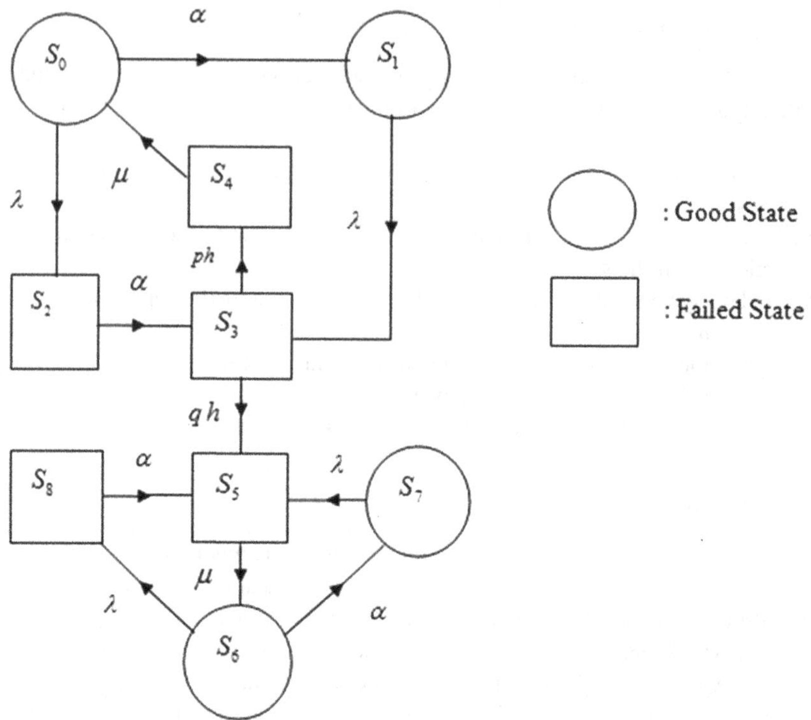

FIGURE 2.1 State–transition diagram of the system.

cost-free to the users; otherwise the repair charges are borne by the users. However, when the failed machine has been repaired, the repairman will go for a vacation again. It has been assumed that the failure, inspection, repair time of a machine, and arrival time of the repairman after his vacations follow a negative exponential distribution (Figure 2.1).

2.4.1 Formulation of the Mathematical Model

Based on the above description, we can formulate the difference-differential equations by using the probabilistic arguments of each state of the system and these are summarized as follows:

$$\left[\frac{d}{dt}+\alpha+\lambda\right]p_0(t)=\mu p_4(t) \tag{2.1}$$

$$\left[\frac{d}{dt}+\lambda\right]p_1(t)=\alpha p_0(t) \tag{2.2}$$

$$\left[\frac{d}{dt}+\alpha\right]p_2(t)=\lambda p_0(t) \tag{2.3}$$

$$\left[\frac{d}{dt}+h\right]p_3(t)=\alpha p_2(t)+\lambda p_1(t) \tag{2.4}$$

$$\left[\frac{d}{dt}+\mu\right]p_4(t)=php_3(t) \tag{2.5}$$

$$\left[\frac{d}{dt}+\mu\right]p_5(t)=qhp_3(t)+\lambda p_7(t)+\alpha p_8(t) \tag{2.6}$$

$$\left[\frac{d}{dt}+\alpha+\lambda\right]p_6(t)=\mu p_5(t) \tag{2.7}$$

$$\left[\frac{d}{dt}+\lambda\right]p_7(t)=\alpha p_6(t) \tag{2.8}$$

$$\left[\frac{d}{dt}+\alpha\right]p_8(t)=\lambda p_6(t) \tag{2.9}$$

The initial conditions are:

$$p_i(0)=\begin{cases}1 & ; i=0\\ 0 & ; i\neq 0\end{cases} \tag{2.10}$$

2.4.2 Solution of the Equations

In order to solve the above-formulated Equations 2.1–2.9, we use the Laplace transforms corresponding to the initial conditions given in Equation 2.10 and we get

$$(s+\lambda+\alpha)p_0(s)=1+\mu p_4(s) \tag{2.11}$$

$$(s+\lambda)p_1(s)=\alpha p_0(s) \tag{2.12}$$

$$(s+\alpha)p_2(s)=\lambda p_0(s) \tag{2.13}$$

$$(s+h)p_3(s)=\alpha p_2(s)+\lambda p_1(s) \tag{2.14}$$

$$(s+\mu)p_4(s)=php_3(s) \tag{2.15}$$

$$(s+\mu)p_5(s)=qhp_3(s)+\lambda p_7(s)+\alpha p_8(s) \tag{2.16}$$

$$(s+\lambda+\alpha)p_6(s) = \mu p_5(s) \tag{2.17}$$

$$(s+\lambda)p_7(s) = \alpha p_6(s) \tag{2.18}$$

$$(s+\alpha)p_8(s) = \lambda p_6(s) \tag{2.19}$$

From Equation 2.12, we get

$$p_1(s) = \frac{\alpha}{(s+\lambda)} p_0(s) \tag{2.20}$$

From Equation 2.13, we get

$$p_2(s) = \frac{\lambda}{(s+\alpha)} p_0(s) \tag{2.21}$$

Using Equations 2.20 and 2.21 in Equation 2.14, we get

$$p_3(s) = A(s)p_0(s) \tag{2.22}$$

where

$$A(s) = \frac{\lambda\alpha(2s+\lambda+\alpha)}{(s+\alpha)(s+\lambda)(s+h)} \tag{2.23}$$

Using Equation 2.22 in Equation 2.15, we get

$$p_4(s) = \frac{ph}{(s+\mu)} A(s)p_0(s) \tag{2.24}$$

Similarly, using Equation 2.24 in Equation 2.11, we get

$$p_0(s) = \frac{1}{M(s)} \tag{2.25}$$

where

$$M(s) = \frac{(s+\lambda+\alpha)(s+\mu) - ph\mu A(s)}{(s+\mu)} \tag{2.26}$$

From Equation 2.17, we get

$$p_6(s) = \frac{\mu}{(s+\alpha+\lambda)} p_5(s) \tag{2.27}$$

Using Equation 2.27 in Equation 2.18, we get

$$p_7(s) = B(s)p_5(s) \tag{2.28}$$

where

$$B(s) = \frac{\alpha\mu}{(s+\lambda)(s+\lambda+\alpha)} \tag{2.29}$$

Similarly, using Equation 2.27 in Equation 2.19, we get

$$p_8(s) = C(s)p_5(s) \tag{2.30}$$

where

$$C(s) = \frac{\lambda\mu}{(s+\alpha)(s+\lambda+\alpha)} \tag{2.31}$$

Using Equations 2.22, 2.28, 2.29 and 2.30 in Equation 2.16, we get

$$p_5(s) = D(s)p_0(s) \tag{2.32}$$

where

$$D(s) = \frac{qhA(s)}{\left[(s+\mu)-\lambda B(s)-\alpha C(s)\right]} \tag{2.33}$$

Using Equation 2.32 in Equations 2.27, 2.28, and 2.30, we get

$$p_6(s) = \frac{\mu D(s)}{(s+\alpha+\lambda)}p_0(s) \tag{2.34}$$

$$p_7(s) = B(s)D(s)p_0(s) \tag{2.35}$$

$$p_8(s) = C(s)D(s)p_0(s) \tag{2.36}$$

It is worth noting that

$$p_0(s)+p_1(s)+p_2(s)+p_3(s)+p_4(s)+p_5(s)+p_6(s)+p_7(s)+p_8(s) = \frac{1}{s}. \tag{2.37}$$

2.4.3 Reliability of the System *R*(*t*)

Reliability *R*(*t*) is the probability that the system functions well in a specified period of time (Ebeling, 2001; Garg et al., 2014; Garg, 2016). Using the method similar to that in Subsection 2.4.1, the difference-differential equations for reliability are:

$$\left[\frac{d}{dt}+\alpha+\lambda\right]p_0(t) = 0 \tag{2.38}$$

$$\left[\frac{d}{dt}+\lambda\right]p_1(t)=\alpha\, p_0(t) \tag{2.39}$$

Using the initial conditions and taking the Laplace transforms of the above equations, the solution can be written as

$$p_0(s)=\frac{1}{(s+\lambda+\alpha)} \tag{2.40}$$

$$p_1(s)=\frac{\alpha}{(s+\lambda+\alpha)(s+\lambda)} \tag{2.41}$$

$$R(s)=p_0(s)+p_1(s)=\frac{1}{(s+\lambda+\alpha)}+\frac{\alpha}{(s+\alpha)(s+\lambda+\alpha)} \tag{2.42}$$

Taking the inverse Laplace transforms, we get

$$R(t)=\exp(-\lambda t) \tag{2.43}$$

Now, based on Equation 2.43, the Mean Time to System Failure (MTSF) is defined as:

$$\mathrm{MTSF}=\int_0^\infty \mathrm{R}(\mathrm{t})\mathrm{dt}$$

$$\mathrm{MTSF}=\frac{1}{(\lambda)}. \tag{2.44}$$

2.4.4 Availability of the System $Av(t)$

Availability $A_v(t)$ is the probability that the system is operating satisfactorily at time t. By using Equations 2.20, 2.25, 2.26, 2.34, and 2.35, the Laplace transforms of $Av(t)$ at time t is as follows:

$$A_v(s)=p_0(s)+p_1(s)+p_6(s)+p_7(s)$$

$$=\frac{\left(s^4+a_3s^3+a_2s^2+a_1s+a_0\right)\left(s^5+c_4s^4+c_3s^3+c_2s^2+c_1s+c_0\right)}{\left(s^5+b_4s^4+b_3s^3+b_2s^2+b_1s+b_0\right)\left(s^5+d_4s^4+d_3s^3+d_2s^2+d_1s+d_0\right)} \tag{2.45}$$

where

$a_3=(\mu+\lambda+2\alpha+h)$, $a_2=\left[(h+\alpha)(\mu+\lambda+\alpha)+\lambda\mu+\alpha\mu\right]$,

$a_1=\left[\left(\mu h\alpha+h\lambda\alpha+h\alpha^2+h\alpha\right)+\left(h\lambda\mu+\alpha h\mu+\lambda\alpha\mu+\alpha^2\mu\right)\right]$, $a_0=\left(h\lambda\mu\alpha+h\mu\alpha^2\right)$,

$b_4=(a_3+\lambda)$, $b_3=(a_2+a_3\lambda)$, $b_2=(a_1+a_2\lambda)$, $b_1=(a_0+a_1\lambda-2\mu\lambda h\alpha p)$,

$b_0=\left(a_0\lambda-\mu\lambda^2h\alpha p-\mu h\lambda\alpha^2p\right)$, $c_4=b_4$, $c_3=b_3$, $c_2=(b_2-2\lambda\alpha\mu)$,

$c_1=\left(b_1-\lambda\alpha\mu(\lambda+\alpha)\right)$, $c_0=b_0$, $d_4=b_4$, $d_3=b_3$, $d_2=(b_2-2\lambda\alpha\mu)$,

$d_1=\left(a_0+a_1\lambda-\lambda\alpha\mu(\lambda+\alpha)-2\lambda\alpha\mu h\right)$ and $d_0=\left(a_0\lambda-h\lambda\alpha\mu(\lambda+\alpha)\right)$

Taking the Laplace transform of Equation 2.45, we get

$$A_v(t)=\sum_{i=1}^{10}A_i\exp(s_it) \tag{2.46}$$

where

$$A_i=\frac{\left(s_i^4+a_3s_i^3+a_2s_i^2+a_1s_i+a_0\right)\left(s_i^5+c_4s_i^4+c_3s_i^3+c_2s_i^2+c_1s_i+c_0\right)}{\prod_{j\neq i=1}^{10}(s_i-s_j)};\quad i=1,2,\ldots,10$$

$s_1, s_2,\ldots,s_5$ are the roots of the Equation $\left(s^5+b_4s^4+b_3s^3+b_2s^2+b_1s+b_0\right)=0$ and

$s_6, s_7,\ldots,s_{10}$ are the roots of the Equation $\left(s^5+d_4s^4+d_3s^3+d_2s^2+d_1s+d_0\right)=0$

2.4.5 Busy Period of the Repairman During the Warranty Period

Since $p_3(t)$ and $p_4(t)$ are the probabilities that the system is in failed state during the warranty period, the repairman remains busy in repairing the failed machine. Suppose that the warranty period of the system is (0, w], the busy period of the repairman during the warranty period $B_1(t)$, in the interval (0, w] is therefore given by

$$B_1(t)=\int_0^w\left(p_3(t)+p_4(t)\right)dt \tag{2.47}$$

Using Equations 2.22, 2.23, 2.24, and 2.25, we get

$$p_3(s)+p_4(s)=\frac{\lambda\alpha(2s+\lambda+\alpha)(s+\mu+ph)}{\left(s^5+b_4s^4+b_3s^3+b_2s^2+b_1s+b_0\right)} \tag{2.48}$$

Taking the inverse Laplace transforms of the above Equation 2.48, we get

$$p_3(t)+p_4(t)=\sum_{i=1}^{5} B_i \exp(s_i t) \tag{2.49}$$

where

$$B_i=\frac{\left[\lambda\alpha\left(2s_i+\lambda+\alpha\right)\left(s_i+\mu+ph\right)\right]}{\prod_{j\neq i=1}^{5}\left(s_i-s_j\right)};\quad i=1,2,\ldots,5$$

$s_1, s_2,\ldots, s_5$ are the roots of the Equation $\left(s^5+b_4s^4+b_3s^3+b_2s^2+b_1s+b_0\right)=0$

Using Equation 2.49 in Equation 2.47, we get

$$B_1(t)=\int_0^w\left(\sum_{i=1}^{5} B_i \exp(s_i t)\right)dt=\sum_{i=1}^{5}\left(\frac{B_i \exp(s_i w-1)}{s_i}\right) \tag{2.50}$$

2.4.6 Busy Period of the Repairman Beyond the Warranty Period

If the warranty period of the system is (0, w], then beyond the warranty period, the repairman remains busy for time (t – w) during the interval $\left(w,t\right]$ (Niwas, 2018); therefore, the busy period of the repairman beyond the warranty period $B_2(t)$, in the interval $\left(w,t\right]$ is given by

$$B_2(t)=\int_w^t p_5(t)dt \tag{2.51}$$

Using Equations 2.25, 2.26, 2.32, and 2.33, we get

$$p_5(s)=\frac{\lambda qh\alpha\left(2s+\lambda+\alpha\right)\left(s^5+e_4s^4+e_3s^3+e_2s^2+e_1s+e_0\right)}{\left(s^5+b_4s^4+b_3s^3+b_2s^2+b_1s+b_0\right)\left(s^5+d_4s^4+d_3s^3+d_2s^2+d_1s+d_0\right)} \tag{2.52}$$

where

$$e_4=b_4,\ e_3=b_3,\ e_2=b_2,\ e_1=\left(a_0+a_1\lambda\right)\text{ and } e_0=a_0\lambda.$$

Taking the inverse Laplace transforms of the Equation 2.52, we get

$$p_5(t)=\sum_{i=1}^{10} C_i \exp(s_i t) \tag{2.53}$$

where

$$C_i = \frac{\left[\lambda qh\alpha\left(2s_i+\lambda+\alpha\right)\left(s_i^5+e_4s_i^4+e_3s_i^3+e_2s_i^2+e_1s_i+e_0\right)\right]}{\prod_{j\neq i=1}^{10}\left(s_i-s_j\right)};\quad i=1,2,\ldots,10$$

$s_1, s_2,\ldots,s_5$ are the roots of the Equation $\left(s^5+b_4s^4+b_3s^3+b_2s^2+b_1s+b_0\right)=0$ and

$s_6, s_7,\ldots,s_{10}$ are the roots of the Equation $\left(s^5+e_4s^4+e_3s^3+e_2s^2+e_1s+e_0\right)=0$

Using Equation 2.53 in Equation 2.51, we get

$$B_2(t)=\int_w^t\left(\sum_{i=1}^{10}C_i\exp(s_it)\right)dt=\sum_{i=1}^{10}\left(\frac{C_i\left\{\exp(s_it)-\exp(s_iw)\right\}}{s_i}\right). \tag{2.54}$$

2.4.7 Vacation Time of the Repairman During the Warranty Period

Since $p_0(t)$, the probability is that the system is in state 0, where the repairman is on his vacation during the warranty period; therefore, the vacation time of the repairman during the warranty period $V_1(t)$ in the interval (0, w] is given by

$$V_1(t)=\int_0^w p_0(t)dt \tag{2.55}$$

From Equations 2.25 and 2.26, we get

$$p_0(s)=\frac{\left(s^4+f_3s^3+f_2s^2+f_1s+f_0\right)}{\left(s^5+b_4s^4+b_3s^3+b_2s^2+b_1s+b_0\right)} \tag{2.56}$$

where

$$f_3=\left(\alpha+\mu+h+\lambda\right),\quad f_2=\left(\left(\alpha+\mu\right)\left(\lambda+h\right)+\alpha\mu+\lambda h\right),$$

$$f_1=\left(\left(\lambda+h\right)\alpha\mu+\left(\alpha+\mu\right)\lambda h\right),\quad f_0=\left(\lambda h\alpha\mu\right)$$

Taking the inverse Laplace transforms of Equation 2.56, we get

$$p_0(t)=\sum_{i=1}^{5}D_i\exp\left(s_it\right) \tag{2.57}$$

where

$$D_i=\frac{\left(s_i^4+f_3s_i^3+f_2s_i^2+f_1s_i+f_0\right)}{\prod_{j\neq i=1}^{5}\left(s_i-s_j\right)};\quad i=1,2,\ldots,5$$

$s_1, s_2,\ldots,s_5$ are the roots of the Equation $\left(s^5 + b_4 s^4 + b_3 s^3 + b_2 s^2 + b_1 s + b_0\right) = 0$

Using Equation 2.57 in Equation 2.55, we get

$$V_1(t) = \sum_{i=1}^{5} \frac{D_i\left(\exp\left(s_i w - 1\right)\right)}{s_i} \tag{2.58}$$

2.4.8 Vacation Time of the Repairman Beyond the Warranty Period

Since $p_6(t)$, the probability is that the system is in state 6, where the repairman is on his vacation beyond the warranty period; therefore, the vacation time of the repairman beyond the warranty period $V_2(t)$ in the interval (w, t) is given by

$$V_2(t) = \int_w^t p_6(t)dt \tag{2.59}$$

From Equations 2.25, 2.26, 2.33, and 2.34, we get

$$p_6(s) = \frac{\mu q h \alpha \lambda\left(2s + \lambda + \alpha\right)\left(s^4 + f_3 s^3 + f_2 s^2 + f_1 s + f_0\right)}{\left(s^5 + b_4 s^4 + b_3 s^3 + b_2 s^2 + b_1 s + b_0\right)\left(s^5 + d_4 s^4 + d_3 s^3 + d_2 s^2 + d_1 s + d_0\right)} \tag{2.60}$$

Taking the inverse Laplace transforms of Equation 2.60, we get

$$p_6(t) = \sum_{i=1}^{10} E_i \exp\left(s_i t\right) \tag{2.61}$$

where

$$E_i = \frac{\mu q h \alpha \lambda\left(2s_i + \lambda + \alpha\right)\left(s_i^4 + f_3 s_i^3 + f_2 s_i^2 + f_1 s_i + f_0\right)}{\prod_{j \neq i=1}^{10}\left(s_i - s_j\right)}; \quad i = 1,2,3,\ldots,10$$

$s_1, s_2,\ldots,s_5$ are the roots of the Equation $\left(s^5 + b_4 s^4 + b_3 s^3 + b_2 s^2 + b_1 s + b_0\right) = 0$ and

$s_6, s_7,\ldots,s_{10}$ are the roots of the Equation $\left(s^5 + d_4 s^4 + d_3 s^3 + d_2 s^2 + d_1 s + d_0\right) = 0$

Using Equation 2.61 in Equation 2.59, we get

$$V_2(t) = \int_w^t \left(\sum_{i=1}^{10} E_i \exp(s_i t)\right) dt = \sum_{i=1}^{10}\left(\frac{E_i\left\{\exp(s_i t) - \exp(s_i w)\right\}}{s_i}\right) \tag{2.62}$$

2.4.9 Idle Time of the Repairman During the Warranty Period

Since state 1 of the system represents the state of the machine running well, the repairman remains idle during the warranty period; therefore, the idle time of the repairman during the warranty period $I_1(t)$ during the interval $(0, w]$ is given by

$$I_1(t) = \int_0^w p_1(t)dt \tag{2.63}$$

From Equations 2.20, 2.25, and 2.26, we get

$$p_1(s) = \frac{\alpha\left(s^3 + g_2 s^2 + g_1 s + g_0\right)}{\left(s^5 + b_4 s^4 + b_3 s^3 + b_2 s^2 + b_1 s + b_0\right)} \tag{2.64}$$

where

$$g_2 = (\alpha + \mu + \lambda),\ g_1 = (\alpha\mu + \alpha h + \mu h) \text{ and } g_0 = (h\alpha\mu)$$

Taking the inverse Laplace transforms of Equation 2.64, we get

$$p_1(t) = \sum_{i=1}^{5} F_i \exp(s_i t) \tag{2.65}$$

where

$$F_i = \frac{\alpha\left(s_i^3 + g_2 s_i^2 + g_1 s_i + g_0\right)}{\prod_{j\neq i=1}^{5}(s_i - s_j)};\quad i = 1,2,\ldots,5$$

$s_1, s_2, \ldots, s_5$ are the roots of the Equation $\left(s^5 + b_4 s^4 + b_3 s^3 + b_2 s^2 + b_1 s + b_0\right) = 0$

Using Equation 2.65 in Equation 2.63, we get

$$I_1(t) = \int_0^w \left(\sum_{i=1}^{5} F_i \exp(s_i t)\right)dt = \sum_{i=1}^{5}\left(\frac{F_i \exp(s_i w - 1)}{s_i}\right). \tag{2.66}$$

2.4.10 Idle Time of the Repairman Beyond the Warranty Period

Since state 7 of the system represents the state of the machine running well, the repairman remains idle beyond the warranty period; therefore, the idle time of the repairman beyond the warranty period $I_2(t)$ during the interval $(w, t]$ is given by

$$I_2(t) = \int_w^t p_7(t)dt \tag{2.67}$$

Using Equations 2.25, 2.26, 2.29, 2.33, and 2.35, we get

$$p_7(s) = \frac{qh\alpha^2\lambda\mu\left(2s+\lambda+\mu\right)\left(s^3+g_2s^2+g_1s+g_0\right)}{\left(s^5+b_4s^4+b_3s^3+b_2s^2+b_1s+b_0\right)\left(s^5+d_4s^4+d_3s^3+d_2s^2+d_1s+d_0\right)} \tag{2.68}$$

Taking the inverse Laplace transforms of Equation 2.68, we get

$$p_7(t) = \sum_{i=1}^{10} G_i \exp\left(s_i t\right) \tag{2.69}$$

where

$$G_i = \frac{qh\alpha^2\lambda\mu\left(2s_i+\lambda+\mu\right)\left(s_i^3+g_2s_i^2+g_1s_i+g_0\right)}{\prod_{j\neq i=1}^{10}\left(s_i-s_j\right)}; \quad i=1,2,3,\ldots,10.$$

$s_1, s_2, \ldots, s_5$ are the roots of the Equation $\left(s^5+b_4s^4+b_3s^3+b_2s^2+b_1s+b_0\right)=0$ and

$s_6, s_7, \ldots, s_{10}$ are the roots of the Equation $\left(s^5+d_4s^4+d_3s^3+d_2s^2+d_1s+d_0\right)=0$

Using Equation 2.69 in Equation 2.67, we get

$$I_2(t) = \int_w^t \left(\sum_{i=1}^{10} G_i \exp\left(s_i t\right)\right)dt = \sum_{i=1}^{10}\left(\frac{G_i\left\{\exp(s_i t)-\exp(s_i w)\right\}}{s_i}\right). \tag{2.70}$$

2.4.11 Profit Analysis of the User

Let K_1 be the revenue per unit time and K_2 be the repair cost per unit time, respectively, then the expected profit $E_p\left(t\right)$ during the interval $\left(0,t\right]$ is given by (Niwas, 2018):

$$E_p\left(t\right) = K_1\int_0^t A_v\left(t\right)dt - K_2\left(t-w\right) \tag{2.71}$$

By using Equation 2.46 and after solving, we get

$$E_p\left(t\right) = K_1\left(\sum_{i=1}^{10}\frac{A_i\left(\exp\left(s_i t\right)-1\right)}{s_i}\right) - K_2\left(t-w\right) \tag{2.72}$$

2.5 NUMERICAL RESULTS

2.5.1 Interpretation of the Numerical Results

To analyze the behavior of the system, we conducted an analysis where we varied the values of the parameters such as failure rate (λ), arrival rate of the repairman after his vacation (α), and the repair cost (K_2) with respect to time (t). Based on this, the values of reliability and profit of the system are computed and depicted in Tables 2.1 and 2.2, respectively. Table 2.1 shows that as we increase the failure rate (λ) from 0.1 to 0.2 and then further to 0.3 at a particular time, say 4 unit, then the reliability of the system decreases by 32.96% every time. However, the complete variation of reliability with λ is summarized in Figure 2.2.

On the other hand, Table 2.2 depicts the behavior of the various parameters on the expected profit $\left(E_p(t)\right)$ and indicates that if we decrease the repair cost (K_2) from 200 to 100 by fixing the other parameters, then the expected profit $\left(E_p(t)\right)$ at a particular time, say 4 unit, increases from 55660.91 to 56010.91.

TABLE 2.1
Effect of the Failure Rate (λ) on the Reliability of the System $R(t)$

Time(t)	$R(t)$ $\lambda = 0.1$	$R(t)$ $\lambda = 0.2$	$R(t)$ $\lambda = 0.3$
1	0.904837	0.818731	0.740818
2	0.818731	0.67032	0.548812
3	0.740818	0.548812	0.40657
4	0.67032	0.449329	0.301194
5	0.606531	0.367879	0.22313
6	0.548812	0.301194	0.165299
7	0.496585	0.246597	0.122456

TABLE 2.2
Effect of the Repair Cost (K_2) and Arrival Rate of Repairman (α) on the Expected Profit $E_p(t)$

	$\lambda = 0.1, \mu = 0.7, \alpha = 0.8, h = 0.9, p = q = 0.5, K_1 = 1000$	$\lambda = 0.1, \mu = 0.7, \alpha = 0.8, h = 0.9, p = q = 0.5, K_1 = 1000$	$\lambda = 0.1, \mu = 0.7, h = 0.9, K_2 = 200, p = q = 0.5, K_1 = 1000$
Time (t)	$E_p(t)$ $K_2 = 200$	$E_p(t)$ $K_2 = 100$	$E_p(t)$ $\alpha = 0.9$
1	1436.647	1486.647	1303.232
2	6253.559	6403.559	5605.296
3	20294.29	20544.29	18183.58
4	55660.91	56010.91	49985.9
5	137618	138068	124069
6	317616.4	318166.4	287809.7
7	698502.8	699152.8	636787.8

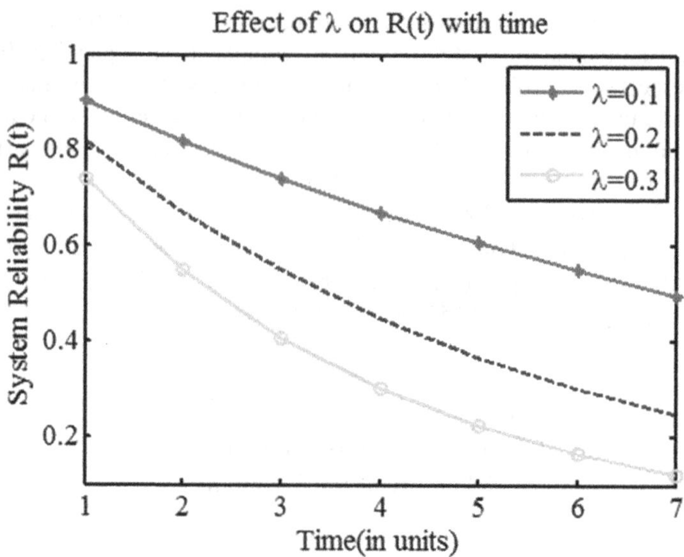

FIGURE 2.2 Effect of 'λ' on system reliability.

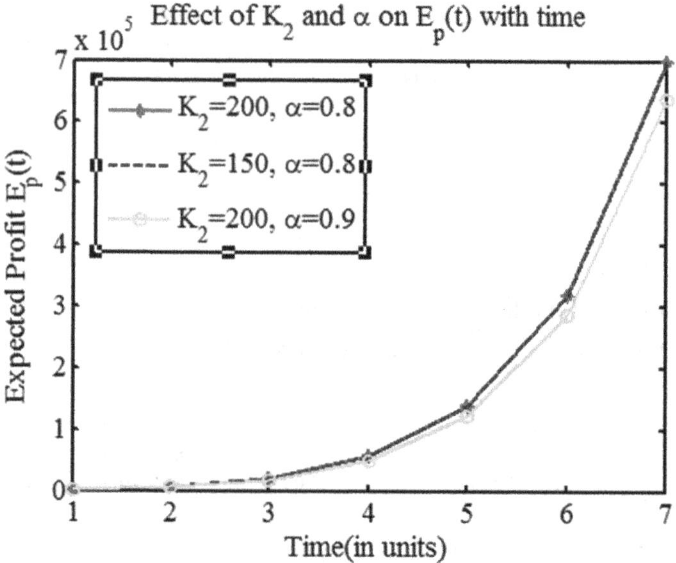

FIGURE 2.3 Effect of K_2 and 'α' on the expected profit.

Similarly, if we increase the arrival rate of the repairman after his vacation (α) from 0.8 to 0.9, then the expected profit $\left(E_p(t)\right)$ at a particular time, say 4 unit, decreases from 55660.91 to 49985.9. The complete variation of the expected profit is summarized in Figure 2.3. Thus, the different parameters have shown their effect on the system, which will be beneficial for the system analyst to increase the productivity of the system by adopting the necessary actions.

2.6 CONCLUSIONS

In the present chapter, we have analyzed the reliability and profit of a single unit system under the cost-free warranty policy with the repairman taking multiple vacations. In this analysis, the unit is repaired free of cost to the users when the failure occurs during the warranty period. However, the users will have to repair the failed unit at their own expenses beyond the warranty period. The repairman might take a sequence of vacations in the idle time. Also, the repairman can contribute to the economic benefit of the system directly or indirectly. Therefore, various expressions such as busyness, vacation, and idle times of the repairman within/beyond the warranty period have been derived, which can affect the performance of the system. Further, the effect of the various parameters on the reliability of the system and the expected profit has been analyzed and found that by varying the repair cost (K_2) and the arrival rate of the repairman after his vacation (α), the expected profit is increased. Based on this observation, the system analyst may focus on the (K_2) and (α) parameters so as to increase the performance and productivity of the system. In Niwas and Grag (2018), a single-unit repairable system with a warranty for repair is analyzed in detail. This study assumes that repair of any failure during the warranty period is cost free to the users, that the users will have to repair the failed unit at their own expenses beyond the warranty, and that the system has a working period followed by a rest period. Single repair facility is always available with the system for its functioning. Also, within the warranty period, the repairman inspects the failed unit to check whether the system is under a warranty policy or not, and during the rest period no unit can fail, but a failed unit can be repaired. To carry out a cost-benefit analysis, the expressions for some measures of system performance such as reliability, MTSF, and availability are derived using the Markov process. The current research model consists of a single-unit repairable system in which there is a single repairman who can take multiple vacations. We not only obtained the busy, vacation, and idle times of the repairman within/beyond the warranty period but also analyzed the reliability and profit function on the basis of various parameters and costs. The current model does not consider the system with periods of working and rest, but the reliability model was built for a single-unit system with a cost-free warranty policy with the repairman taking multiple vacations. In the study of Niwas and Grag (2018), when $a = b = 0$, i.e., the system does not go into rest. Also, in the current model, if the repairman always remains with the system and does not go for vacations, then the results agree with the results of Niwas and Grag's study (2018). The main objective of our study is to investigate the influence of the repairman's vacations or the arrival rate of the repairman after his vacation on the profit function as well as on the performance of the system, which will be economically beneficial for the users.

Hence, the study reveals that a single-unit repairable system with a cost-free warranty policy in which the repairman goes for vacations in his idle time for doing other assigned tasks will be economically beneficial. So, our studied model is more economically viable and advance than the existing models.

2.7 FUTURE RESEARCH DIRECTIONS

Although we have focused on a single-unit repairable system with a cost-free warranty policy and the repairman taking multiple vacations, the general idea presented here can also be applicable to many systems such as the series system, parallel system, standby system, k-out-of-n system, multi-state system, and so on. Also, the presented methodology can be further extended and improved using other optimization tools, parameters, and techniques. Moreover, we can also extend our research in the future for analyzing the reliability and performance of multi-state repairable systems in which the repairman does not always remain with the system and repairs are not perfect.

REFERENCES

Alqahtani A. Y., Gupta S. M. "Warranty as a marketing strategy for remanufactured products", *Journal of Cleaner Production*, 161, 1294–1307 (2017).

Bai J., Pham H. "Discounted warranty cost for minimally repaired series systems", *IEEE Transactions on Reliability*, 53(1), 37–42 (2004).

Blischke W. R. "Mathematical models for analysis of warranty policies", *Mathematical and Computer Modelling*, 13(7), 1–16 (1990).

Bura G. S. "Transient solution of an queue with catastrophes", *Communication in Statistics-Theory and Methods*, 48(14), 1–12 (2018).

Bura G. S., Gupta S. "Time dependent analysis of an queue with catastrophes", *Journal of Reliability: Theory and Applications*, 14, 79–86 (2019).

Chao X., Zhao Y. "Analysis of multi-server queues with station and server vacations", *European Journal of Operational Research*, 110(2), 392–406 (1998).

Chukova S., Johnston M. R. "Two-dimensional warranty repair strategy based on minimal and complete repairs", *Mathematical and Computer Modelling*, 44(11–12), 1133–1143 (2006).

Chun Y. H. "Optimal number of periodic preventive maintenance operations under warranty", *Reliability Engineering and Systems Safety*, 37(3), 223–225 (1992).

Dongliang Y., Hu Tao, Tong C. "Reliability analysis of series system with a repairman taking single vacation", in *29th Chinese Control and Decision Conference (CCDC)* May 28–30, Chongqing, China, 3818–3822, (2017).

Ebeling C. *An Introduction to Reliability and Maintainability Engineering*. Tata McGraw-Hill Company Ltd, New York, (2001).

Garg H. "Reliability, availability and maintainability analysis of industrial systems using PSO and fuzzy methodology", *MAPAN- Journal of Metrology Society of India*, 29(2), 115–129 (2014).

Garg H. "A novel approach for analyzing the reliability of series-parallel system using credibility theory and different types of intuitionistic fuzzy numbers", *Journal of Brazilian Society of Mechanical Sciences and Engineering*, 38(3), 1021–1035 (2016).

Garg H., Rani M., Sharma S. P. "Preventive maintenance scheduling of the pulping unit in a paper plant", *Japan Journal of Industrial and Applied Mathematics*, 30(2), 397–414 (2013).

Garg H., Rani M., Sharma S. P., Vishwakarma Y. "Bi-objective optimization of the reliability-redundancy allocation problem for series-parallel system", *Journal of Manufacturing Systems*, 33(3), 335–347 (2014).

He G., Wu W., Zhang Y. "Analysis of a multi-component system with failure dependency, N-policy and vacations", *Operations Research Perspectives*, 5, 191–198 (2018).

Hu L., Yue D., Li J. "Probabilistic analysis of a series-parallel repairable system with three units and vacation", *Applied Mathematical Modelling*, 34(10), 2711–2721 (2010).

Huang Y., Huang C., Ho J. "A customized two-dimensional extended warranty with preventive maintenance", *European Journal of Operational Research*, 257(3), 971–978 (2017).

Huang Y., Yen C. "A study of two-dimensional warranty policies with preventive maintenance", *IIE Transactions*, 41, 299–308 (2009).

Ja S., Kulkarni V., Mitra A., Partaker G. "A renewable minimal-repair warranty policy with time-dependent costs", *IEEE Transactions on Reliability*, 50(4), 346–352 (2001).

Jahromi A. E., Vahdani H. "Replacement-repair policy based on a simulation model for multi-state deteriorating products under warranty", *Transaction E: Industrial Engineering*, 16(1), 26–35 (2009).

Jain N. K., Bura G. S. " queue subject to modified binomially distributed catastrophic intensity with restoration time", *Journal of the Indian Statistical Association*, 49(2), 135–147 (2011).

Jia J., Wu S. "A replacement policy for a repairable system with its repairman having multiple vacations", *Computers and Industrial Engineering*, 57(1), 156–160 (2009).

Kadyan M. S., Ramniwas. "Cost benefit analysis of a single-unit system with warranty for repair", *Applied Mathematics and Computation*, 223, 346–353 (2013).

Ke J. C. "The optimal control of an M/G/1 queueing system with server startup and two vacation types", *Applied Mathematical Modelling*, 27(6), 437–450 (2003).

Ke J. C., Wang K. H. "Vacation policies for machine repair problem with two type spares", *Applied Mathematical Modelling*, 31(5), 880–894 (2007).

Ke J. C., Wu C. H., Liou C. H., Wang T. Y. "Cost analysis of a vacation machine repair model", *Procedia - Social and Behavioral Sciences*, 25, 246–256 (2011).

Kim H. G., Rao B. M. "Expected warranty cost of two-attribute free-replacement warranties based on a bivariate exponential distribution, *Computers and Industrial Engineering*, 38(4), 425–434 (2000).

Levy Y., Yechiali U. "Utilization of the idle time in an M/G/1 queue", *Management Science*, 22(2), 202–211 (1975).

Mo S., Zeng J., Xu W. "A new warranty policy based on a buyer's preventive maintenance investment", *Computers and Industrial Engineering*, 111, 433–444 (2017).

Monga A., Zuo M. J. "Optimal system design considering maintenance and warranty", *Computers and Operations Research*, 25(9), 691–705 (1998).

Murthy D. N. P. "A usage dependent model for warranty costing", *European Journal of Operational Research*, 57(1), 89–99 (1991).

Murthy D. N. P., Djamaludin I. "New product warranty: A literature review", *International Journal of Production Economics*, 79(3), 231–260 (2002).

Murthy D. N. P., Iskandar B. P., Wilson R. J. "Two dimensional failure-free warranty policies: Two dimensional point process models, *Operations Research*, 43(2), 356–366 (1995).

Niwas R. "Reliability analysis of a maintenance scheduling model under failure free warranty policy", *Journal of Reliability: Theory and Applications*, 13(3), 49–65 (2018).

Niwas R., Garg H. "An approach for analyzing the reliability and profit of an industrial system based on the cost free warranty policy", *Journal of Brazilian Society of Mechanical Sciences and Engineering*, 40(5), 1–9 (2018).

Niwas R., Kadyan M. S. "Stochastic analysis of a single-unit system with repairman having multiple vacations", *International Journal of Computers and Applications*, 8(1), 137–147 (2018).

Niwas R., Kadyan M. S., Kumar J. "Probabilistic analysis of two reliability models of a single-unit system with preventive maintenance beyond warranty and degradation", *Eksploatacja i Niezawodnosc-Maintenance and Reliability*, 17(4), 535–543 (2015).

Niwas R., Kadyan M. S., Kumar J. "MTSF (mean time to system failure) and profit analysis of a single-unit system with inspection for feasibility of repair beyond warranty", *International Journal of Systems Assurance Engineering and Management*, 7, 198–204 (2016).

Pan Y., Thomas M. U. "Repair and replacement decisions for warranted products under Markov deterioration", *IEEE Transactions on Reliability*, 59(2), 368–373 (2010).

Podolyakina N. "Estimation of the relationship between the products reliability, period of their warranty service and the value of the enterprise cost", *Procedia Engineering*, 178, 558–568 (2017).

Qiao J., Guo L. "A repairable system with warning device and repairman vacation", in *Proceedings of the 27th Chinese Control and Decision Conference (CCDC)*, 23–25 May, Qingdao, China. 3941–3947, (2015).

Rahman A. "Modeling analysis of reliability and costs for lifetime warranty and service contract policies", Ph. D. Thesis, Queensland University of Technology, Australia (2007).

Sahin I., Polatoglu H. "Maintenance strategies following the expiration of warranty", *IEEE Transactions on Reliability*, 45(2), 221–228 (1996).

Su B. H., Shi D. H. "Reliability analysis of n-unit series systems with multiple vacations of a repairman", *Mathematical Statistics and Applied Probability*, , 10, 78–82 (1995).

Sun H., Ma L. "Modeling of a deteriorating system with single vacation and reliability analysis", in *Proceedings of the 33rd Chinese Control Conference*, July 28–30, Nanjing, Jiangsu Province, China 3141–3146, (2014).

Takagi H. *Queueing Analysis: A Foundation of Performance Evaluation Vacation and Priority Systems*, Elsevier Science Ltd, North-Holland, Amsterdam, (1991).

Tian N., Zhang Z. G. *Vacation Queueing Models-Theory and Applications.* Springer, New York, (2006).

Vahdani H., Chukova S., Mahlooji H. "On optimal replacement-repair policy for multi-state deteriorating products under renewing free replacement warranty", *Computers and Mathematics with Applications*, 61(4), 840–850 (2011).

Varnosafaderani S., Chukova S. "A two-dimensional warranty servicing strategy based on reduction in product failure intensity", *Computers and Mathematics with Applications*, 63(1), 201–213 (2012).

Wu W. Q., Tang Y. H., Jiang Y. "Study on a k-out-of-n: G repairable system with multiple vacations and one replaceable repair facility", *System Engineering Theory and Practice*, 33(10), 2604–2614 (2013).

Yang D. Y., Tsao C. L. "Reliability and availability analysis of standby systems with working vacations and retrial of failed components", *Reliability Engineering and Systems Safety*, 182, 46–55 (2019).

Yuan L., Cui Z. D. "Reliability analysis for the consecutive-k-out-of-n: F system with repairmen taking multiple vacations", *Applied Mathematical Modelling*, 37(7), 4685–4697 (2013).

Yuan L., Xu J. "A deteriorating system with its repairman having multiple vacations", *Applied Mathematics and Computation*, 217(10), 4980–4989 (2011).

Zhang M. Y., Wu Y. F. "Reliability analysis of warm maintenance repairable system with continuous switch and multiple vacations for repairman", *Journal of Lanzhou University of Technology* , 41(5), 152–157 (2015).

3 A Bayesian Approach for Parameter Estimation of Ball Bearing Failure Data

Jitendra Kumar, Srikant Gupta, and Sachin Chaudhary

CONTENTS

3.1 INTRODUCTION

Rolling-element bearings are widely used in machinery and play a significant role in reliability engineering. Over the last decades, rolling-element bearings were referred to as anti-friction bearings, since they have much lower friction in comparison to sliding bearings. Many types of rolling-element bearings are available in a large variety of designs, which can be applied for most arrangements in machinery for supporting radial and thrust loads. It is well known that the rolling motion has lower friction when compared with that of sliding. In addition to friction, the rolling action causes much less wear in comparison to sliding. In most cases, rolling-element bearings require less maintenance than hydrodynamic bearings. The service life of a ball-bearing is the period that it works under actual operating conditions before it is replaced or fails. It depends on the friction between the outer and inner surfaces of the bearing and their operating environments. Their operational reliability is the basis for devising optimization and improvement strategies and implementing failure factor analysis, which directly relates to the operational safety of the product during service time. Failure analysis involves the inspection of the quality and investigation of the causes for the failure of the engineering equipment.

There are some primary failure causes such as design deficiency, material defects, manufacturing or installation defects, and operating life anomalies, but in this study we considered the failure time of the ball bearing equipment. Therefore, bearing manufacturers are engaged continuously in fatigue-testing operations in order to obtain information relating to fatigue life and load (pressure) as well as other related factors. Several manufacturers have recently pooled their test data in a cooperative effort to set up a uniform and standardized ball-bearing application formula, which would be benefited by many users in the field of anti-friction bearings. Lieblein and Zelen (1956) report on the ball bearing data of four companies named A, B, C, and D (number of test groups = 213 and the total number of bearings in the test groups = 4948). These tests were conducted by the four companies mentioned above. The authors also obtained the Weibull slope and analyzed the reliability measure for the ball bearing data. The study reported 23 bearing failure data out of 25 because the company representatives said that two failures were not part of the given experiment. Zaretsky (2013) discussed the fitting of the bearing life data for Weibull distribution and predict its characteristics. Dowson and Hamrock (1981) discussed some of the fundamental elements, features, and quality of ball bearing equipment in their NASA technical report. There are various types of ball bearing data, but in this study we have used the ball bearing data that was discussed by Lieblein and Zelen (1956). Caroni (2002a;2002b) discussed the modeling of ball bearing data. In the statistical literature, many authors have used the ball bearing data for lifetime analysis of ball bearings.

3.2 LITERATURE REVIEW AND METHODOLOGY

Generally, in reliability analysis life-testing experiments are performed to analyze the expected life of the experimental units/systems. Often some units are removed or lost during the execution of the experiment due to shortage of time and scarcity of funds or due to some unavoidable circumstances. In such situations, the censoring scheme is commonly used, and it is an essential feature of life-testing experiments. Various types of censoring schemes have been discussed in the statistical literature for life-testing experiments, namely time censoring (Type I), failure censoring (Type II), and progressive censoring. Numerous censoring schemes have been developed for lifetime experiments in the past decade [see Lawless (2003) and Sinha (1986) for a brief preview]. Here we consider the progressive type-II censoring scheme, which is very popular among academicians working in the field of reliability studies. Balakrishnan and Sandhu (1995) discussed the algorithm of type-II progressive sample generation from a lifetime distribution.Garg (2013) proposed the confidence interval-based fuzzy Lambda-Tau (CIBFLT) methodology and a methodology for analyzing the behavior of real-life complex repairable industrial systems. Garg et al. (2013)stochastically analyzed the behavior of an industrial system using imprecise, vague, and uncertain data. Since fuzzy numbers require the expert opinions on uncertainty, operating conditions, and imprecision in reliability information, they obtained repair time, failure rate, reliability, mean time between failures (MTBF), maintainability, and availability using the Weibull distribution [see more for fuzzy reliability Garg(2014a) and Garg (2014b)]. Singh et al. (2013) discussed the Bayesian estimation procedure for the generalized Lindley distribution under progressively type-II censored beta-binomial removals and calculated the expected total test time. Dey and Dey (2014) used progressively type-II censoring binomial removals for estimation and discussed the statistical inference for the Rayleigh distribution (Balakrishnan, 2007; Balakrishnan & Aggarwala, 2000; Balakrishnan & Cramer, 2014; Kumar et al., 2014; Shanker et al., 2015). The progressively type-II censoring method is described as follows. Let the random variable X denote the lifetime of a unit. Suppose that n identical units are put to test and non-negative integers $R_1, R_2, \ldots, R_m$ are also random variables and follow a binomial distribution with parameter p and satisfying $R_1 + R_2 + \ldots + R_m = n - m$. At the time of the first failure, R_1 units out of the remaining $n-1$ units are randomly removed. At the time of the second failure, R_2 units out of the remaining $n-2-R_1$ units are randomly removed, and so on. Finally, at the time of mth failure, the experiment is terminated by removing all the remaining $R_m = n - R_1 + R_2 + \ldots + R_{m-1} - m + 1$ units. Table 3.1 presents the data derived from the sampling procedure under progressively type-II censored beta-binomial removals for a lifetime experiment.

TABLE 3.1
Sampling Procedure for a Life-Test Under Progressive Type-II Censored Beta-Binomial Removals

Stage	Failed unit	Removed units	Survived unit
1	x_1	$r_1 = BB(n-m, \alpha, \beta)$	$n-1-r_1$
2	x_2	$r_2 = BB(n-m-r_1, \alpha, \beta)$	$n-2-r_1-r_2$
i	x_i	$r_i = BB(n-m-\sum_{j=1}^{i-1} r_j, \alpha, \beta)$	$n-i-\sum_{j=1}^{i} r_j$
m-1	x_{m-1}	$r_{m-1} = BB(n-m-\sum_{j=1}^{m-2} r_j, \alpha, \beta)$	$n-m+1-\sum_{j=1}^{m-1} r_j$
m	x_m	All remaining units are failed	0

3.3 LINDLEY DISTRIBUTION

In the statistical literature, there are many lifetime distributions used in reliability analysis, including exponential, normal, gamma, and Weibull distributions. The Weibull distribution is very popular for lifetime analysis, and Lieblein and Zelen (1956) used it to fit ball bearing data, but we introduced one parameter distribution known as the Lindley distribution, and it is used as an alternative model for existing statistical distributions. The Lindley distribution has several real applications where the data show the non-monotone shape for their hazard rates. The Lindley distribution belongs to an exponential family because it is a mixture of exponential and gamma distributions. The probability density function of the Lindley distribution (Ghitany et al., 2011) is

$$f(x;\theta) = \frac{\theta^2}{\theta+1}(x+1)\,e^{-\theta x}, \qquad \theta > 0, x > 0 \tag{3.1}$$

where θ is the shape parameter.

The Lindley distribution was first introduced by Lindley (1958) to analyze failure lifetime data, and he applied it to estimate the stress–strength reliability model. The model for the failure time data set has increasing, decreasing, unimodal, and bathtub-shaped hazard rates. The exponential distribution has a constant hazard rate, whereas the Lindley distribution has a monotonically increasing hazard rate. Ghitany et al. (2008) derived some statistical properties and the parameter estimation of the Lindley distribution and showed that the Lindley distribution is superior to exponential distribution. Ghitany et al. (2011) showed that the Lindley distribution is especially useful for modeling in mortality studies. Tomy (2018) discussed some of the various forms of the Lindley distribution such as quasi-Lindley, power Lindley, generalized Lindley, truncated Lindley, etc., and showed some statistical properties for all the given distributions. Mazucheli and Achcar (2011) proposed the Lindley distribution as a possible alternative to exponential and Weibull distributions. Aslam and Feroze (2019) obtained Bayes estimates of the parameters of a mixture of generalized exponential distributions using Lindley's approximation and the MCMC method with progressively censored multimodal data. Abdi et al. (2019) discussed some characteristics of the gamma and Lindley lifetime distributions and obtained the maximum likelihood estimate of the reliability and stress-strength parameters. Bai et al. (2019) used the EM and Gibbs algorithm to obtain the system reliability of the stress–strength model with progressively type-I censoring data from a finite mixture distribution. Nie and Gui (2019) discussed the parameter estimation of the Lindley distribution for competing risk data based on the progressively type-II censoring method with binomial distribution (Chaturvedi et al., 2018; Kumar et al., 2014; Valiollahi et al., 2018).

3.4 BASIC DEFINITION

Reliability Function: The reliability function is the probability that the time to failure of a unit is later than some specified time x_o. The reliability function must be non-increasing, i.e., $S(x_o+t) > S(x_o)$, if $x_o+t > x_o$, with $S(0) = 1$. This property follows directly from $S(x_o)$ being the integral of a non-negative function. The reliability function is usually assumed to approach zero as age increases without bound, i.e., $S(x_o) \to 0$ as $x_o \to \infty$.

$$S(x_o;\theta) = P[X > x_o] = \int_{x_o}^{\infty} f(x,\theta)\, dx.$$

The reliability function of the Lindley distribution at a given time $x_0 > 0$ is given by

$$S(x_o;\theta) = \frac{1+\theta+x_o\theta}{\theta+1}\, e^{-\theta x_o} \tag{3.2}$$

Hazard Rate: It may be noted that $h(x)dt$ represents the probability that a device which has survived up to time t will fail in the small interval of time x to $x+dx$. Let X be a continuous random variable with probability density function $f(x;\theta)$. Then the hazard rate or failure rate $h(x_o,\theta)$, at a time x_o, is defined to be

$$H(x_0:\theta) = \lim_{\Delta x_o \to 0} \frac{P\left(x_0 \le X \le x_0 + \Delta x_0 \mid X > x_0\right)}{\Delta x_0}$$

For the Lindley distribution, the hazard function at a given time $x_0 > 0$ is given by

$$h(x_o;\theta) = \frac{\theta^2}{1+\theta+x_o\theta}\,(1+x_o) \tag{3.3}$$

The failure rate $h(x_o;\theta)$, as a function of x_o, plays an important role in the choice of a lifetime distribution model. There are distributions which may have (i) a constant failure rate, (ii) an increasing failure rate, (iii) a decreasing failure rate, and (iv) a bath-tub shaped failure rate.

3.5 PROPOSED MODEL

Since we put n items in a life testing experiment and n-m items out of n are removed from the experiment by the progressively censoring scheme, the progressively censoring removals are the random variables and follow the beta-binomial distribution. Here, we assume that progressive removal $R_i = \left(r_1, r_2, \ldots, r_i\right)^i$ at ith failure, i.e., r_i units are removed from the experiment at follow-up times $t_i; i = 1,2,\ldots m,$ follows a binomial distribution with parameter $\left(n - m - \sum_{j=1}^{i} r_j\right)$ and p. At the first failure, $P(r_1)$ is defined as:

$$P\left(R_1 = r_1\right) = \binom{n-m}{r_1} p^{r_1}\left(1-p\right)^{n-m-r_1}, \qquad 0 \le r_1 \le n, \quad m \le n.$$

and

$$P(R_2 = r_2 \mid R_1 = r_1) = \binom{n-m-r_1}{r_2} p^{r_2} (1-p)^{n-m-r_1-r_2}, \qquad 0 \le r_2 \le n.$$

So, in general, the term for $i = 3,4,\ldots, m$ can be written as

$$P(R_i = r_i \mid R_{i-1} = r_{i-1}) = \binom{n-m-\sum_{j=1}^{i-1} r_j}{r_i} p^{r_j} (1-p)^{n-m-\sum_{j=1}^{i} r_j}, \text{ where } 0 \le r_i \le n-m-\sum_{j=1}^{i-1} r_j.$$

Further, as mentioned above, the binomial probability of removals is not fixed at every stage, and it follows the beta distribution having a probability density function as given below:

$$g(p \mid \alpha,\beta) = \frac{1}{\mathrm{B}(\alpha,\beta)} p^{\alpha-1}(1-p)^{\beta-1}, \ \alpha, \beta > 0, \ 0 > p > 1.$$

So the distribution of r_i is given by

$$P(R_i = r_i, \alpha, \beta) = \frac{1}{\mathrm{B}(\alpha,\beta)} \binom{n-m-\sum_{j=1}^{i-1} r_j}{r_i} \int_0^1 p^{r_j+\alpha-1}(1-p)^{n+\beta-m-\sum_{j=1}^{i-1} r_j-1}\, dp.$$

To solve the above equation, we get

$$P(R_i = r_i, \alpha, \beta) = \binom{n-m-\sum_{j=1}^{i-1} r_j}{r_i} \frac{\mathrm{B}\left(r_j+\alpha,\ n+\beta-m-\sum_{j=1}^{i-1} r_j\right)}{\mathrm{B}(\alpha,\beta)}$$

where $\mathrm{B}(\alpha,\beta) = \dfrac{\Gamma\alpha\Gamma\beta}{\Gamma(\alpha,\beta)}$ and $r_i = 0,1,\ldots\ldots,n-m-\sum_{j=1}^{i} r_j; i = 1,2,\ldots\ldots,(m-1)$. This is the probability mass function of the beta-binomial distribution.

The joint probability of $R_1 = r_1,\ R_2 = r_2,\ldots, R_m = r_m$ is given by

$$P(R = r,\alpha,\beta) = P(R_m = r_m \mid R_{m-1} = r_{m-1},\ldots R_1 = r_1) \times \ldots\ldots \times P(R_2 = r_2 \mid R_1 = r_1) \times P(R_1 = r_1)$$

Also, supposing further that R_i is independent of X_i for all i, then the joint likelihood function of the observed data is

$$L(\theta,\alpha,\beta,x) \propto L(\theta, x \mid R = r)\ P(R = r,\alpha,\beta) \tag{3.4}$$

Using Equations 3.1, 3.2, and 3.4, the likelihood function is as follows:

$$L(\theta,\alpha,\beta\mid x)\propto\frac{\theta^{2m}}{(1+\theta)^{m+\sum_{i=1}^{m}r_i}}\prod_{i=1}^{m}\left(1+\theta+\theta x_i\right)^{r_i}\left(1+x_i\right)\prod_{i=1}^{m-1}\binom{n-m-\sum_{j=1}^{i-1}r_j}{r_i} \tag{3.5}$$

$$\times\frac{\mathrm{B}\left(r_j+\alpha,\,n+\beta-m-\sum_{j=1}^{i-1}r_j\right)}{\mathrm{B}(\alpha,\,\beta)}$$

We rewrite the above likelihood function in the form

$$L(\theta,\alpha,\beta\mid x)\propto L_1(\theta,R\mid x)L_2(\alpha,\beta\mid R)$$

where

$$L_1(\theta\mid R,\,x)=\theta^{2m}(1+\theta)^{-\left(m+\sum_{i=1}^{m}r_i\right)}\prod_{i=1}^{m}\left(1+\theta+\theta x_i\right)^{r_i}\left(1+x_i\right)e^{-\theta\sum_{i=1}^{m}(r_i+1)x_i} \tag{3.6}$$

$$L_2(\alpha,\,\beta\mid R)=\prod_{i=1}^{m-1}\binom{n-m-\sum_{j=1}^{i-1}r_j}{r_i}\frac{\mathrm{B}\left(r_j+\alpha,\,n+\beta-m-\sum_{j=1}^{i-1}r_j\right)}{\mathrm{B}(\alpha,\,\beta)} \tag{3.7}$$

3.6 MAXIMUM LIKELIHOOD ESTIMATION

It may be noted that we have obtained the MLEs of the θ, β, α, density function, hazard function, and reliability function based on progressively type-II censored data with beta-binomial removals. We observe that the joint likelihood function is proportional to two functions, namely, $L_1(\theta\mid x,R)$, and $L_2(\alpha,\,\beta\mid R)$. $L_1(\theta\mid x,R)$ is a function of θ only, whereas $L_2(\alpha,\,\beta\mid R)$ involves α and β only. Therefore, the MLEs of θ can be derived by maximizing $L_1(\theta\mid x,R)$. According to the previous section, the maximum likelihood function of θ is obtained by maximizing the logarithm of the likelihood function $L_1(\theta\mid x,R)$, which is given in Equation 3.6. So, the log-likelihood functions of $L_1(\theta\mid x,R)$ and $L_2(\alpha,\,\beta\mid R)$ are given by

$$\log L_1\propto 2m\ln(\theta)-\left(m+\sum_{i=1}^{m}r_i\right)\ln(1+\theta)+\sum_{i=1}^{m}r_i\ln\left(1+\theta+\theta x_i\right)+\sum_{i=1}^{m}\left(1+x_i\right)-\theta\sum_{i=1}^{m}\left(r_i+1\right)x_i \tag{3.8}$$

and

$$Log\,L_2(\alpha,\,\beta\mid R)\propto\sum_{i=1}^{m-1}\ \log B\left(\alpha+r_i,\,n+\beta-m-\sum_{ji=1}^{i-1}r_j\right)-(m-1)\log B(\alpha,\,\beta) \tag{3.9}$$

To obtain the maximum likelihood estimate of the θ, β, and α, we differentiate Equations 3.8 and 3.9 partially with respect to the parameters θ, α, and β, respectively. Let $\hat{\theta}$, $\hat{\alpha}$, and $\hat{\beta}$ be the maximum likelihood estimates of θ, α, and β, and these are defined in the following equations, respectively,

$$\frac{\partial L_1(\theta, R, x)}{\partial \theta} = \frac{2m}{\theta} - \frac{m + \sum_{i=1}^{m} r_i}{1+\theta} - \sum_{i=1}^{m}(r_i+1)x_i + \sum_{i=1}^{m} \frac{r_i\,(1+x_i)}{(1+\theta+\theta\,x_i)} \tag{3.10}$$

$$\frac{\partial L_2(\alpha,\beta \mid R)}{\partial \alpha} = (m-1)\left[\psi(\alpha+\beta)-\psi(\alpha)\right] + \sum_{i=1}^{m-1}\psi(\alpha+r_i) - \sum_{i=1}^{m-1}\psi\left(\alpha+n+\beta-m-\sum_{ji=1}^{i} r_j\right) \tag{3.11}$$

$$\frac{\partial L_2(\alpha,\beta \mid R)}{\partial \beta} = (m-1)\left[\psi(\alpha+\beta)-\psi(\beta)\right] + \sum_{i=1}^{m-1}\psi\left(n+\beta-m-\sum_{ji=1}^{i-1} r_j\right) + \sum_{i=1}^{m-1}\psi\left(\alpha+n+\beta-m-\sum_{ji=1}^{i} r_j\right)$$

where

$$\psi(\zeta) = \frac{\partial}{\partial \zeta}\log(\Gamma\zeta) \text{ and } \psi(\zeta+c) = \sum_{l=0}^{c}\frac{1}{\zeta+l} + \frac{\partial}{\partial \zeta}\log(\Gamma\zeta). \tag{3.12}$$

In Equation 3.10, MLE of θ is not in closed form. So, we have used here the numerical method, say the Newton-Raphson method, for calculating the ML estimate. The MLE of α and β are obtained by optimizing the standard equations3.11 and 3.12. Note that the $r(t)$, $f(t)$, and $h(t)$ is a function of θ, and so by using the invariance property of MLE, the MLEs of $r(t)$, $f(t)$, and $h(t)$ are given by

$$\hat{r}(x_0) = \frac{1+\hat{\theta}+x_0\hat{\theta}}{\hat{\theta}+1}e^{-\hat{\theta}x_0},\ \hat{f}(x;\hat{\theta}) = \frac{\hat{\theta}^2}{\hat{\theta}+1}(x_0+1)e^{-\hat{\theta}x_0} \text{ and } \hat{h}(x_o;\hat{\theta}) = \frac{\theta^2}{1+\theta+x_o\theta}(1+x_o).$$

In the next section, we obtain the asymptotic confidence interval (ACI) for the parameters and the parametric function.

3.7 CONFIDENCE INTERVALS

3.7.1 Asymptotic Confidence Intervals

Since the MLEs of $\hat{\Theta} = \{\theta, \alpha, \beta\}$ obtained in Equations 3.10, 3.11, and 3.12 are not in closed form, it is not possible to derive their exact distributions. Thus, we evaluate the ACIs for the parameters using the asymptotic normality of MLEs. Therefore, we first obtain the approximate Fisher's information matrix given by

$$I(\hat{\Theta}) = \begin{bmatrix} \frac{\partial^2 L_1}{\partial \theta^2} & \frac{\partial^2 L_1}{\partial \theta\, \partial \alpha} & \frac{\partial^2 L_1}{\partial \theta\, \partial \beta} \\ \frac{\partial^2 L_2}{\partial \alpha\, \partial \theta} & \frac{\partial^2 L_2}{\partial \alpha^2} & \frac{\partial^2 L_2}{\partial \alpha\, \partial \beta} \\ \frac{\partial^2 L_2}{\partial \beta\, \partial \theta} & \frac{\partial^2 L_2}{\partial \beta\, \partial \alpha} & \frac{\partial^2 L_2}{\partial \beta^2} \end{bmatrix}\Bigg|_{(\theta,\alpha,\beta)=(\hat{\theta},\hat{\alpha},\hat{\beta})} \tag{3.13}$$

where $\frac{\partial^2 L_1}{\partial \theta^2} = -\frac{2m}{\theta^2} - \frac{m+\sum_{i=1}^{m} r_i}{(1+\theta)^2} - \sum_{i=1}^{m} \frac{r_i\,(1+x_i)^2}{(1+\theta+\theta x_i)^2}$, $\frac{\partial^2 L_1}{\partial \theta\, \partial \alpha} = \frac{\partial^2 L_1}{\partial \theta\, \partial \beta} = \frac{\partial^2 L_2}{\partial \alpha\, \partial \theta} = \frac{\partial^2 L_2}{\partial \beta\, \partial \theta} = 0$

$$\frac{\partial^2 L_2}{\partial \alpha^2} = (m-1)\left[\psi'(\alpha+\beta) - \psi'(\alpha)\right] + \sum_{i=1}^{m-1} \psi'(\alpha + r_i) - \sum_{i=1}^{m-1} \psi'\left(\alpha + n + \beta - m - \sum_{ji=1}^{i} r_j\right)$$

$$\frac{\partial^2 L_2}{\partial \alpha\, \partial \beta} = (m-1)\left[\psi'(\alpha+\beta)\right] - \sum_{i=1}^{m-1} \psi'\left(\alpha + n + \beta - m - \sum_{ji=1}^{i} r_j\right),$$

$$\frac{\partial L_2}{\partial \beta\, \partial \alpha} = -(m-1)\left[\psi'(\alpha+\beta)\right] - \sum_{i=1}^{m-1} \psi'\left(\alpha + n + \beta - m - \sum_{ji=1}^{i} r_j\right)$$

$$\frac{\partial^2 L_2}{\partial \beta^2} = -(m-1)\left[\psi'(\alpha+\beta) - \psi'(\beta)\right] + \sum_{i=1}^{m-1} \psi'\left(n + \beta - m - \sum_{ji=1}^{i-1} r_j\right)$$
$$+ \sum_{i=1}^{m-1} \psi'\left(\alpha + n + \beta - m - \sum_{ji=1}^{i} r_j\right)$$

where $\psi'(\zeta) = \frac{\partial}{\partial \zeta}\psi(\zeta)$. Thus, using Equation 3.13, we get the 100(1 – α)% confidence limits for Θ, which is given by $\hat{\Theta} \pm z_{\alpha/2}\, S.E.(\hat{\Theta})$, where, $z_{\alpha/2}$ is the upper 100(α/2)th percentile of the standard normal variate.

Now we use the delta method to obtain the asymptotic confidence interval of $r(t)$, $f(t)$, and $h(t)$. The delta method (Qehlert, 1992) allows a normal approximation for a continuous and differentiable function of a sequence of random variables that already has a normal limit in distribution. According to the delta method, the variance of $\gamma(\theta)$ is estimated by

$$\sqrt{n}\left[\gamma(X) - \gamma(\theta)\right] \to N\left(0,\ V(\theta)\left[\gamma'(\theta)\right]^2\right)$$

So the ACI of $\gamma(\theta)$ is obtained as follows:

$$\left[\hat{\gamma}(\theta) + z_{\alpha/2}\ S.E.\ (\hat{\gamma}(\theta)),\ \hat{\gamma}(\theta) - z_{\alpha/2}\ S.E.(\hat{\gamma}(\theta))\right].$$

3.7.2 Bootstrap Confidence Interval

Here, we have obtained the parametric bootstrap confidence intervals proposed by Efron and Tibshirani (1994). The bootstrap method is very useful when the sample observations may not be sufficiently large. In such cases, an assumption regarding the normality is invalid. Consequently, the ACI may not be an approximate choice. In what follows, we consider the bootstrap CI for the parameter based on the bootstrapping method. In order to obtain boot-p confidence intervals, the computational algorithm is given as follows.

i. Compute the MLE $\hat{\Theta}$ by using Equations 3.10, 3.11, and 3.12.
ii. Use $\hat{\Theta}$ to generate a failure censored sample $\left\{x_1^*, x_2^*, \ldots, x_m^*\right\}$ of size m from $f(x;\hat{\theta})$.
iii. Using the sample obtained in step (ii), compute the bootstrap estimate of Θ, say $\hat{\Theta}^*$.
iv. Repeat steps (ii–iii), B times, to get the set of bootstrap estimators $(\hat{\Theta}_j^*; j = 1,2,\ldots B)$.
v. Arrange $(\hat{\Theta}_j^*; j = 1,2,\ldots B)$ in ascending order and obtain $\left(\hat{\Theta}_1^*, \hat{\Theta}_2^*, \ldots, \hat{\Theta}_B^*\right)$.
vi. A two-sided $100(1-\alpha)\%$ boot-p confidence interval is given by,

$$\left(\hat{\Theta}_L^*, \hat{\Theta}_U^*\right) = \left(\hat{\Theta}^*_{[B(\alpha/2)]}, \hat{\Theta}^*_{[B(1-\alpha/2)]}\right)$$

where $[q]$ denotes the integrated part of q.

3.8 BAYESIAN ESTIMATION

In this section, we obtain the Bayes estimates of the parameters θ based on progressively type-II censored data with binomial removals. We must assume that the parameter θ is a random variable. The random variable θ has prior distribution PDFs.

$$q(\theta) = \frac{\kappa^\eta}{\Gamma\eta}\theta^{\eta-1}\, e^{-\kappa\theta}; \quad \eta, \kappa > 0, \quad \theta > 0. \tag{3.14}$$

It may be noted that the gamma prior of θ is $q_1(\theta)$, and $q_1(\theta)$ is flexible enough to cover a wide variety of the prior beliefs. Combining the likelihood function, $L(x,\theta)$, with the prior distribution of θ, we obtain the posterior density of θ for the given data as follows:

$$\Pi\left(\theta, p \mid x\right) = \frac{L(\theta, \alpha, \beta, x, R)\, q(\theta)}{\int_\Theta L(\theta, \alpha, \beta, x, R) q(\theta) d\Theta}$$

$$
=\frac{\dfrac{\theta^{2m+\eta-1}}{(1+\theta)^{m+\sum_{i=1}^{m} r_i}}\prod_{i=1}^{m}\left(1+\theta+\theta x_i\right)^{r_i} e^{-\theta\left(\kappa+\sum_{i=1}^{m}(r_i+1)x_i\right)}\prod_{i=1}^{m-1}\binom{n-m-\sum_{j=1}^{i-1} r_j}{r_i}\dfrac{\mathrm{B}\left(r_j+\alpha,\, n+\beta-m-\sum_{j=1}^{i-1} r_j\right)}{\mathrm{B}(\alpha,\,\beta)}}{\iiint_{\Theta}\dfrac{\theta^{2m+\eta-1}}{(1+\theta)^{m+\sum_{i=1}^{m} r_i}}\prod_{i=1}^{m}\left(1+\theta+\theta x_i\right)^{r_i} e^{-\theta\left(\kappa+\sum_{i=1}^{m}(r_i+1)x_i\right)}\prod_{i=1}^{m-1}\binom{n-m-\sum_{j=1}^{i-1} r_j}{r_i}\dfrac{\mathrm{B}\left(r_j+\alpha,\, n+\beta-m-\sum_{j=1}^{i-1} r_j\right)}{\mathrm{B}(\alpha,\,\beta)}\,d\Theta}
\tag{3.15}
$$

Here the denominator of the above equation is the normalizing constant of the posterior distribution. The Bayes estimate of ω (the function of θ) under the square error loss function can be obtained by

$$
E\left(\omega,\, \tilde{\omega}\right)=\in\left(\omega-\tilde{\omega}\right)^2, \qquad \in>0,
$$

where $\tilde{\omega}$ is the estimate of the parameter ω and the Bayes estimator $\tilde{\omega}$ of ω comes out to be $E\left[\omega\right]$, where $E\left[\ \omega\right]$ denotes the posterior expectation of ω. However, the squared error loss function is a symmetric loss function and can only be justified if overestimation and underestimation of equal magnitudes are of equal seriousness. Now the Bayes estimate of $\varphi(\theta)$ under SELF is given by

$$
\tilde{\xi}(\theta)=\int_{\Theta}\frac{K^{-1}\varphi(\theta)\ \theta^{2m+\eta-1}}{(1+\theta)^{m+\sum_{i=1}^{m} r_i}}\frac{\prod_{i=1}^{m}\left(1+\theta+\theta x_i\right)^{r_i}}{e^{\theta\left(\kappa+\sum_{i=1}^{m}(r_i+1)x_i\right)}}\prod_{i=1}^{m-1}\binom{n-m-\sum_{j=1}^{i-1} r_j}{r_i}\frac{\mathrm{B}\left(r_j+\alpha,\, n+\beta-m-\sum_{j=1}^{i-1} r_j\right)}{\mathrm{B}(\alpha,\,\beta)}\,d\,\Theta
\tag{3.16}
$$

$$
k=\int_{\Theta}\frac{\theta^{2m+\eta-1}}{(1+\theta)^{m+\sum_{i=1}^{m} r_i}}\prod_{i=1}^{m}\left(1+\theta+\theta x_i\right)^{r_i} e^{-\theta\left(\kappa+\sum_{i=1}^{m}(r_i+1)x_i\right)}\prod_{i=1}^{m-1}\binom{n-m-\sum_{j=1}^{i-1} r_j}{r_i}\frac{\mathrm{B}\left(r_j+\alpha,\, n+\beta-m-\sum_{j=1}^{i-1} r_j\right)}{\mathrm{B}(\alpha,\,\beta)}\,d\Theta
$$

where K is the normalized constant and $\varphi\left(\theta\right)$ is a parametric function of the parameters. It may be noted that Equation 3.16 does not have any closed form and does not have any distribution form; therefore, here, one needs numerical techniques for computations. We, therefore, propose to use Markov Chain Monte Carlo (MCMC) methods (Chen et al., 2012; Metropolis & Ulam,1949). Using MCMC techniques, we considered the Metropolis–Hastings algorithm to generate samples from posterior distributions, and these samples are used to compute Bayes estimates. The Gibbs sampling is an algorithm for simulating from the fully conditional posterior distributions, while the Metropolis–Hastings algorithm generates samples from an arbitrary proposal distribution. The conditional posterior distributions of the parameters θ, $\propto$, and β can be written as

$$\pi_1\left(\theta|\alpha,\beta,x\right)\propto\theta^{2m+\eta-1}(1+\theta)^{-\left(m+\sum_{i=1}^{m}r_i\right)}\prod_{i=1}^{m}\left(1+\theta+\theta x_i\right)^{r_i}e^{-\theta\left(\kappa+\sum_{i=1}^{m}(r_i+1)x_i\right)} \tag{3.17}$$

$$\pi_2\left(\alpha|\theta,\beta,x\right)\propto\prod_{i=1}^{m-1}\frac{\Gamma(\alpha+\beta)}{\Gamma\alpha}\frac{\Gamma\left(\alpha+r_i\right)}{\Gamma\left(n+\alpha+\beta-m-\sum_{j=1}^{i-1}r_j\right)} \tag{3.18}$$

and

$$\pi_3\left(\beta|\alpha,\theta,x\right)\propto\prod_{i=1}^{m-1}\frac{\Gamma(\alpha+\beta)}{\Gamma\beta}\frac{\Gamma\left(n+\beta-m-\sum_{j=1}^{i}r_j\right)}{\Gamma\left(n+\alpha+\beta-m-\sum_{j=1}^{i-1}r_j\right)} \tag{3.19}$$

respectively. Now using Equations 3.17, 3.18, and 3.19, we generate the Gibbs sequence $(\theta^0, \alpha^0, \beta^0)$, $(\theta^1, \alpha^1, \beta^1)$, ..., $(\theta^h, \alpha^h, \beta^h)$ as follows.

1. Choose an initial value of (θ,α,β), say $(\theta^0, \alpha^0, \beta^0)$.
2. Set $i = 1$.
3. Generate θ^1 via $\pi_1\left(\theta^0|\alpha^0,\beta^0\right)$.
4. Generate α^1 via $\pi_2\left(\alpha^0|\theta^1,\beta^0\right)$.
5. Generate β^1 via $\pi_3\left(\beta^0|\alpha^1,\theta^1\right)$.
6. Repeat steps 2–5, h times and obtain $(\theta^1,\alpha^1,\beta^1)(\theta^2,\alpha^2,\beta^2)$, ...,$(\theta^h,\alpha^h,\beta^h)$.

where h is a sufficiently large value. After discarding the burn-in-process samples, we obtain the samples from the posterior distributions of θ, α, and β. With these generated samples, we can evaluate the Bayes estimate of the parameters or the parametric function.

3.8.1 Bayesian Intervals

Once the samples are generated from the posterior of Θ via the MCMC technique, we obtain the credible intervals for the parameters on the basis of these samples. In this section, we provide the procedure to obtain the BCIs, based on the algorithm of Chen and Shao (1999), to evaluate credible intervals and the highest posterior density (HPD) intervals. The steps of the algorithm are as follows.

(a) **Credible interval**

(i) Order the observed sample,

$$\Theta_{(1)} \leq \Theta_{(2)} \leq \ldots\ldots \leq \Theta_{(h)}.$$

(ii) The $100(1-\alpha)\%$ Bayesian credible interval for Θ is given by

$$\left(\Theta_{[(\alpha/2)h]}, \Theta_{[(1-\alpha/2)h]}\right),$$

where $[q]$is an integral part of q.

(b) **Highest posterior density interval**

(i) Find all possible $100(1-\alpha)\%$ credible intervals with their respective lengths as follows:

$$\left(\Theta_{(j)},\ \Theta_{(j+(1-\alpha/2)h)}\right),\quad l_j^{\Theta} = \left(j+\left(1-\lambda/2\right)h\right)-\left(j\right);\ \ j=1,2,\ldots.,\alpha h$$

(ii) Search for the credible intervals having the smallest length for l_j^{Θ}. We picked the interval that has the smallest length credible interval, which is the HPD interval for Θ.

The same process calculates the Bayesian credible HPD interval for the other parameters or the parametric function.

3.9 DATA STUDY

In this section, we have considered simulated and real data to show the application of the Lindley distribution. The real data fit the Kolmogorov-Smirnov (K-S) test and cumulative distribution plotting. All data sets show how one can use the results to a real-life problem.

3.9.1 Simulated Data

We consider the analysis of a simulated data set to show how one can use the results to a real-life problem. We generate a simulated sample from the Lindley population having a parameter θ. Here, removals are removed from the experiment under the progressively type-II beta-binomial censoring scheme. The observed samples based on different parametric values are shown in Table 3.2. We assume that m observations are failed and the remaining(n-m) observations are removed from the experiment by progressively type-II beta-binomial removals with parameters α and β. For this study, α and β takes different values and are also presented in Table 3.2. The $f(t)$, $h(t)$, and $s(t)$ are evaluated at a given time, which is defined in the respective tables. The ML estimate of the parametric function and their respective asymptotic

TABLE 3.2
Failed Observation Based on Progressively Type-II Beta-Binomial Removals

Sample 1:n=50; m=30; $\theta=1.5$; $\alpha=2$; $\beta=3$ $t_r=0.6$; $t_f=1.2$; $t_h=1.5$									
0.1049	0.1210	0.1221	0.1336	0.1431	0.1607	0.2117	0.2275	0.2348	0.2389
0.2531	0.2684	0.2739	0.2756	0.2868	0.3077	0.3253	0.3761	0.5107	0.5284
0.5696	0.6999	0.8087	0.9272	0.9876	1.0080	1.1280	1.3976	1.6272	2.1954
Sample 1:n=50; m=40; $\theta=1.5$; $\alpha=2$; $\beta=3$ $t_r=0.6$; $t_f=1.2$; $t_h=1.5$									
0.0254	0.0338	0.0384	0.0795	0.1147	0.1171	0.1205	0.1524	0.1780	0.2163
0.2282	0.3663	0.3780	0.3995	0.4780	0.5317	0.5840	0.6060	0.6442	0.7135
0.7378	0.7620	0.8824	0.9121	0.9620	0.9836	1.0211	1.0260	1.1945	1.2041
1.2994	1.6602	1.6786	1.8253	1.8338	1.8642	1.9262	1.9706	2.1152	2.1152
Sample 2:n=40, m=25, $\theta=1.2$; $\alpha=1.5$; $\beta=5$ $t_r=0.7$; $t_f=1.2$; $t_h=1.3$									
0.0128	0.0471	0.0649	0.0848	0.1301	0.1423	0.2239	0.2386	0.2819	0.4593
0.4691	0.4928	0.5195	0.6287	0.6719	0.8527	1.3606	1.4563	1.5068	1.5273
2.3250	2.3250	2.3250	2.3250	2.3250					
Sample 2:n=60, m=40, $\theta=2$; $\alpha=3$; $\beta=4$ $t_r=0.5$; $t_f=1$; $t_h=1.4$									
0.0068	0.0264	0.0328	0.0565	0.0668	0.0714	0.0729	0.0782	0.0794	0.0847
0.1043	0.1052	0.1073	0.1444	0.2561	0.2669	0.3408	0.4578	0.5582	0.5829
0.6426	0.6454	0.6907	0.7207	0.7306	0.7314	0.7791	0.7802	0.7887	0.9874
1.1916	1.2240	1.2339	1.2673	1.4265	1.4539	1.6152	1.8019	2.4175	2.4175
Sample 5:n=60, m=50, $\theta=2.5$; $\alpha=4$; $\beta=8$ $t_r=0.8$; $t_f=1.5$; $t_h=2$									
0.0041	0.0123	0.0151	0.0156	0.0166	0.0455	0.0633	0.0983	0.1189	0.1203
0.1409	0.1469	0.1942	0.2121	0.2309	0.2515	0.2551	0.2929	0.2957	0.3404
0.3692	0.3695	0.3978	0.4353	0.4487	0.4509	0.4767	0.4789	0.4973	0.5273
0.5348	0.5717	0.6371	0.6389	0.6433	0.6520	0.7192	0.7474	0.7840	0.7853
0.7865	0.8450	0.8754	0.9501	1.0940	1.3480	1.7029	1.8240	2.1750	2.1750

confidence interval and boot-p interval are presented in Table 3.3. In the Bayesian paradigm, we assume that the arbitrary value of the hyper-parameter of the gamma distribution is $\kappa = 3$ and $\eta = 2$. The Bayes estimate of θ, $\propto$, β, $f(t)$, $h(t)$, and $r(t)$ with their HPD and credible Bayesian intervals are presented in Table 3.4. In the Bayesian paradigm, two pairs of intervals of the given parametric values are shown in Table 3.4, the first is the HPD interval and the second is the Bayesian credible intervals. From the study, we see that the length of boot-*p* is greater than ACI. Similarly, the length of the HPD intervals is greater than the Bayesian credible intervals.

3.9.2 Real Example

The data given below are related to tests on the endurance of deep groove ball bearings (Lawless, 2003). The data are the number of million revolutions before the failure for each of the 23 ball bearings in the life-tests and they are:

17.88, 28.92, 33.00, 41.52, 42.12, 45.60, 48.80, 51.84, 51.96, 54.12, 55.56, 67.80, 68.44, 68.64, 68.88, 84.12, 93.12, 98.64, 105.12, 105.84, 127.92, 128.04, 173.40.

First, we have used the K-S test to check whether the data fit the distribution or not. We applied the K-S test using *R* software, and the *D*-statistic and *p*-value for the given data are 0.19281 and 0.3175, respectively, and the tabulated *D*-statistic is 0.27490, which shows that the distribution is quite a good fit for the data.

We also show the empirical cumulative distribution function (ECDF) plot for these data in Figure 3.1. The estimated value of θ is 0.0273213. Using these data, we fixed $m = \{10, 15, 20\}$ and obtained different samples according to various choices of the removal patterns of observations which is based on $\propto$ and β and are defined in Table 3.5. We have five different samples based on the choice of some combinations of m, $\propto$, and β. We present the MLEs and their intervals of parameters, the density function, hazard function, and reliability function based on the given samples in Tables 3.6 and 3.7. Since we do not have any prior information about the given parameters, we use a non-informative prior for the Bayesian estimation. It should be mentioned here that the non-informative prior $g_1(\theta)$ for θ can be obtained by choosing the values of the hyper-parameter to be $\eta = 1$ and $\kappa = 0$. For the given samples, the values of Bayes estimates of θ, $\propto$, and β with their confidence intervals are presented in Table 3.8. In Table 3.9, the Bayes estimates of $f(t)$, $h(t)$, and $r(t)$ and their HPD and credible Bayesian intervals are discussed. The MCMC cumulative sum and the iteration plots of θ for different samples (ball bearing data) are presented in Figures 3.2 through 3.6.

TABLE 3.3
ML Estimate of θ, α, and β, $r(t)$ and $h(t)$Based on the Given Samples with Their ACI and Boot-p Confidence Interval for Simulated Data

		$\hat{\theta}$	$\hat{\alpha}$	$\hat{\beta}$	$\hat{r}(t)$	$\hat{f}(t)$	$\hat{h}(t)$
Sample 1	MLE	1.7658	2.0215	3.0152	0.4794	0.4669	1.4397
	ACI	(1.2312, 2.3004)	(1.9504, 2.1231)	(2.9252, 3.0952)	(0.4597, 0.4991)	(0.4572, 0.4765)	(1.3372, 1.5422)
	Boot-p	(1.1952, 2.3251)	(1.9254, 2.1452)	(2.8954, 3.1045)	(0.4418,0.5140)	(0.4489, 0.4601)	(1.3215, 1.5568)
Sample 2	MLE	1.6043	2.1251	3.1152	0.5231	0.5581	1.2841
	ACI	(1.2025, 2.0061)	(2.0124, 2.1821)	(3.0541, 3.1752)	(0.5071, 0.5390)	(0.5520, 0.5641)	(1.2023, 1.3659)
	Boot-p	(1.1154, 2.0244)	(1.9807, 2.1978)	(3.0421, 3.1812)	(0.5011, 0.5420)	(0.5429, 0.5578)	(1.1987, 1.4217)
Sample 3	MLE	1.2810	1.9850	2.8941	0.5683	0.7917	0.9564
	ACI	(0.8782, 1.6837)	(1.8751, 2.0785)	(2.7921, 2.9875)	(0.5464, 0.5901)	(0.7901, 0.7934)	(0.8729, 1.0398)
	Boot-p	(0.8421, 1.7012)	(1.8247, 2.0954)	(2.7142, 2.9424)	(0.5502, 0.6012)	(0.7624, 0.8014)	(0.8800, 1.0458)
Sample 4	MLE	1.8678	2.1115	3.0548	0.5210	0.5101	1.5271
	ACI	(1.3757, 2.3598)	(1.954, 2.1641)	(2.9514, 3.1472)	(0.5060, 0.5360)	(0.5035, 0.5168)	(1.4527, 1.6015)
	Boot-p	(1.2851, 2.3921)	(1.9418, 2.1120)	(2.7514, 3.1485)	(0.4982, .5278)	(0.4952, 0.5214)	(1.4409, 1.6248)
Sample 5	MLE	2.2785	2.1715	3.1642	0.5514	0.5674	1.9877
	ACI	(1.7501, 2.8069)	(2.1140, 2.2014)	(3.1045, 3.2241)	(0.5385, 0.5643)	(0.5592, 0.5756)	(1.8629, 2.1126)
	Boot-p	(1.7028, 2.8357)	(2.1025, 2.2271)	(3.1145, 3.2409)	(0.5244, 0.5688)	(0.5500, 0.5810)	(1.8694, 2.1199)

TABLE 3.4
Bayes Estimate of θ, α, and β, $r(t)$ and $h(t)$ based on the Given Samples with Their HPD and BC Interval for Simulated Data

		$\tilde{\theta}$	$\tilde{\alpha}$	$\tilde{\beta}$	$\tilde{r}(t)$	$\tilde{f}(t)$	$\tilde{h}(t)$
Sap-1	Bayes	1.5008	2.0541	3.0352	0.5527	0.6250	1.1850
	HPDI	(1.4894, 1.5134)	(1.9985, 2.0325)	(2.9598, 3.0647)	(0.5490, 0.5561)	(0.6164, 0.6328)	(1.1740, 1.1970)
	BCI	(1.4887, 1.5130)	(1.9254, 2.0655)	(2.9504, 3.1002)	(0.5491, 0.5562)	(0.6167, 0.6332)	(1.1734, 1.1966)
Sap-2	Bayes	1.7025	2.0954	3.0452	0.5352	0.5485	1.3451
	HPDI	(1.5285, 1.8985)	(2.0124, 2.1821)	(3.0024, 3.0525)	(0.5259, 0.5589)	(0.5152, 0.5754)	(1.2895, 1.3825)
	BCI	(1.6525, 1.9850)	(1.9807, 2.1978)	(3.0321, 3.0941)	(0.5158, 0.5685)	(0.5089, 0.5798)	(1.3651, 1.4015)
Sap-3	Bayes	1.2007	1.9845	2.9542	0.5963	0.8618	0.8815
	HPDI	(1.1879,1.2123)	(1.9245, 2.0421)	(2.9141, 2.9800)	(0.5922, 0.6009)	(0.8514, 0.8735)	(0.8696, 0.8923)
	BCI	(1.1884, 1.2131)	(1.9345, 2.0678)	(2.8541, 3.4244)	(0.5919, 0.6007)	(0.8507, 0.8729)	(0.8701, 0.8930)
Sap-4	Bayes	2.0008	2.1025	3.0642	0.4903	0.4508	1.6560
	HPDI	(1.9859, 2.0154)	(2.0452, 2.1351)	(2.9924, 3.1015)	(0.4871, 0.4937)	(0.4447, 0.4571)	(1.6415, 1.6701)
	BCI	(1.9862, 2.0159)	(2.0245, 2.1420)	(2.9548,3.1509)	(0.4869, 0.4936)	(0.4445, 0.4570)	(1.6418, 1.6706)
Sap-5	Bayes	2.5009	2.1105	3.1025	0.2125	0.1216	2.2068
	HPDI	(2.4849, 2.5164)	(2.0751, 2.1584)	(3.0584, 3.1345)	(0.2100, 0.2151)	(0.1189, 0.1245)	(2.1910, 2.2221)
	BCI	(2.4849, 2.5164)	(2.0618, 2.1625)	(3.0355, 3.1489)	(0.2100, 0.2151)	(0.1189, 0.1245)	(2.1910, 2.2220)

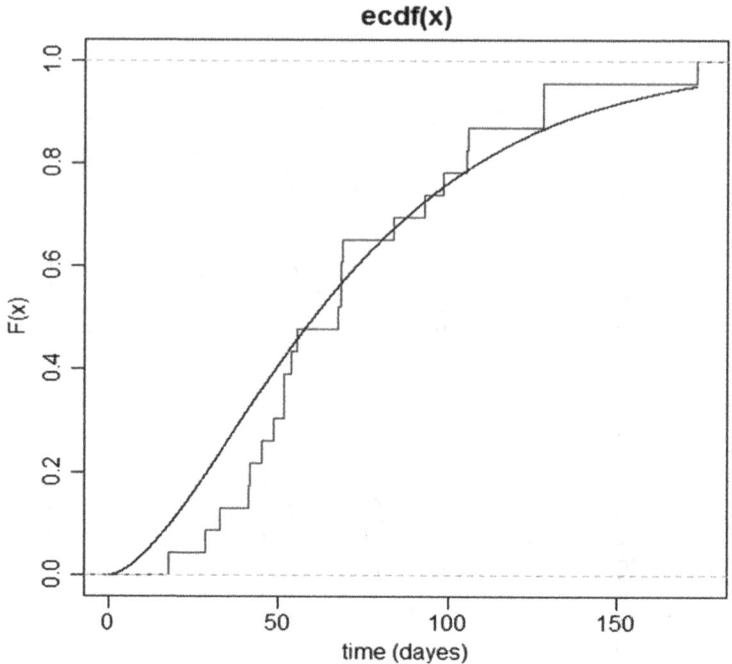

FIGURE 3.1 ECDF and CDF plot for ball bearing data.

TABLE 3.5
Ball Bearing Failed Observation Under Progressively Type-II Beta-Binomial Censored Data

Sample 1: Failed observation based on $n=23$, $m=10$, $\alpha=2$, $\beta=4$

17.88, 28.92, 45.6, 48.8, 51.84, 51.96, 54.12, 68.44, 84.12, 93.12

Sample 2: Failed observation based on $n=23$, $m=15$, $\alpha=2$, $\beta=8$

17.88, 28.92, 33, 41.52, 42.12, 45.6, 48.8, 55.56, 67.8, 68.44, 68.88, 84.12, 105.12, 105.84, 127.92.

Sample 3: Failed observation based on $n=23$, $m=15$, $\alpha=3$, $\beta=5$

17.88, 28.92, 41.52, 45.6, 48.8, 51.84, 51.96, 54.12, 55.56, 67.8, 68.44, 68.64, 68.88, 84.12, 93.12.

Sample 4: Failed observation based on $n=23$, $m=15$, $\alpha=8$, $\beta=14$

17.88, 28.92, 33, 41.52, 42.12, 48.8, 51.96, 54.12, 55.56, 67.8, 68.44, 68.64, 84.12, 98.64, 105.12.

Sample 5: Failed observation based on $n=23$, $m=20$, $\alpha=4$, $\beta=10$

17.88, 28.92, 33, 41.52, 42.12, 45.6, 48.8, 51.84, 51.96, 54.12, 55.56, 67.8, 68.44, 68.64, 68.88, 93.12, 105.12, 105.84,128.04, 173.4.

TABLE 3.6
ML Estimate of θ, α, and β Based on the Given Samples with Their ACI and Boot-p Confidence Interval for Ball Bearing Data

		$\hat{\theta}$	$\hat{\alpha}$	$\hat{\beta}$
Sample 1	MLE	0.02905	2.04152	4.12541
	ACI	(0.01698, 0.04111)	(2.02584, 2.05741)	(4.01581, 4.25684)
	Boot-p	(0.01455, 0.04756)	(2.02351, 2.05904)	(3.97542, 4.28525)
Sample 2	MLE	0.02749	2.10251	8.06541
	ACI	(0.01793, 0.03704)	(1.98414, 2.22054)	(7.85420, 8.12450)
	Boot-p	(0.01547, 0.04251)	(1.92575, 2.25547)	(7.76224, 8.14289)
Sample 3	MLE	0.03175	3.13524	5.08245
	ACI	(0.02067, 0.04284)	(3.01854, 3.28410)	(4.86751, 5.24841)
	Boot-p	(0.01753, 0.04755)	(2.97152, 3.34015)	(4.71058, 5.34021)
Sample 4	MLE	0.02897	7.89542	13.87542
	ACI	(0.01895, 0.03899)	(7.68452, 8.15582)	(13.58542, 14.15421)
	Boot-p	(0.01563, 0.04632)	(7.50821, 8.11083)	(13.42185, 14.21455)
Sample 5	MLE	0.02810	4.14521	10.34521
	ACI	(0.01947, 0.03673)	(3.87120, 4.23484)	(10.01542, 10.68247)
	Boot-p	(0.01745, 0.0434)	(3.70251, 4.35552)	(10.01455, 10.55980)

TABLE 3.7
ML Estimate of $r(t)$, $f(t)$, and $h(t)$ Based on the Given Samples with their ACI and Boot-p Intervals for Ball Bearing Data

		$\hat{r}(t=60)$	$\hat{f}(t=100)$	$\hat{h}(t=50)$
Sample 1	MLE	0.47149	21.14737	0.01734
	ACI	(0.24985, 0.69312)	(21.14561, 21.14913)	(0.00723, 0.02744)
	Boot-p	(0.25485, 0.75124)	(21.14175, 21.15140)	(0.00624, 0.02156)
Sample 2	MLE	0.50074	23.76374	0.01604
	ACI	(0.31798, 0.68349)	(23.76245, 23.76502)	(0.00814, 0.02394)
	Boot-p	(0.25481, 0.75140)	(23.71912, 23765420)	(0.00500, 0.00291)
Sample 3	MLE	0.42357	17.20800	0.01963
	ACI	(0.23481, 0.61232)	(17.20626, 17.20973)	(0.01016, 0.02910)
	Boot-p	(0.18241, 0.72145)	(17.20526, 17.21452)	(0.00945, 0.03501)
Sample 4	MLE	0.47289	21.26886	0.01728
	ACI	(0.28837, 0.65740)	(21.26741, 21.27032)	(0.00889, 0.02567)
	Boot-p	(0.24157, 0.71254)	(21.26040, 21.27185)	(0.00726, 0.02155)
Sample 5	MLE	0.48902	22.69568	0.01655
	ACI	(0.32651, 0.65152)	(22.69448, 22.69688)	(0.00938, 0.02373)
	Boot-p	(0.29423, 0.72422)	(22.69381, 22.69752)	(0.00853, 0.02670)

TABLE 3.8
Bayes Estimate of θ, α, and β Based on the Given Samples with Their HPD and BC Interval for Ball Bearing Data

		$\tilde{\theta}$	$\tilde{\alpha}$	$\tilde{\beta}$
Sample 1	Bayes	0.0273227	2.052514	4.082101
	HPDI	(0.0269516, 0.0276802)	(2.031521, 2.075214)	(4.051453, 4.092148)
	BCI	(0.0269633, 0.0276975)	(2.030151, 2.080154)	(4.058252, 4.110285)
Sample 2	Bayes	0.0273230	2.625152	8.045415
	HPDI	(0.0270403, 0.0276074)	(2.622542, 2.622054)	(8.040155, 8.102558)
	BCI	(0.0270423, 0.0276098)	(2.632514, 2.625892)	(8.038218, 8.110836)
Sample 3	Bayes	0.0273260	3.102152	5.132255
	HPDI	(0.0269998, 0.0276639)	(3.100128, 3.114525)	(5.105641, 5.165248)
	BCI	(0.0270009, 0.0276683)	(3.1125211, 3.125401)	(5.097550, 5.169421)
Sample 4	Bayes	0.0273238	8.045415	14.064852
	HPDI	(0.0270280, 0.0276202)	(8.040155, 8.102558)	(14.045851, 14.092574)
	BCI	(0.0270299, 0.0276244)	(8.038218, 8.110836)	(14.049810, 14.081891)
Sample 5	Bayes	0.0273235	4.154892	10.5625232
	HPDI	(0.0270652, 0.0275746)	(4.125148, 4.182154)	(10.524582, 10.594101)
	BCI	(0.0270659, 0.0275762)	(4.110825, 4.189252)	(10.510845, 10.597245)

TABLE 3.9
Bayes Estimate of *r*(*t*), *f*(*t*), and *h*(*t*) Based on the Given Samples with Their HPD and BC Intervals for Ball Bearing Data

		$\tilde{r}(t = 60)$	$\tilde{f}(t = 100)$	$\tilde{h}(t = 50)$
Sample 1	Bayes	0.5038550	24.0536172	0.0159073
	HPDI	(0.4970156, 0.5110035)	(23.4218359, 24.7211434)	(0.0156013, 0.0162027)
	BCI	(0.4966842, 0.5107687)	(23.3915062, 24.6990167)	(0.0156110, 0.0162170)
Sample 2	Bayes	0.5038474	24.0524594	0.0159074
	HPDI	(0.4984010, 0.5092858)	(23.5488356, 24.5595510)	(0.0156744, 0.0161424)
	BCI	(0.4983437, 0.5092372)	(23.5435760, 24.5549821)	(0.0156759, 0.0161444)
Sample 3	Bayes	0.5037900	24.0473924	0.0159100
	HPDI	(0.4973247, 0.5100696)	(23.4501384, 24.6332175)	(0.0156410, 0.0161892)
	BCI	(0.4972287, 0.5100464)	(23.4413449, 24.6310339)	(0.0156419, 0.0161928)
Sample 4	Bayes	0.5038319	24.0510864	0.0159081
	HPDI	(0.4981568, 0.5095247)	(23.5264156, 24.5819925)	(0.0156642, 0.0161530)
	BCI	(0.4980670, 0.5094863)	(23.5181835, 24.5783803)	(0.0156658, 0.0161565)
Sample 5	Bayes	0.5038353	24.0511922	0.0159079
	HPDI	(0.4990263, 0.5088059)	(23.6062772, 24.5145004)	(0.0156948, 0.0161153)
	BCI	(0.4989858, 0.5087889)	(23.6025546, 24.5129045)	(0.0156954, 0.0161166)

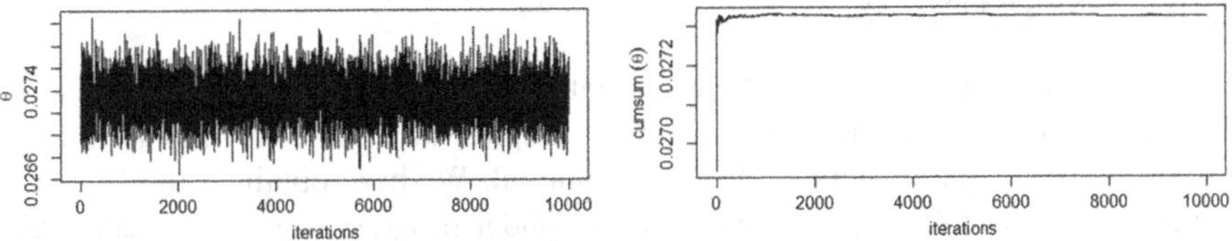

FIGURE 3.2 MCMC cumulative sum (CUSUM) and iteration plots of θ for sample 1 (ball bearing data).

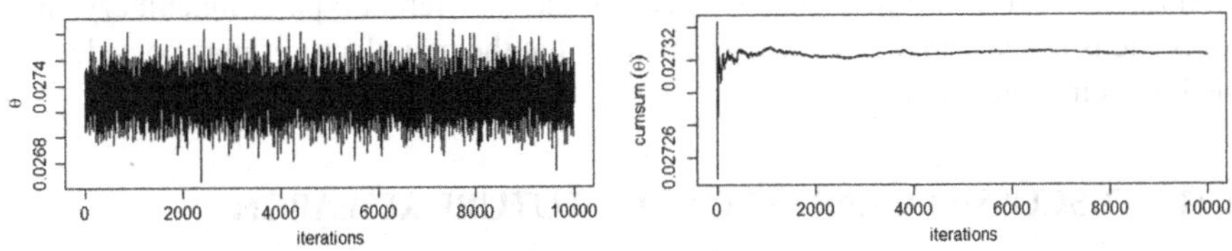

FIGURE 3.3 MCMC CUSUM and iteration plots of θ for sample 2 (ball bearing data).

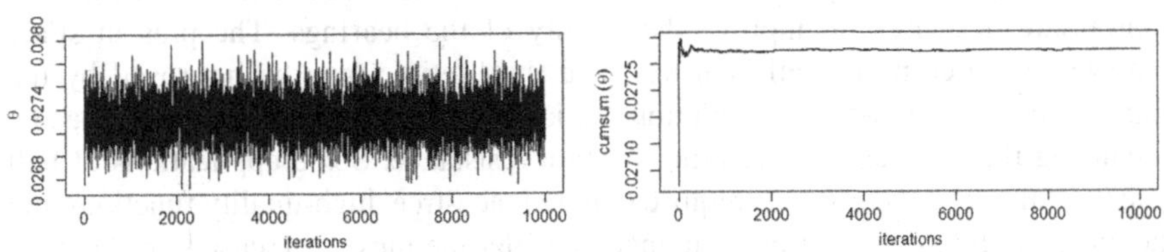

FIGURE 3.4 MCMC CUSUM and iteration plots of θ for sample 3 (ball bearing data).

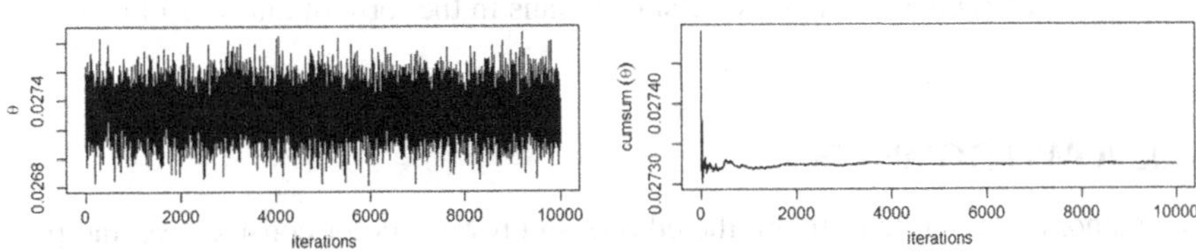

FIGURE 3.5 MCMC CUSUM and iteration plots of θ for sample 4 (ball bearing data).

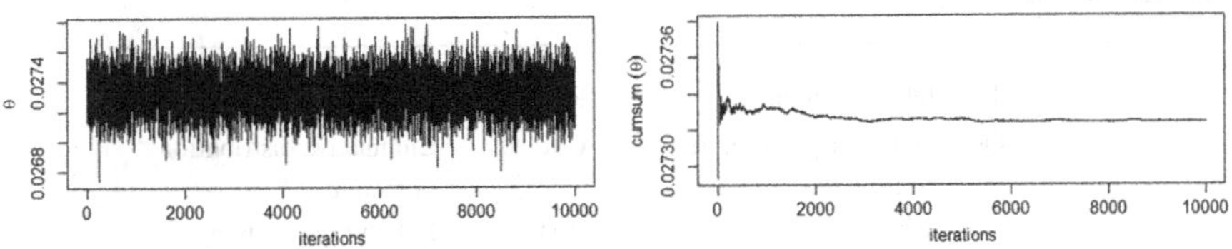

FIGURE 3.6 MCMC CUSUM and iteration plots of θ for sample 5 (ball bearing data).

3.10 CONCLUSION

Since the ball bearing is a useful component in engineering or reliability analysis, the estimation procedure is based on failed samples, which are obtained by different values of the parameters of the beta-binomial. We have obtained the ML and Bayes estimates of θ, $\propto$, β, $f(t)$, $h(t)$, and $r(t)$ under progressively type-II censored beta-binomial removals. The confidence intervals of the parametric function are also discussed in this study. The Bayes estimates of the parametric functions are obtained by using the MCMC method. Since, in this study, we considered that some sample observation is removed from the experiments under progressively type-II beta-binomial removals, the ball bearing has been used as a different type of machinery, and so its life is based on many characteristics so that the manufacturer may consider all the basic environments.

3.11 DISCUSSION AND SCOPE OF FUTURE RESEARCH

This study is helpful to improve fatigue life, quality, and lifetime of ball bearing products. We finally conclude that the discussed methodology can be used in medical, engineering, and in other areas where such type of life-tests is needed. This bearing study can be implemented in real-life applications; manufacturers can take appropriate measures to improve the quality of the bearings. The present study analyzes the lifetime of ball bearings, and the lifetime may be improved by the manufacturer for high-speed revolutions, axial movement, weight carrying capacity, enhancing the lifecycle, and refining friction resistance. Some applications of ball bearings such as wind turbines require maintenance-free, high-quality functions and therefore predictions about performance and lifetime must be focused on these. In future, we intend to consider some problems that involve analysis of various kinds of censored data under the competing risk model. These problems are untouched in reliability analysis for masked data. The advancements in various fields of reliability produce simultaneous challenges for statisticians in the form of analysis of complex data sets.

ACKNOWLEDGMENTS

The authors would like to thank the editors and reviewers for constructive and pertinent comments and the referees for their valuable suggestions that improvised the original version of the manuscript.

ABBREVIATIONS

X	The lifetime of items	Pdf	Probability density function
x_i	Observed lifetime of ith item	Cdf	Cumulative distribution function.
n	Sample size	R(t)	Reliability function
m	Total failed items out of n	h(t)	Hazard function

R_i	Removals at ith time.	θ, α, β	Scale parameters
MLE	Maximum-likelihood estimate	t_r	Reliability at the given time
MCMC	Markov Chain Monte Carlo	t_f	Pdf at a given time
ACI	Asymptotic confidence interval	t_h	Hazard rate at a given time
BCI	Bayesian credible interval	CUSUM	Cumulative sum
HPD	Highest probability density	p	Binomial probability

REFERENCES

Abdi, M., Asgharzadeh, A., Bakouch, H. S., & Alipour, Z. 2019. A new compound gamma and Lindley distribution with application to failure data. *Austrian Journal of Statistics*, 48(3), 54–75.

Aslam, M., & Feroze, N. 2019. Optimal Bayesian reliability estimation from progressively censored multimodal data. *Iranian Journal of Science and Technology, Transactions A: Science*, 2407–2422.

Bai, X., Shi, Y., Liu, Y., & Liu, B. 2019. Reliability estimation of stress–strength model using finite mixture distributions under progressively interval censoring. *Journal of Computational and Applied Mathematics*, 348, 509–524.

Balakrishana, N. 2007. Progressive censoring methodology: An appraisal (with discussions). *Test*, 16(2), 211–296.

Balakrishanan, N., & Aggarwala, R. 2000. *Progressively Censoring: Theory, Methods, and Application*. Birkhauser, Boston, MA.

Balakrishnan, N., & Cramer, E. 2014. *The Art of Progressive Censoring*. Springer, NewYork.

Balakrishnan, N., & Sandhu, R. A. 1995. A simple simulational algorithm for generating progressive type-ii censored samples. *The American Statistician*, 49(2), 229–230.

Caroni, C. 2002a. Modeling the reliability of ball bearings. *Journal of Statistics Education*, 10(3), 1–8.

Caroni, C. 2002b. The correct "ball bearings" data. *Lifetime Data Analysis*, 8(4), 395–399.

Chaturvedi, A., Kumar, N., & Kumar, K. 2018. Statistical inference for the reliability functions of a family of lifetime distributions based on progressive type II right censoring. *Statistica*, 78(1), 81–101.

Chen, M. H., & Shao, Q. M. 1999. Monte Carlo estimation of Bayesian credible and HPD intervals. *Journal of Computational and Graphical Statistics*, 8(1), 69–92.

Chen, M.-H., Shao, Q.-M., & Ibrahim, J. G. 2012. *Monte Carlo Methods in Bayesian Computation*. Springer Science & Business Media.

Dey, S., & Dey, T. 2014. Statistical inference for the Rayleigh distribution under progressively type-II censoring with binomial removal. *Applied Mathematical Modelling*, 38(3), 974–982.

Dowson, D., & Hamrock, B. J. 1981. History of ball bearings. NASA Technical Report.

Efron, B., & Tibshirani, R. J. 1994. *An Introduction to the Bootstrap*. Chapman and Hall/CRC, Taylor & Francis.

Garg, H. 2013. Performance analysis of complex repairable industrial systems using PSO and fuzzy confidence interval based methodology. *ISA Transactions*, 52(2), 171–183.

Garg, H. 2014a. Analyzing the behavior of an industrial system using fuzzy confidence interval based methodology. *National Academy Science Letters*, 37(4), 359–370.

Garg, H. 2014b. Reliability, availability and maintainability analysis of industrial systems using PSO and fuzzy methodology. *MAPAN*, 29(2), 115–129.

Garg, H., Sharma, S. P., & Rani, M. 2013. Weibull fuzzy probability distribution for analysing the behaviour of pulping unit in a paper industry. *International Journal of Industrial and Systems Engineering*, 14(4), 395–413.

Ghitany, M. E., Alqallaf, F., Al-Mutairi, D. K., & Husain, H. A. 2011. A two parameter weighted Lindley distribution and its applications to survival data. *Mathematics and Computers in Simulation*, 81(6), 1190–1201.

Ghitany, M. E., Atieh, B., & Nadarajah, S. 2008. Lindley distribution and its applications. *Mathematics and Computers in Simulation*, 78(4), 493–506.

Kumar, J., Panwar, M. S., & Tomer, S. K. 2014. Bayesian estimation of component reliability using progressively censored masked system lifetime data from Rayleigh distribution. *Journal of Reliability and Statistical Studies*, 7(2), 37–52.

Lawless, J. F. 2003. *Statistical Models and Methods for Lifetime Data*. Wiley, New York.

Lieblein, J., & Zelen, M. 1956. Statistical investigation of the fatigue life of deep-groove ball bearings. *Journal of Research of the National Bureau of Standards*, 57(5), 273–316.

Lindley, D. V. 1958. Fiducial distributions and Bayes theorem. *Journal of the Royal Statistical Society: Series A*, 20(1), 102–107.

Mazucheli, J., & Achcar, J. A. 2011. The Lindley distribution applied to competing risks lifetime data. *Computer Methods and Programs in Biomedicine*, 104(2), 188–192.

Metropolis, N., & Ulam, S. 1949. The Monte Carlo method. *Journal of the American Statistical Association*, 44(247), 335–341.

Nie, J., & Gui, W. 2019. Parameter estimation of Lindley distribution based on progressive type-II censored competing risks data with binomial removals. *Mathematics*, 7(7), 646.

Oehlert, G. W. 1992. A note on the delta method. *The American Statistician*, 46(1), 27–29.

Shanker, R., Hagos, F., & Sujatha, S. 2015. On modeling of lifetimes data using exponential and Lindley distributions. *Biometrics and Biostatistics International Journal*, 2(5), 1–9.

Singh, S. K., Singh, U., & Sharma, V. K. 2013. Expected total test time and Bayesian estimation for generalized Lindley distribution under progressively type-II censored sample where removals follow the beta-binomial probability law. *Applied Mathematics and Computation*, 222, 402–419.

Sinha, S. K. 1986. *Reliability and Life Testing*. Wiley Eastern Ltd., New Delhi.

Tomy, L. 2018. A retrospective study on Lindley distribution. *Biometrics and Biostatistics International Journal*, 7(3), 163–169.

Valiollahi, R., Raqab, M. Z., Asgharzadeh, A., & Alqallaf, F. A. 2018. Estimation and prediction for power Lindley distribution under progressively type II right censored samples. *Mathematics and Computers in Simulation*, 149, 32–47.

Zaretsky, E. V. 2013. Rolling bearing life prediction, theory, and application. NASA Technical Report.

4 Dynamic Fault Tree Analysis

State-of-the-Art in Modeling, Analysis, and Tools

Koorosh Aslansefat, Sohag Kabir, Youcef Gheraibia, and Yiannis Papadopoulos

CONTENTS

4.1 INTRODUCTION

Safety critical systems have become an integral part of our life. Over the years, new functionalities and capabilities have been added to such systems, and information and communication technologies are increasingly used to make them more sophisticated. While additional features and sophistication bring significant benefits, this might lead to additional complexity, making system development, safety assurance, and reliability properties more challenging. Safety is the avoidance of harm to people and the environment, and reliability is the ability to perform the intended function uninterrupted by a failure, which is often a precondition for safety. Both properties are crucial, and as systems become more complex, their prediction via analysis plays a vital role in the successful design and development of the system. Safety and reliability analysis are important tasks performed throughout the system's lifecycle, which systematically explores the potential safety-related issues in a system to verify whether or not a system is safe to use.

Over the years, several methodologies have been developed to facilitate safety and reliability analysis of systems. Among them, the FTA is one of the oldest and most popular techniques widely used to perform safety and reliability analysis of systems. In the traditional FTA, systems and their components are usually considered to have two states: the *working* state and the *failed* state. To model the logical interaction between different failure events Boolean AND OR gates are used, and the causes of system failure are determined in the form of combinations of events. To facilitate reliability analysis, each of the components can have its probability of failure, failure rate, distribution of time of failure, or steady-state or instantaneous (un)availability defined. At the same time, if the component can be repaired then a repair rate is defined. However, modern, large-scale complex systems have the capacity to work in different states, and they can have a complex repair process. A component in such a system can work as a primary component at a particular point in time, and in another instance the same component can work as a secondary component. Moreover, if a component acts as a spare component in a system, it can be in different modes such as cold, warm, and hot spares.

Such a multi-modal operational capability of systems and the complex interactions between their components give rise to different dynamic failure characteristics such as priorities among events and functionally dependent events. However, using a classical fault tree approach it is not possible to explicitly consider system dynamics and sequencing/timing of events while performing analyses, which may produce inaccurate results (Kabir, 2017). The limitations of the classical analysis techniques have not gone unnoticed, and it was recognized that methodologies with more powerful modeling capabilities are required to take into account the dynamic behavior of systems for a comprehensive and accurate analysis of complex systems.

Several attempts have been reported in the literature to improve the modeling power of SFTs through augmentation to include different types of temporal and statistical dependencies in the FT model. The concept of the Priority-AND (PAND) gate was introduced by Fussell et al. (1976). Later, several extensions to the SFTs such as the DFT (Dugan et al., 1992; 2000), temporal fault trees (Palshikar, 2002; Walker, 2009), and state/event fault trees (Kaiser et al., 2007) have been proposed. Among these extensions, DFT is the most popular dynamic extension of SFTs. The DFT retains the PAND gate and it additionally introduces new dynamic gates like the Functional Dependency (FDEP) gate and the SPARE and Sequence Enforcing (SEQ) gate.

Over the years, significant advancements have been made in the area of dynamic system analysis using DFTs. In this chapter, we review a variety of such developments in DFT analysis, which include both qualitative and quantitative analyses. Development in qualitative analysis started with the extension of the concept of minimal cut sets of SFTs to the minimal cut sequences (MCSQs) of DFTs. This was followed by the introduction of approaches for the determination of MCSQs from the structure of DFTs. On the other hand, the development in the area of quantitative analysis mainly focuses on the quantitative evaluation of the top event of the DFT based on the quantitative failure behavior-related information, e.g. failure rate or probability of the basic events. To accomplish this task, a number of existing approaches such as Markov models, Bayesian networks, Petri nets, mathematical formulations, and simulations have been utilized. In addition, uncertainty quantification of DFT analysis through criticality and sensitivity analysis is another important area that received considerable attention from both academia and industries. Regarding uncertainty handling in DFT analysis, the application of fuzzy set theory has been reported in the literature; a brief review of these methods is also be provided. Moreover, the developed software and the applications capable of handling DFTs are reviewed. The capabilities and limitations of these applications are also briefly addressed.

In addition to reviewing the above developments in DFT analysis as the state of art technology in reliability analysis, a combination of machine learning and reliability models, especially DFTs, will be discussed. The combination of machine learning with reliability models can be classified into five main categories: (i) using the reliability model as a core in machine learning with the aim of fault detection and diagnosis, (ii) selection of predefined and co-evaluated models through machine learning, (iii) updating the value of failure rates through machine learning-based algorithms, (iv) reconfiguration of DFTs through process mining, and (v) updating the membership functions of fuzzy models via machine learning. The chapter concludes with a concise discussion on the current challenges and potential future trends in DFT-related research.

4.2 OVERVIEW OF DFT

DFT has a logical structure similar to its static counterpart. The event at the top of the tree is known as the top event (TE), which almost always represents a system failure. This top event is decomposed into a combination of intermediate events (IEs). Unlike the static fault tree, DFT uses both Boolean and dynamic gates to specify logical relationships among events to represent the IEs. IEs are further decomposed down to lowest level events, which are known as basic events (BEs).

To allow the fault tree to model the sequence/time-dependent failure behavior of systems, several dynamic gates have been introduced. Figure 4.1 shows the commonly used DFT gates. The Priority-AND (PAND) gate is a special version of the AND gate. It delineates the priority behavior in a dynamic system. In this gate, the output will be true when both inputs occur and the first input (event A) occurs sooner than the second input (event B). In other words, the occurrence time of event A should be less than the occurrence time of event B and both should fail to have the failure as the output of this gate. Like the PAND gate, the Priority-OR (POR) gate also delineates a sequence; however, it defines an ordered disjunction rather than an ordered conjunction. In this gate, the first input (event A) has priority over other inputs. This event must happen first for the POR gate output to be true but does not require the occurrence of all other events (Walker, 2009). If other non-priority events

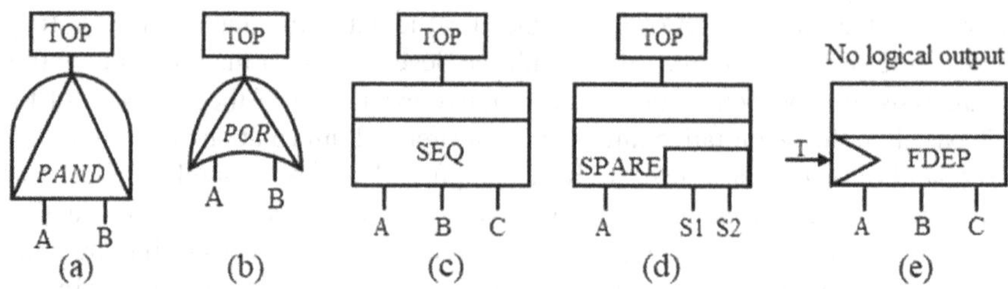

FIGURE 4.1 DFT logic gates.

occur, they must occur after the priority input. The Sequence Enforcing gate (SEQ) gate represents the sequential failure behavior of events A, B, and C, respectively. It means that events B and C cannot fail before the failure of event A. Also, event C cannot fail before the failure of event B.

The SPARE gate is used to model redundancy in the system design. The inputs to the SPARE are all BEs. The leftmost of the input corresponds to a primary event, and other inputs represent spare components. In the SPARE gate of Figure 4.1(d), the input A is the primary component, and S1 and S2 are two spare components. The behavior of this gate is defined as such that when the primary component (A) fails, the first spare (S1) will be activated; and if S1 fails, then S2 will be activated. Finally, the outcome of the gate will become true when all of its inputs become true. A SPARE gate can represent three different types of dynamic redundancies: (i) CSP: Cold Standby Spare in which the spare parts will be activated to be replaced when the primary unit (A) fails. That means in the cold spare mode the spare components are deactivated until they are required. (ii) HSP: Hot Standby Spare in which the spare parts start to work in parallel with the primary unit and when it fails the spare parts will be replaced immediately. (iii) WSP: Warm Standby Spare in which the spare parts partially work in parallel with the primary unit to be replaced when needed. In other words, the spare components are neither ON nor OFF; instead they are kept in-between these two states, i.e., components are kept in a reduced readiness state until required.

The Functional Dependency (FDEP) gate represents the functional dependency of some events on another trigger event. This gate helps to design a scenario when the operations of some components of a system are dependent on the operation of another component of the system. For example, when many components of a system receive power from a single source of supply, then the interruption in the power supply would cause all the dependent components to fail. In the FDEP gate, there is only one trigger event (either a basic event or an intermediate event) but there could be multiple functionally dependent events. As illustrated, the event T is the trigger event and the events A, B, and C are the dependent events, and they will fail if T occurs. In other words, these events (A, B, and C) are functionally dependent on event T. However, they can have their own individual failures, which will not affect the occurrence of the trigger events. The FDEP gate is particularly useful for modeling networked systems, where communication between the connected components takes place through a common network element, and failure of the common element isolates other connected components. This type of gate can also model interdependencies, which would otherwise introduce loops in fault trees.

4.3 METHODOLOGIES OF DFT ANALYSIS

As DFTs introduce dynamic gates in classical fault trees, the typical combinatorial analysis techniques available for classical fault tree analysis cannot be directly applied to analyze DFTs. Several methodologies have been developed for both qualitative and quantitative analyses of DFTs. Qualitative analysis mainly focuses on determining the cut sequences from DFTs. On the other hand, quantitative analysis aims at determining the probability of the top event given the failure rate or failure probability or failure probability distribution of the basic events of the DFTs. Additionally, criticality analysis of events is also performed as part of the quantitative evaluation of DFTs. The approaches used for developing methodologies for DFT analysis include, but are not limited to, Markov models, Petri nets, Bayesian networks, analytical solutions, and Monte Carlo simulation. In the following sections, we briefly discuss the qualitative and quantitative analysis approaches of DFTs.

4.3.1 Qualitative Analysis of DFTs

In a qualitative analysis of traditional static fault trees, minimal cut sets (MCSs) are determined from the fault tree structure. An MCS represents the minimal combination of events that can cause the top event of the fault tree. An MCS-based qualitative analysis of a DFT is possible if the dynamic gates of the DFT are replaced by static gates, for instance, by replacing the FDEP gates by OR gates and replacing PAND and SPARE gates by AND gates. However, in this case, the temporal dependencies between events would not be retained. Xiang et al. (2012) proposed a method to allow the combinatorial analysis of DFT with the Priority-AND gate only. In their work, the PAND gate was transformed into an AND gate by adding some conditioning events and the new gate was called CAND. The work was later extended in the study by Xiang et al. (2013).

To capture the temporal dependencies between events, the concept of minimal cut sequences (MCSQ) was proposed by Tang and Dugan (2004). An MCSQ is the minimal sequence of events that is sufficient and necessary to cause the top event of the DFT. To generate the cut sequences for a DFT, the zero-suppressed binary decision diagrams (ZSBDDs) (Minato, 2001) were used. It was shown that the dynamic gates can be replaced by the static gates to determine the cut sets, and the cut sequences can then be obtained by adding the necessary sequencing information into the cut sets. Later, for cut sequence generation, Liu et al. (2007a) proposed an algorithm called the cut sequence set algorithm (CSSA) using the notion of sequential failure symbol (SFS). The SFS is a mechanism that describes sequential failure between two independent events. Later, the concept of the extended cut sequence was proposed based on the general cut sequence by Zhang et al. (2011). In the above approaches, the concept of the cut sequence was based on the assumption of the non-repairability of system components. In the work by Chaux et al. (2013), a new definition of cut sequences was provided for binary systems, i.e., the system can either be in working or in failed states, with repairable components.

Walker (2009) proposed a qualitative analysis approach for the Pandora temporal fault tree. He also provided temporal laws to facilitate the minimization of the temporal sequences of events. One year later, Merle (2010) introduced an algebraic method for determining and expressing cut sequences of dynamic fault trees. This approach was based on the extension of the structure function used for classical static fault tree analysis. Rauzy (2011) introduced a variant of the ZSBDD approach proposed in the work by Minato (2001) to include sequencing information. This variant can be used for the determination of the cut sequences of DFT. In the study by Kabir et al. (2017), a model-based approach was proposed for the qualitative analysis of the dynamic failure behavior of systems. Elderhalli et al. (2017) integrated theorem proving and model checking to propose a comprehensive approach for qualitative and quantitative analysis of DFTs. Most recently, Piriou et al. (2019) provided a new definition of MCSQ for dynamic, repairable, and reconfigurable systems. Afterward, an algorithm was proposed to derive the MCSQs from the generalized Boolean logic driven Markov processes (GBDMP) (Piriou et al., 2017).

4.3.2 Quantitative Analysis of DFTs

A brief taxonomy of the DFT quantitative solution techniques reviewed in this chapter is shown in Figure 4.2. The meaning of each sign has been explained at the bottom of the figure. As an example in this figure, 'R' stands for the ability to model and solve the repairable DFTs, 't' refers to a time-consuming procedure, and 'D' means the solution is applicable for on-demand safety analysis.

4.3.2.1 Algebraic Solutions for DFTs

In SFTs, mathematical formulas are often used to quantify the probability of the Boolean gates, thus evaluating the probability of the MCSs and the top event. However, in DFTs, the logic gates not only model the effects of a combination of events, but also model the effects of the order of the failure. By taking the sequencing into account, at first, Fussell et al. (1976) provided an algebraic method to find an approximate solution to the Priority-AND gate. In 2000, Long et al. (2000) provided a solution for DFT with the Priority-AND gate. In their work, they used sequential failure logic (SFL) to model the behavior of the PAND gate, and mathematical equations with multiple integration were proposed to quantify the SFL model.

There are algebraic approaches that utilize the inclusion–exclusion method to determine the MCSQs of the DFT first. Subsequently, the inclusion–exclusion principle is used to quantify the MCSQs to determine the probability of the top event. One such approach is applied by Liu et al. (2007b), which used a technique similar to that of Long et al. (2000). In this work, MCSQs are expressed as sequential failure expressions (SFEs) and SFEs are solved using different multi-integration formulas. Finally, the top event probability is evaluated by summing the probabilities of different SFEs. At the same time, Yuge and Yanagi (2008) proposed an algebraic method for computing the exact top event probability of the DFT containing the PAND gate and the repeated events. This approach also assumed that the basic events are

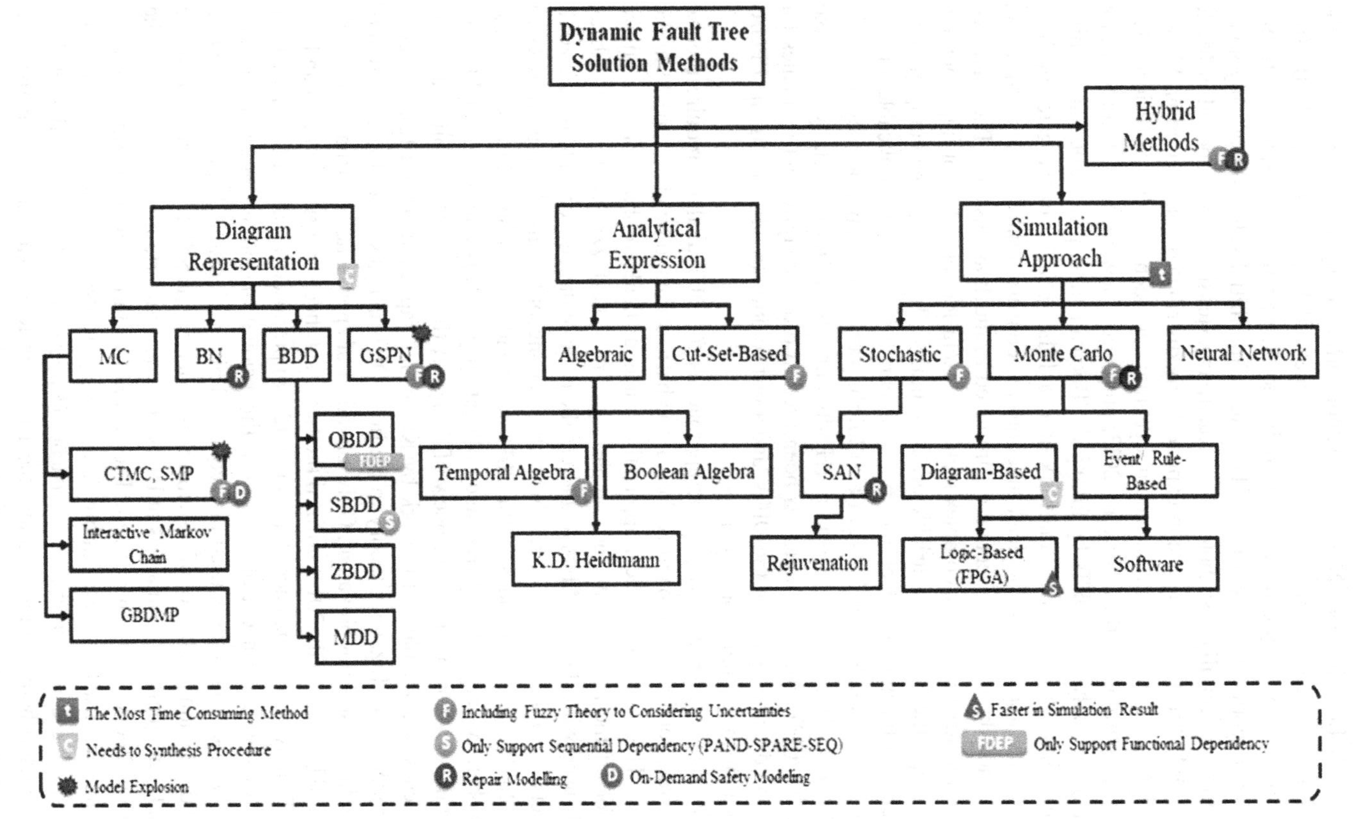

FIGURE 4.2 Taxonomy of solutions for DFTs.

statistically independent, exponentially distributed, and the components associated with the events are non-repairable. Note that the abovementioned methods can quantify DFTs with the PAND gate only, not with other dynamic gates such as the SPARE gate. By considering all the dynamic gates of DFTs, Merle et al. (2010; 2011) determined the structure function of DFTs and then proposed an algebraic framework for algebraically modeling DFTs' gates. This initial solution was only applicable to exponentially distributed data. Later, an extension was proposed in Merle et al. (2014; 2016) to consider non-exponentially distributed data. Based on Merle's work, Edifor et al. (2012) proposed an algebraic approach to solve the Priority-OR gate of temporal fault trees. Ni et al. (2013) proposed a new algebraic framework for quantitative analysis of DFTs by taking the Boolean state, probability, and timing of events into account. The framework modeled the behavior of the gates in three steps. In the first step, the Boolean functions are converted into the sum-of-product forms. Then the repeated events are eliminated as much as possible. The final step reduces the structure of the complex inclusion–exclusion formula. For implementing the concept, they used the variable array definition. To improve the computational efficiency of the existing algebraic approaches and to make them applicable for analyzing highly coupled DFTs, Ge et al. (2015b) proposed an approach by using the adapted K.D. Heidtmann's algorithm (Heidtmann, 1989). In the work by Aliee and Zarandi (2013), stochastic logic has been used to propose equivalent templates for static and dynamic gates of the DFTs, and a fast solution to DFTs using the FPGA-based implementation is provided.

Note that all the above algebraic approaches for DFT analysis require precise failure rates or failure probability data of the basic events to be able to perform the analysis. However, in practical applications, it is difficult to obtain precise failure data for all the basic events for complex systems. The fuzzy set theory has been widely used with classical SFTs to address the issue of data uncertainty (Garg, 2014; Garg et al., 2014a). The fuzzy set theory-based concept has also been used by Garg et al. (2014b) and Garg and Sharma (2011) to provide a solution to bi-objective and multi-objective reliability–redundancy allocation problems under the condition of uncertainty. A comprehensive review of fuzzy set theory-based reliability analysis approaches is available in the work by Kabir and Papadopoulos (2018). However, the application of fuzzy set theory to facilitate DFT analysis under the condition of uncertainty is still not prevalent. Only a handful of approaches (Verma et al., 2006; Ping, 2011; Jiang et al., 2018) utilized the fuzzy set theory for uncertainty handling in DFT analysis. For instance, Jiang et al. (2018) proposed a method for fuzzy DFT analysis using the concept of weakest n-dimensional t-norm arithmetic operations on fuzzy sets. The authors used the sequential binary decision diagram (SBDD) to model the dynamic behavior of systems. Subsequently, the SBDD is transformed into DFTs. In the quantitative analysis of DFTs, to handle uncertainty in failure data, the fuzzy set theory has been utilized. The weakest n-dimensional t-norm arithmetic operations are used on fuzzy failure data of basic events, thus reducing fuzzy accumulation. There are a couple of approaches (Kabir et al., 2014a, 2016) that utilized the fuzzy set theory to handle data uncertainty in temporal fault tree analysis. In these approaches, fuzzy operators for the temporal gates have been developed first. Subsequently, fuzzy failure rates of basic events of DFTs were used in the fuzzy operators of the logic gates to evaluate the top event probability of the DFTs. Most recently, intuitionistic fuzzy set theory has been combined with expert elicitation by Kabir et al. (2020) to quantify temporal fault trees with uncertain data.

4.3.2.2 Markov Models for Quantifying DFTs

Solving the DFTs using the continuous time Markov chain (CTMC) is regarded as one of the first and most important solution methods developed for the quantitative evaluation of DFTs. This method has been employed in the development of software tools such as Galileo, DIFtree, and HiRel (Dugan et al., 1997; Bavuso et al., 1994).

As shown in Figure 4.3, Markov models can be categorized into five types: (i) Homogenous Continuous Time Markov Chain (HCTMC) known as a traditional CMTC can model failures with exponential probability distribution with

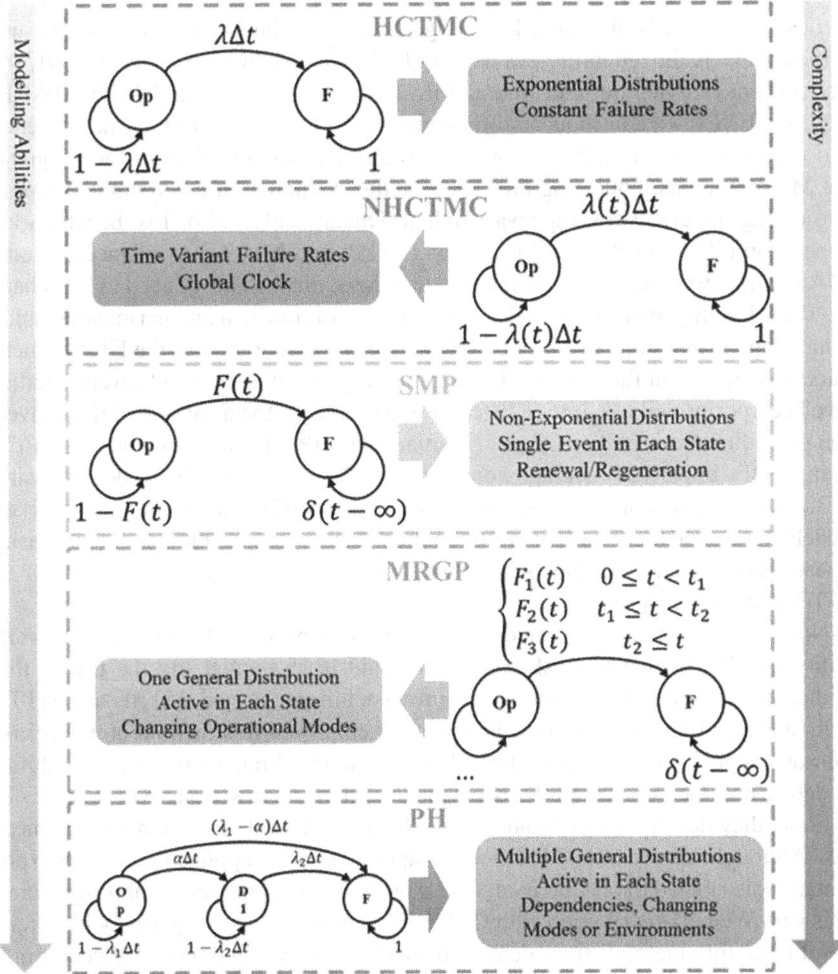

FIGURE 4.3 Classification of Markov Models.

constant failure rates. (ii) The Non-Homogenous Continuous Time Markov Chain (NHCTMC) can model global clock and exponential type failures with time-variant failure rates. (iii) The Semi-Markov Process (SMP) is capable of considering non-exponential probability distributions and renewal processes. (iv) The Markov Regenerative Process (MRGP) is capable of considering operational mode changes in one transition. (v) The Phased Type Markov Process (PH) can model multiple general distributions by dividing some of the systems' states into degraded states (the more the degraded states, the more accuracy) (Trivedi & Bobbio, 2017). It should be noted that there are some other extensions of Markov models such as input/output interactive Markov chains and generalized Boolean logic driven Markov processes (GBDMP) which are obtained from the combination of the Markov theorem and Automata. In fact, each of those introduced Markov types can be merged with Automata or similar theories to generate the extended versions. In Figure 4.3, the modeling capability is shown to increase from top to bottom while the complexity of computation is also raised. Having categorized the Markov models, the use of these models for reliability evaluation of DFTs is briefly discussed. In 1991, the first concept of dynamic fault tree and its dynamic gates such as PAND, SPARE, SEQ, and FDEP were introduced through their CTMCs (Boyd, 1992). The reference also recommended an automatic way for the conversion of DFTs to their equivalent Markov chains. Following this, in 1993, the evaluation of the system behaviors considering imperfect coverage was studied (Dugan et al., 1993). Two benchmarks, named Fault Tolerant Parallel Processors (FTPP) and Mission Avoidance Systems (MAS) which were used later by many researchers, are also introduced in this chapter. The reliability analysis of DFT in the presence of transient and permanent faults, failure dependencies, recovery of a system, and reconfiguration of the FTPP benchmark was studied in the work by Dugan (1993). From 1993 to 2009 several studies have been performed to address different issues such as the accuracy of the conversion procedure from DFT to CTMC (Manian et al., 1999), uncertainty analysis (Yin et al., 2001), imperfect coverage consideration (Vesely et al., 2002), decomposing DFTs into independent modules (Huang & Chang, 2007), introducing new Markov models for components' failures (Dominguez-Garcia et al., 2008), and considering repeated events and their effects in state-space modeling (Yuge & Yanagi, 2008) in DFT-based reliability analysis.

Norberg et al. (2009) presented a model for merging static fault trees with availability CTMC, so that the risk parameter could be evaluated. By the use of this method, reliability, risk, availability, failure rate, failure interval, MTBF, and MTTF were induced from the fault tree. This chapter employed this method on the drinking water supply system. Verma et al. (2010) studied different methods for reliability modeling and then discussed the behavior of dynamic gates along with CTMC. In addition, they described DFT solutions by the use of CTMC and Monte Carlo theories. Although, in general, CTMC-based approaches are applicable only to exponentially distributed data, Guo et al. (2011) proposed an approach combining failure rates with Weibull distribution with CTMC. Zixian et al. (2011) reported widespread use of reliability methods in evaluating the risk of surgery, and they evaluated time-independent risk and time-dependent risk by merging CTMC and static fault tree analysis. By calculating the failure rate of medical facilities, they evaluated surgery frequency, rescue timeliness, and the risk of gastric-esophageal surgery using the fault tree analysis. Then by using sensitivity analysis, the effect of the retrieval time factor and rescue timeliness was measured. A power factor correction (PFC) using CTMC in the DFT of power systems has been presented in the work by Ranjbar et al. (2011).

Fuzzy-CTMC models were proposed by Li et al. (2012) to solve fuzzy DFTs and their reliability was evaluated under the condition of uncertainty. They presented an example of an automatic hydraulic system cutting machine (CNC). Their study only considered a dynamic fault tree example with an FDEP gate, and the fuzzy evaluation of other gates was left unaddressed. This fuzzy approach was also used in another study for the reliability evaluation of a driver in an array of solar cells (Huang et al., 2013). A year later, the statistical reliability evaluation of a dynamic fault tree with the PAND gate was proposed by Xiang et al. (2013) in which the conversion of the PAND gate into an AND gate was introduced, while considering some dependent conditional events. Moreover, the newly introduced AND gate called CAND was assumed to be dependent upon conditional events. In this study, CTMCs for PAND and CAND gates were provided with a discussion on their differences and were used in the reliability evaluation of FTPP's benchmark. The combination of Binary Decision Diagrams (BDDs) and CTMC for reliability evaluation of DFTs was studied by Hao et al. (2014).

The use of Shannon's decomposition theory was proposed by Ge and Yang (2015) to solve DFTs. The proposed method increased the computational efficiency. However, the study only considered the PAND gate and the method was not generalized for other dynamic gates. Brameret et al. (2015) proposed a framework called "AltaRica" to reduce the state explosion by combining the Dijkstra's algorithm and the notion of the distance factor for the DFT solution. An approximate solution for DFT by truncating Markov chain states was presented by Yevkin (2016). The method was appropriate for both repairable and non-repairable systems. In 2017, the research work of Ge and Yang (2015) was extended and published in Ge and Yang (2017). The research covered spare and sequence gates using the De Morgan theorem, and for negating a generalized cut sequence, they provided an improved explicit formula. A new state-space generation approach for solving the DFTs was also proposed (Volk et al., 2018). The presented method enabled model reduction through model checking theories. A hierarchical and approximate solution for availability analysis in DFTs based on equivalent two-state Markov models was proposed by Ramezani et al. (2016). Their approach was only tailored for exponential failure distribution-based events. An automated tool for the evaluation of repairable DFT was presented by Manno et al. (2014). The study proposed a mapping from the DFT entity to the adaptive transition system entity, and a conception of failure gates for the evaluation of both reliability and availability was illustrated. This study used the SMP for reliability evaluation of DFTs. A novel hierarchical SMP-based solution for reliability assessment of DFTs was also proposed by Aslansefat (2014) in which the computational complexity and the state explosion of the SMP have decreased significantly.

As mentioned before, input/output interactive Markov chain (I/O IMC) is an extension of CTMC which is used for DFT solutions (refer to Hermanns, 2002; Crouzen, 2006; Boudali et al., 2007a,b, 2010; Arnold et al., 2013a,b). The use of I/O

IMCs can reduce the state-space explosion. In addition, these models enable us to consider the standby spare behaviors in the basic events. Generalized Boolean logic driven Markov processes (GBDMP), another extension of Markov process, have also been used for qualitative and quantitative analysis of the DFT by Piriou et al. (2017). Moreover, the sequential binary decision diagram (SBDD) and its extensions have been used in some studies (Xing et al., 2011; Xing et al., 2012; Tannous et al., 2011; Ge et al., 2015a, 2016) for quantitative evaluation of DFTs. The Markov process was used by Niwas and Garg (2018) to propose an approach to evaluate the reliability and availability of an industrial system under the cost-free warranty policy, where the working period of a system was followed by a rest period. To address the issue of uncertain failure data in the Markov chain-based reliability evaluation, Garg (2015) used a fuzzy Markov model of a repairable system to develop the nth-order fuzzy Kolmogorov's differential equations. Later the fuzzy reliability of the system both in the transient and steady states was evaluated using the Runge-Kutta method. Aslansefat and Latif-Shabgahi's article (2019) is one of the recent research works that proposed a novel hierarchical SMP-based approach as a solution for reliability evaluation of DFTs. The study presented a number of hypothetical and industrial examples. It also presents an example related to the repair consideration in DFTs and its SMP-based solution.

4.3.2.3 Petri Nets for Quantifying DFTs

Petri nets are formal graphical and mathematical modeling schemes used widely for the specification and analysis of complex, distributed, and concurrent systems. Graphically, a PN model is represented by a directed bipartite graph composed of a set of places, a set of transitions, and a set of directed arcs. PNs have been widely used in system safety and the reliability analysis domain, and Kabir and Papadopoulos (2019) present a review of PN-based safety, reliability, and risk assessment approaches .

As evident in the literature, previously researchers used the application of PNs to evaluate static fault trees Hura & Atwood, 1988; Malhotra & Trivedi, 1995; Liu & Chiou, 1997; Bobbio et al., 2003). The underlying reachability graph of a PN model is isomorphic to CTMC, and there are established approaches for mapping between CTMCs and PNs. As a result of this, similar to Markov chains, PNs are also used to solve DFTs. In the literature, readers can find many extensions of PNs that can model transitions governed by both exponentially and non-exponentially distributed rates. For instance, the use of the Weibull distribution in PN was shown in the studies of Fecarotti et al. (2016) and Le and Andrews (2016) and, in addition to the Weibull distribution, the use of other types of distributions such as normal and lognormal distribution was shown in Bernardi et al. (2011) and Volovoi (2004). A number of Petri net tools that can offer the abovementioned modeling capabilities are reported by Longo et al. (2016). Therefore, while the Markov chain-based approaches are applicable only to systems with exponentially distributed lifetime, the PN-based approaches can be used for the analysis of systems with both exponentially and non-exponentially distributed lifetimes. Moreover, other approaches had been developed (Knezevic & Odoom, 2001; Garg, 2013) to address the issue of uncertainty in failure data in PN-based reliability analysis.

The first approach to evaluate DFTs via Petri nets was provided by Codetta-Raiteri (2005). In her approach, she provided graph transformation rules to translate dynamic gates of DFT to Petri nets. Similar approaches for evaluating DFTs and temporal fault trees were proposed (Zhang et al., 2009; Herscheid & Tröger, 2014; Kabir et al., 2015; Junges et al., 2018) as well. In the DFT to PN transformation process, each of the basic events and logic gates of a DFT are translated into a sub-net and then all the sub-nets are combined together to form the PN model of the DFT. Figure 4.4 shows the PN model of a BE of a DFT. A token (the black dot) in the place *x.up* represents that at the beginning of system operation the component associated with the BE x is fully functional, i.e., BE has not occurred. The firing rate of the timed transition *x.f* is determined according to the failure rate of the component represented by this BE. If the component has an exponentially distributed failure rate λ, then the probability of the transition *x.f* firing at time t is $1 - e^{-\lambda t}$. As mentioned earlier, in a PN model, this kind of timed transition can be characterized by a non-exponentially distributed failure rate as well. On firing the transition *x.f*, the place *x.dn* will get a token, which will mark the occurrence of the basic event, i.e., failure of the corresponding component.

Figure 4.5 shows the PN models of the Boolean and dynamic gates used in DFTs. As seen in the PN model of the AND gate in Figure 4.5(a), all input places: $X_1.dn$ to $X_n.dn$ are connected to the single immediate transition called *AND*. That means when all the input places get a token each, then the transition *AND* will fire to make the AND gate output true by depositing a token to the place *X.dn*. On the contrary, the PN model of the OR gate in Figure 4.5(b) models a disjunctive behavior. In this model, each of the input places, *Xi.dn*, is connected to a distinct immediate transition. This will ensure that whenever any of the input places gets a token the respective transition will fire to make the OR gate output true by depositing a token to the place *X.dn*. The PN model of the PAND gate is shown in Figure 4.5(c). This PN model is designed in such a way that it will ensure that the place (*X.dn*) representing the output of the PAND gate will get a token if and only if the input places get tokens in sequential order, i.e., the occurrence of the BEs obey the required sequencing. If the order of occurrence of the BEs is violated, the place *X.ok* will get a token, which will eventually prohibit the transition T_n from firing, thus forcing the PAND output to be false.

As seen in Section 4.2, the FDEP gate does not have a logical output, but the occurrence of the trigger event would force the dependent events to fail. In the PN model of the FDEP gate in Figure 4.5(d), the place *T.dn* would get a token if the trigger event occurs, and in the presence of a token *T.dn* will cause all the immediate transitions to fire to deposit tokens to the places $D_i.dn$, thus forcing the dependent

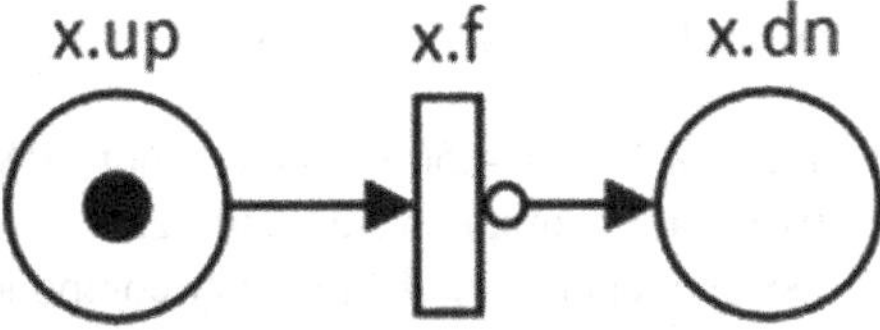

FIGURE 4.4 PN of a BE.

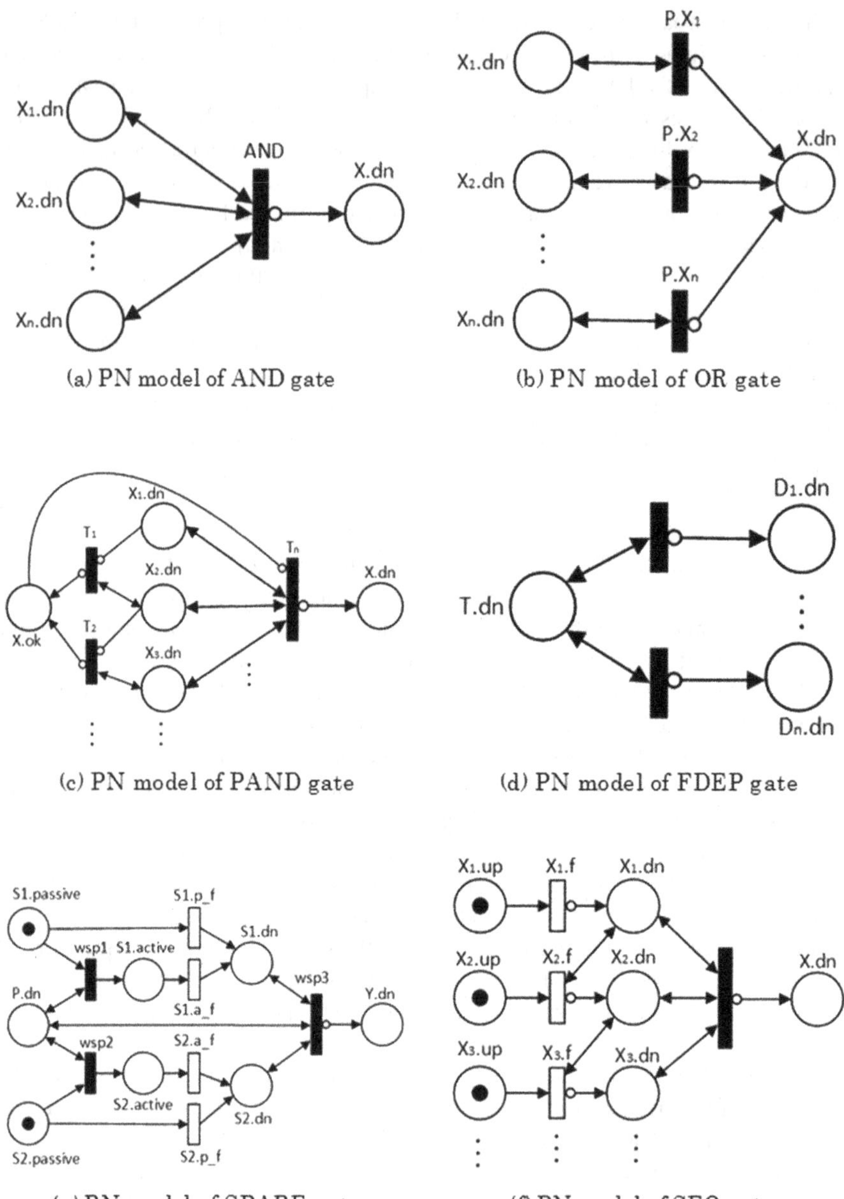

FIGURE 4.5 PN models of Boolean and dynamic gates.

events to fail. The PN model of Figure 4.5(e) models a hot SPARE gate with a primary component *P* and two spare components *S1* and *S2*.

At the initial stages of system operation, the primary component acts as the active component and the spare components are in passive mode, which is represented by the tokens in places *S1.passive* and *S2.passive*. The places *S1.dn* and *S2.dn* represent the failed state of the two spare components. It can be seen in the figure that as the spare components are in the hot spare mode, their failed states can be reached in two different ways. First, they can reach the S_i.*dn* state from their passive mode through the firing of transitions S_i.*p-f*, which are the failure rates of the components in the passive mode. Second, the spare components will reach their active states (represented by places *Si.active*) due to the failure of the primary component, and then their failed state can be reached from the active state through the firing of the transitions S_i.*a_f*.

The PN model of the SEQ gate is presented in Figure 4.5(f). This model ensures that the input events of the SEQ gate will occur in a predefined sequence. For instance, in this model, the place X_1.*dn* will get a token when the transition X_1.*f* fires. However, the transition X_2.f would not fire to deposit a token to X_2.*dn* until X_1.*dn* gets a token. This means that event X_2 cannot occur until event X_1 occurs. This way the model ensures that all the events in the SEQ gate will occur in a sequence and the occurrence of all the events will put a token in the place *X.dn*, denoting the occurrence of the output of the gate. Given the transformation rules for the basic event and the DFT's gates, Figure 4.6 shows the pseudocode of a function that converts a DFT to GSPN in the course of a depth-first traversal of the DFT.

After the PN model is formed, it can be evaluated in many different ways to perform a different analysis. For instance, in the PN model, if all the timed transitions are exponentially distributed, then the PN model can be evaluated using an underlying Markov model. In this case, as the analysis is performed based on the Markov model, it is clear that its application will also be limited only to exponentially distributed failure data. On the other hand, if the PN model contains non-exponentially distributed timed transitions, then simulations like the Monte Carlo simulation can be used for evaluation. Due to the use of simulation, this type of analysis could be

```
dftTOgspn (dft node) {

dft rn

  if (node is basic event) {
        translate node to GSPN module
        add new GSPN module to list of GSPN modules
  }
  else {                                   //node is a gate
     if (child of node has GSPN translation) { //gate inputs have GSPN translations
              translate node to GSPN module      //create GSPN module for gate
              add new GSPN module to list of GSPN modules

       }
       else {                                //gate inputs not translated
              for (rn = child of node and all siblings) {     //all gate inputs
                     dftTOgspn (rn)               //recursive call

              }

       }
  }
}
```

FIGURE 4.6 Pseudocode to convert DFT to GSPN (Kabir et al., 2018b).

computationally time-consuming. Although PNs provide more flexibility in terms of using different types of distributions, it has many features in common with the Markov model. For instance, like Markov model-based approaches, PN-based approaches have to generate the state space of the system for analysis; as a result, they face a state-space explosion problem while analyzing moderately complex systems.

4.3.2.4 Bayesian Networks for Quantifying DFTs

Bayesian networks (BNs), as a probabilistic graphical model, have a flexible architecture, which can make decisions under uncertainty, and can provide a global assessment about different dependability properties such as reliability and availability by combining local-level information from different sources. Widespread use of BNs for system dependability assessment have been reported (Weber et al., 2012; Kabir & Papadopoulos, 2019; Yazdi & Kabir, 2017, 2018). In their pioneering work, Bobbio et al. (2001) showed how a classical static fault tree can be evaluated by translating it to BN. Afterward, inspired by this approach, the modeling capability of Bayesian networks has been utilized in different methods for evaluating DFTs.

At first, in Boudali and Dugan (2005), a method was introduced for quantitative analysis of DFTs by translating them into discrete-time BNs. The general idea of the translation process is shown in Figure 4.7. The translation is performed in two steps: qualitative and quantitative. Qualitative translation involves translating the basic events, the logic gates, and the top event of a DFT to root nodes, intermediate events, and the leaf node of a BN, respectively. On the other hand, quantitative translation requires generating prior probabilities of the root nodes based on the failure probabilities of the basic events, conditional probability tables for the intermediate nodes and leaf node based on the logical specification of the DFT gates. As the approach in Boudali and Dugan (2005) uses a discrete-time BN, the granularity

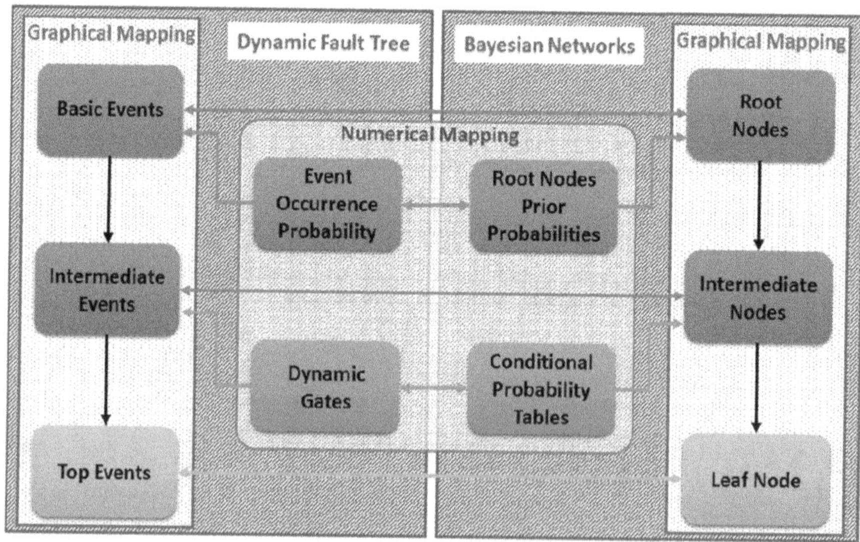

FIGURE 4.7 DFT to BN conversion process.

of time-discretization before the translation process should be decided. Similar to this approach, Montani et al. (2005) proposed a dynamic Bayesian network-based DFT analysis method, which is also considered to be a discretized model. Later, they performed further research (Montani et al., 2006b) to automate the DFT based on the BN generation process. In Montani et al. (2006a) a tool named RADYBAN was presented for automatic conversion of DFTs to 2-time-slice BNs (2TBNs)(Weber and Jouffe, 2003). Other discrete-time BN-based approaches for DFT analysis could be found in Kabir et al. (2014b) and Kabir et al. 2018a). All the above approaches consider system components to be non-repairable. However, to allow the modeling of repairable systems, the concept of a repair box gate was introduced (Portinale et al., 2010).

In addition to the discrete-time model, continuous time BNs (CTBN) have also been used for the quantitative analysis of DFTs. For instance, Boudali and Dugan (2006) introduced a CTBN-based DFT analysis method. In this approach, due to the use of a continuous time model, probability density functions and joint probability density functions were used instead of prior and conditional probability tables. One advantage of such an approach over discrete-time BN-based approaches is that it can provide an exact closed-form solution to DFTs. However, analysis using such approaches may face state-space explosion problems.

In Marquez et al. (2008; 2010) both discrete and continuous nodes were used in the same BN for DFT evaluation. As a result, it was possible to use both empirical and parametric distributions for the time-to-failure of system components. Recently, Codetta-Raiteri (2015), Codetta-Raiteri and Portinale (2017), and Li et al. (2015) have studied CTBN-based DFT analysis, in which a generalized continuous time Bayesian network (GCTBN) was used for the quantification of DFT (2017). In order to analyze GCTBN models, they have to be converted to GSPN models, which leads to the state-space explosion problem. In the study by Li et al. (2015), CTBN was used under a fuzzy environment to quantify DFTs. Some of the applications of the BN-based DFT analysis for fault detection, identification, and analysis could be found in Codetta-Raiteri and Portinale (2015) and Mi et al. (2016).

From the above discussion, it is clear that the discrete-time BN-based approaches for DFT evaluation can provide a fast non-exact solution to DFTs. However, the accuracy of the results can be improved significantly by increasing the number of discretized time intervals, but at the cost of higher computation time. On the other hand, CTBN-based approaches can readily provide exact solutions to DFTs but may suffer from the state-space explosion problem. To alleviate this problem, approximate algorithms instead of exact algorithms can be used for analyzing BN models. However, based on the nature of the application, the users can always make an informed decision by making a trade-off between the computing time and the precision of the results required.

4.3.2.5 Simulation Approaches for Quantifying DFTs

Simulation approaches can be used when a system is too complex and an approximate result is acceptable. The idea behind the simulation approach is that one (i) simulates the behavior of a system, (ii) determines the mission time, (iii) repeats the simulation for a huge number of iterations (e.g. 10e+8) and for each iteration checks whether the system failed before the mission time or not, and finally, (iv) evaluates the system reliability by dividing the number of failures to the number of iterations. It is also possible to (i) decompose the system into its components, (ii) determine the mission time, (iii) check whether each component fails during this mission time, (iv) provide a rule-set, or some similar logical models such as DFT, Petri Nets, Automata, etc., (v) repeat the simulation procedure for a huge number of iterations (e.g. 10e+8) and for each iteration check whether the overall system failed before the mission time or not, and (vi) evaluate, in the last step, the system reliability by dividing the number of failures of the overall system by the number of iterations. These two types of simulation approaches are common for the reliability evaluation of a complex system. Figure 4.8 illustrates these two procedures as (a) and (b), respectively.

Marseguerra et al. (1998) outlined the concepts and principles of the methods used for evaluating dynamic reliability and mentioned some Monte Carlo simulating algorithms. In order to decrease the calculation time and also provide a practical simulation method for evaluating dynamic reliability, this chapter introduces memory possessing methods and effective estimators. Using the "Time-to-failure" (TTF) tree, which is a tree that shows the time relation between the system failure time and each component's failure time, Ejlali and Miremadi (2004) solved the dynamic fault tree. In this tree, the AND gate is converted into MAX, the OR and FDEP gates are converted to MIN and the PAND and SEQ gates are converted to ADDER and spare gates are converted into a selector, all of which are convertible to logic circuits (Ejlali and Miremadi, 2003). In this study, after designing the time-to-failure tree, the logic circuits of the new tree are synthesized using VHDL language on the FPGA programmable chip, and Monte Carlo simulations are performed on this chip. Eventually, a comparison between the efficiency and velocity of this method is performed using computer simulation which indicates that evaluation using the FPGA chip is almost 471 times faster than a computer simulation.

Zonouz and Miremadi (2006) suggested the fuzzy Monte Carlo method for evaluating the dynamic fuzzy fault tree (only spare gate). In this study, the Weibull

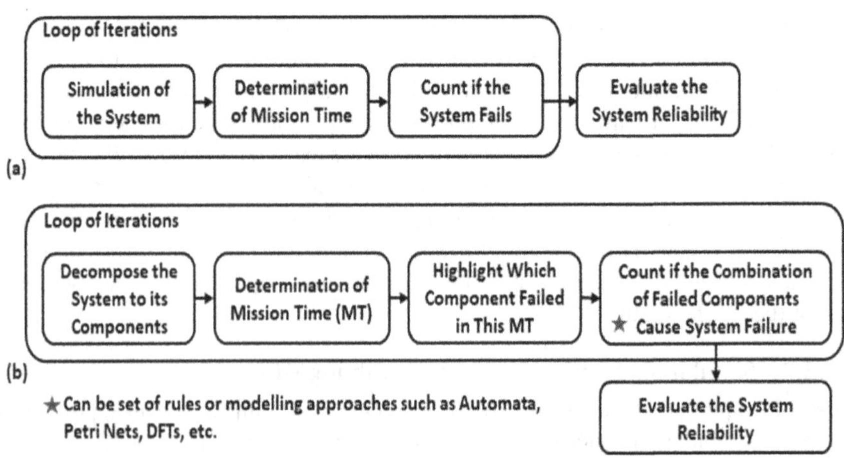

FIGURE 4.8 Two common simulation approaches.

distribution is used for components' failure, and the fault tree is solved after being converted into a time-to-failure tree. The comparison between the simulation time in the two studies shows that the fuzzy Monte Carlo simulation takes as much time as the typical Monte Carlo simulation, and thus this method makes problem solving much slower. Kara-Zaitri and Ever (2009) dealt with evaluating the fault tree with repairable components, and the implemented simulation on FPGA chips has hastened its evaluation process. In fact, in this study, a semi-analytical method is employed, since failure rates and repair rates are achieved through the analytical method, and later by using the Monte Carlo method the failure probability of the final event is calculated. This method is introduced as a less costly and flexible method with the ability of modeling more complex scenarios. However, in this study only exponential distribution is considered for failure, and different failure distributions such as Weibull and normal are not considered, which may be regarded as a disadvantage of this study.

Rao et al. (2009) proposed a solution process for each of the dynamic gates using Monte Carlo simulation. In this study, time curves are considered for each of the dynamic gates, so that the time-dependent failure of each gate becomes tangible. First of all, this study deals with solving and validating the proposed method for evaluation of the non-repairable dynamic fault tree and also compares the results with integral methods. Afterward, an example of an electricity supply system with spare components in a nuclear power station is considered and solved considering repairability. The results achieved from the simulations are then compared with results obtained from the analysis of the Markov model. It has been shown that in some cases the solution of the simulation is somehow similar to the analytical solution and in other cases significant differences exist. In addition to the reliability evaluation, this study also evaluated system availability and performed its simulations using the DRSIM tool.

Yevkin (2010) provided various methods for improving the efficiency of Monte Carlo simulations for both static and dynamic fault trees. In this study, variance reduction, parallel processing, and enhancement based on the structural information of the tree have been used, and the results are validated on the basis of an industrial benchmark. A year later, Chiacchio et al. (2011) presented a MATLAB-based open source software for reliability evaluation using DFTs. By considering the four common benchmarks for validation, the results are compared with respect to accuracy and simulation time using Relex commercial software, Galileo commercial-research software, and DFTSIM research software. This software has appropriate relative accuracy when compared with other software and possesses an acceptable computing speed. Following Chiacchio et al.'s study (2011), Manno et al. (2012a) proposed a toolbox, named MatCarloRe, in the Simulink environment (MATLAB) which allows the users to solve a dynamic fault tree and evaluate reliability in a specific mission time using the Monte Carlo simulation method. In this study, for each of the dynamic and static gates a block is considered and the output of each gate consists of a failure time, and a failure signal is attached to the inputs of the top gates in the tree, and eventually, a point-to-point solution of the dynamic fault tree is achieved through simulation. Among the advantages this toolbox offers, we can refer to a block, named basic event, which allows the users to consider various failure distribution functions for system components. Also, for the dynamic gate of a spare component, a general block is considered which is able to model all (hot, cold, and medium) spares. At the end of the study, the fault tree benchmark of the hypothetical cardiac assist system is used for validating the toolbox.

Lindhe et al. (2012) solved the dynamic fault tree through two methods such as the estimated Markov model and the Monte Carlo simulation for evaluating the risk of drinking water supply. In this method, the failure rate and the average time of system failure are evaluated at each level of the tree. This method is employed on three water supply scenarios and the results achieved from them are compared. In 2012, Aghassi and Aghassi (2012) presented a software based on the Monte Carlo simulation for evaluating the reliability of a dynamic fault tree. This software manages to make a fault tree evaluation 310 times faster than when using parallel processing in the GPU. It should be noted that this software converts a fault tree into time-to-failure in the first place and then performs the simulations. A new quantitative reliability evaluation of DFTs by means of event-driven Monte Carlo simulations has been proposed by Gascard and Simeu-Abazi (2018). The approach has been implemented in a Java-based framework called DFTEDS. Chiacchio et al. (2019) focused on MATLAB Simulink-based modeling and proposed a new solution as a library called Stochastic Hybrid Fault Tree Object Oriented (SHyFTOO). The differences between analytical-based approaches like CTMC and simulation-based methods like Monte Carlo simulations are summarized in Table 4.1.

4.3.2.6 Modularization Approaches for Quantifying DFTs

Analyzing large fault trees, including DFTs, is a big challenge for safety analysis experts. Modularization has been proved to be a powerful method to improve the computational performance of approaches while solving large fault trees (Patterson-Hine & Dugan, 1992). Modularization techniques are also known as hierarchical or compositional approaches. In modularization techniques, a divide-and-conquer strategy is followed. Under this strategy, a large DFT is divided into smaller independent static and dynamic modules. These independent modules are then solved using appropriate approaches depending on their types.

Dugan et al. (1997) proposed the first modularization technique called DIFtree for DFT analysis. They used BDDs and Markov chains to solve the static and dynamic

TABLE 4.1
Comparison between Simulation and Analytical Methods

Attributes	Simulation-Based Methods	Analytical Methods
Computation Time	Time-consuming (depends on number of iterations)	Usually faster than simulation
Final Results	Approximate.	Exact/Precise
Complexity Consideration	It would be easier to evaluate complex systems through simulation	Simplification is needed for large-scale systems
Further Results Extraction	Only statistical calculation	It is possible to drive other factors like MTTF/MTBF from results directly

sub-trees of the DFT. Solutions of smaller sub-trees are combined together to obtain solutions for larger trees. This work was later extended in Gulati and Dugan (1997). A similar solution to DFT was also proposed in Anand and Somani (1998). As the DIFtree approach is only applicable to exponentially distributed failure data, Manian et al. (1998) extended it by including a Monte Carlo simulation to allow the use of non-exponential distributions.

The limitations of the abovementioned approaches are that they cannot perform sensitivity analysis due to modularization and if a module is dynamic then no further modularization is performed in that module. To address these issues, Huang and Chang (2007) proposed a hierarchical approach by modularizing the fault tree, which allows the decomposition of independent sub-trees of a dynamic module. Yevkin (2011) also proposed an improved modular approach for the dynamic fault tree by considering systems without repairable components. He evaluated five different scenarios with the possibility of separating dynamic fault tree modules, and the approach was employed on three DFT benchmarks.

Manno et al. (2012a) proposed a modular method where each independent subtree of a fault tree is detected and solved hierarchically; similar to other approaches, a dynamic sub-tree is replaced by a single basic event where the probability of occurrence of the basic event is the probability of occurrence of the sub-tree. Chiacchio et al. (2013) proposed an algorithm based on a hierarchical approach for the reliability evaluation of dynamic fault trees. The approach used a parametric function (a 4-parameters Weibull) to solve the problem of dynamic evolution of the cumulative distribution function with different gates including the PAND gate. The proposed approach used the MatcarloRE (Manno et al., 2012a) tool to reduce the least square error fitting. In the study by Amari et al. (2003), a new modular approach was proposed in which both static and dynamic sub-modules of a DFT were solved using algebraic formulas, i.e., using multi-level integrals. This approach is different from other modularization techniques in the sense that it did evaluate the dynamic modules without converting them into CTMCs.

4.3.2.7 Application of Machine Learning with DFTs

The use of machine learning algorithms alongside safety models is one of the cutting-edge research topics in the safety and reliability analysis area (Simen et al., 2018). Figure 4.9 presents five categories of solutions using machine learning combined with safety models: (i) using the reliability model as a core in machine learning with the aim of fault detection and diagnosis, (ii) selection of predefined and co-evaluated models through machine learning, (iii) updating the value of failure rates through machine learning-based algorithms, (iv) reconfiguration of the safety model through process mining, and (v) updating the membership functions of fuzzy models via machine learning. Regarding the third category, in 2017, Aizpurua et al. (2017a,b) proposed a method for combining the failure rate and the Remaining Useful Life (RUL) as the basic event in DFTs. We know that the RUL can be estimated through machine learning approaches (Sikorska et al., 2011). For the other categories, the works by Lampis and Andrews (2009), Askarian et al. (2016), Chen and Ge (2018), Getir et al. (2018), and Cheng et al. (2019) can be consulted. However, none of the current research studies on those categories consider DFT as a safety model.

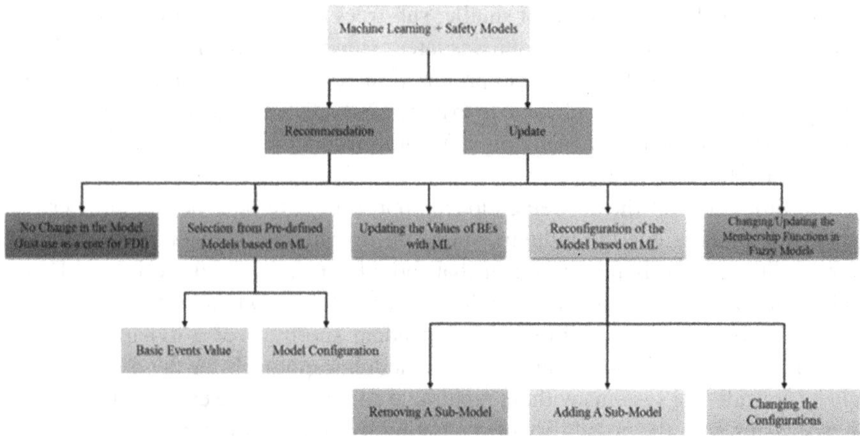

FIGURE 4.9 A classification of different approaches where machine learning can be combined with DFTs.

In addition to the abovementioned works, there exist some other works where machine learning has been used with DFTs. For instance, Zhou et al. (2006) proposed an approach for designing dynamic systems using recursive neural networks. The reliability of the system is designed based on DFTs and neural networks. The DFT is mapped into the recursive neural network with the feed-forwards technique. Raptodimos and Lazakis (2017) proposed a new approach for predictive maintenance of ship machinery. The proposed approach combined DFT with machine learning to facilitate forecasting of the health of selected system components to optimize the maintenance task. The approach aims to predict and monitor the future values of different components' physical parameters by using an autoregressive dynamic time series neural network modeling approach. The neural network model was trained in real time when no bugs or faults occurred in the system. Yassmeen Elderhalli and Tahar (2019) used machine learning to facilitate automating the proof of the subgoals. The verification of subgoals is performed via two steps. The first step is evaluating the existing subgoals, and the second step involves real-time reasoning to verify the remaining subgoals. From the proposed techniques we can understand that the use of machine learning with DFT is to identify a tool that can analyze the dynamic system with minimum user intervention using the formal DFT analysis. Linard et al. (2019) have provided a new evolutionary-based approach to generate a fault tree from observational data. A novel idea to merge the machine learning algorithm and update the fault tree has been provided by Gheraibia et al. (2019). In this research, a one-class support vector machine with a decision tree has been used to update the fault tree of safety critical systems.

4.4 SENSITIVITY ANALYSIS IN DFTS

Given that reliability evaluation is an important aspect in designing safety critical and fault-tolerant systems, the evaluation of the importance factor and the sensitivity can be a useful tool for analyzing this parameter. Considering the factors of importance and sensitivity, the following can be achieved:

- Finding out the part of the system that has contributed most to system failure (failure bottleneck(s)).
- Finding a path (a set of sequential events in a tree) that mostly contributes to the system failure.
- Finding the uncertainty of the tree's results regarding the accuracy of the input parameters (for example, the estimated failure rate).
- Finding the most affordable way to increase the system's reliability.
- Determining the components for which investments in maintenance and repairs have the most impact on system performance (Xing, 2004).
- Assessing the impact of the mission of a subsystem on the entire mission risk (Zixian et al., 2011).

Table 4.2 summarizes the sensitivity and the importance analysis approaches found in the literature.

4.5 TOOL SUPPORT

As we have already seen that the development made with DFT-based safety and reliability is not just theoretical, they have practical applications as well. Moreover, several tools have been developed as outcomes of successful research. A list of such DFT analysis tools including the methodologies used in these tools and what dependability parameters can be measured by them is summarized in Table 4.3.

In parallel with the existing academic DFT tool, there are some useful commercial tools such as ReliaSoft (ReliaSoft, 2016), OpenFTA (Auvation, 2016), Isograph Fault Tree+ (Isograph, 2016), ITEM Toolkit (ITEM Software, 2016), and EPRI CAFTA (EPRI, 2013). However, the abovementioned commercial tools have limited functionality related to the DFT analysis. They have PAND and POR gates, but they cannot support the sequence gate, repair action gate or spare gate. For more details regarding the existing commercial tools, readers are referred to check Section 2.7 in the work of Ruijters and Stoelinga (2015).

4.6 DISCUSSION AND FUTURE RESEARCH DIRECTIONS

Reliability analysis is an integral part of dependable system design and development. Over the years, several approaches have been developed for reliability evaluation. FTA is one of the most widely used approaches for reliability and risk assessment. To meet the increasing demand of the society, different new and complex functionalities have been continuously added to modern systems, thus making it difficult to analyze the failure behavior of such systems using classical FTA. The issue of the inability of a fault tree to model the complex time-dependent dynamic failure behavior and different types of redundancy has not gone unnoticed. Researchers have extended the modeling power of the static fault tree by proposing a dynamic fault tree by introducing new dynamic logic gates. While the inclusion of dynamic gates improves the

TABLE 4.2
Sensitivity and Importance Analysis Approaches

Fault Tree Analysis							
Dynamic				Static			
Stochastic Equation	CutSet	BDD	MC	Simulation	Noncoherent	Coherent	Measured Factor
Petkov and Pekov (2009)		Xing and Dugan (2002)	Ou and Dugan (2000); Chiacchio et al. (2018); Noroozian et al. (2019); Ou (2003); Ou and Dugan (2003); Lo et al. (2005)	Marseguerra et al. (2005)	Beeson (2002)	Iman (1987); Dugan and Lyu (1995); Contini et al. (2010); László (2011); Aslansefat and Latif-Shabgahi (2015)	Sensitivity
Liping and Fuzheng (2009)	Liu et (2012)	Dutuit and Rauzy (2001); Chang et al. (2004)	Ou (2003)	Marseguerra and Zio (2004); Zio et al. (2004)	Andrews and Beeson (2003); Beeson and Andrews (2003); Xing (2004); Lu and Jiang (2007); Vaurio (2010); Contini and Matuzas (2011)	Lu and Jiang (2007); Contini and Matuzas (2011); Contini et al. (2010)	Importance

TABLE 4.3
Different DFT Analysis Tools

Software	Method	Measurements Types	Release Year	Remarks
Radyban (Montani et al., 2008)	Bayesian networks	Reliability, Availability	2008	BE Dependencies Not Exact
Galileo (Sullivan et al., 1999)	BDD and Markov methods	Reliability, Availability, Sensitivity, On-demand Safety	1990	Exact Solution State-space explosion
SHARPE (Sahner and Trivedi, 1987)	Modularization	Reliability, Availability, Sensitivity	1987	Non-Exponential Complex BE
MatCarloRe (Manno et al., 2012a)	Monte Carlo	Reliability, Availability, Sensitivity	2012	Non-Exponential Not Exact
PANDORA (Walker, 2009)	Algebraic	Reliability	2009	Non-Exponential Temporal Gates (POR, pSAND)
RAATSS (Manno et al., 2012b)	Adaptive Transitions System	Reliability, Availability, Sensitivity	2012	Non-Exponential Complex Systems
DFTCalc (Arnold et al., 2013a)	Stochastic model checking	Reliability, Availability, Sensitivity	2013	No State-Space Explosion
DFTSim (Boudali et al., 2009)	Monte Carlo	Reliability, Availability, Sensitivity	2009	Repair Modeling Accept Non-Markovian Properties
DIFtree (Dugan et al., 1997)	Modularization	Reliability	1997	Complex BE
SHADE Tree™ (Pullum and Dugan, 1996)	Modularization	Reliability	1996	Simple Solution
The AltaRica (Boiteau et al., 2006)	Markov chain	Reliability	2006	Model Simplification Complex BE

expressiveness of the fault tree model, it complicates the evaluation of the fault tree. Many approaches have been considered in the literature, including algebraic formulas, Markov models, Petri nets, Bayesian networks, and so on, to develop different methodologies for qualitative and quantitative analyses of DFTs. In this chapter, we have provided a high-level overview of these methodologies.

From the overview, it is observed that the DFT analysis approaches have their own strengths and weaknesses. Although the approaches have some shared capabilities, they do have their distinct capabilities, and one approach may perform relatively better than the others in some specific situations. For instance, if the components of a system have an exponentially distributed lifetime, then the classical Markov chain-based approaches can be used to analyze the DFT of such systems. However, a system with non-exponentially distributed system cannot be analyzed using the Markov chain-based approaches. On the other hand, semi-Markov processes, Petri nets, Bayesian networks, and Monte Carlo simulation-based approaches can analyze DFTs having basic events with both exponentially and non-exponentially distributed lifetimes. From a different point of view, the Markov chain and Petri nets based approaches suffer from state-space explosion while evaluating DFTs of complex systems. Therefore, the application of these approaches is limited to small-scale systems. As algebraic solutions to DFTs use mathematical formulas, they do not have the issue of state-space explosion problem. However, for a DFT of a large and complex system, defining mathematical expressions would be a difficult task. With regard to state-space explosion, BN-based approaches show better performance as they can avoid the state-space explosion problem by avoiding the state-space generation by exploiting the local dependencies between variables while modeling complex behavior. Another strength of BN-based approaches over other DFT analysis approaches is that they can perform diagnostic analysis in addition to predictive analysis. In diagnostic analysis, BN-based approaches can propagate new evidence through the network to obtain new beliefs about the failure probability of the events and update prior beliefs. Unlike other DFT analysis approaches, BNs are therefore able to adapt and refine their diagnostic ability over time. However, if a continuous time model is used for BNs, then for an internal node with many parents, with a probability density function, expressing the joint probability distribution would be tedious.

Although extensive research has been performed, there exist some challenges that need additional research. For instance, most of the DFT analysis approaches reviewed in this chapter perform the analysis under the assumption that the DFT of a system is already available, and in most cases, the DFT of a system is derived from a predefined fixed architecture of the system. However, the advancement of technologies has brought loosely connected systems. Typical examples of such systems are cyber-physical systems and the internet of things. These are systems where temporary system architectures/configurations are formed during operation by combining several smaller systems, and these architectures may cease to exist after a certain period of time to form a new architecture. As a result, there may exist infinite possible configurations of such a system, and it is difficult to ensure certainty about a particular system architecture during safety and reliability assessment. This will also affect the structure of the DFT. Therefore, the assessment of such open systems using DFT would require taking into account the uncertainty of the system architecture at a certain point in time. This opens new research avenues to investigate how a meaningful safety and reliability assessment can be performed for open and adaptive systems by taking into account the architectural uncertainty.

Similar to architectural uncertainty, data uncertainty is an important, but less researched, topic in DFT analysis. In SFTs, this issue has been addressed in many different ways. One of the prominent ways is to use the fuzzy set theory to address data uncertainty. From the literature review, we have noticed that there is much less research in DFT analysis involving the fuzzy set theory. Therefore, in the future, it would be worthwhile to perform more research in this area to address the data uncertainty issue.

Another important issue worth mentioning is that even though there exists tool support for creating and analyzing DFTs, it requires a lot of manual effort. This could also introduce errors into the analysis. Model-based safety analysis (Sharvia et al., 2015), which attracted significant interest from industry and academia, can be used to automate the static fault tree generation process from system models. This offers important advantages, not least the reduction in both effort and potential for error, and supports a more iterative design process via automatic synthesis of fault trees. Although the auto generation of static fault trees as part of MBSA and the availability of tool supports for this task was reported in the literature, to the best of the authors' knowledge, limited effort has been made to replicate the same for the automatic generation of DFTs from system models by considering dynamic behavior. Therefore, future research could be directed to address this issue.

In terms of the application of DFTs, there is some potential for the use of DFT for alarm management in different industries. For instance, Simeu-Abazi et al. (2011) used a combination of Petri nets and DFT for alarm filtering and showed its capabilities for alarm nuisance reduction. In addition, the Automata models of dynamic gates have been introduced by Gascard et al. (2011) for alarm modeling. Recently, Aslansefat et al. (2019) and Bahar-Gogani et al. (2017) used the Priority-AND gate for performance evaluation of the alarm system with a variable threshold. Therefore, there is still some potential to use other dynamic or temporal gates such as SEQ, SPARE, and pSAND for the performance evaluation of alarm systems. Moreover, Xu et al. (2011) and Taheri Kalani et al. (2017) used Markov models to evaluate n-sample delay timers for only one component of alarm systems. Based on the idea provided in Kabir et al. (2019) and Papadopoulos et al. (2019), in the future, such Markov models can be used as a complex BE in DFTs to combine and obtain the performance of large-scale dynamic alarm systems.

As mentioned in Section 4.3.2.7, in the safety and reliability engineering domain, the utilization of machine learning (ML) algorithms with safety models has become an emerging research topic. Therefore, there is a large scope to use DFTs as a safety model together with ML for the safety assurance and reliability management of complex, modern systems. One potential way of using DFT with ML algorithms would be to use it as a core model within an ML algorithm for fault detection and diagnosis of systems under the condition of uncertainty. Another way an ML could be used with DFTs to help update the DFT models and failure probability of the basic events within the DFT models is by learning from the emerging behavior and the changing operating environment of the systems. These future researches have the potential to revolutionize performance assessment for many future generations of autonomous self-adaptive systems.

4.7 CONCLUSION

Reliability engineering and the management of complex and dynamic systems is a complicated task. There are many dynamic interactions between system components and multiple temporal and stochastic dependencies between them that need to be taken into account, and not all the existing safety and reliability analysis formalisms are able to capture these dependencies. DFT is one of the popular mechanisms widely used to model dynamic failure behavior systems to evaluate the dynamic reliability of systems. Many new developments and an upward trend are observed in the application of DFTs in reliability engineering and the management of complex systems. This chapter reviews many such developments in DFT-based reliability analysis. The review provided insights into the working mechanism, applicability, strengths, and challenges of different DFT analysis approaches. A discussion is provided on the reviewed methodologies, and directions on potential future research have been provided based on the identified challenges.

ACKNOWLEDGMENTS

This work was partly supported by the DEIS H2020 Project under Grant 732242. The authors would like to thank the EDF Energy Research and Development UK Centre, AURA Innovation Centre and the University of Hull for their support.

BIBLIOGRAPHY

Aghassi, H., Aghassi, F., 2012. Fault tree analysis speed-up with GPU parallel computing. *Computer Information Systems and Industrial Management Applications, Tehran* 5, 106–114.

Aizpurua, J.I., Catterson, V.M., Papadopoulos, Y., Chiacchio, F., D'Urso, D., 2017a. Supporting group maintenance through prognostics-enhanced dynamic dependability prediction. *Reliability Engineering and System Safety* 168, 171–188.

Aizpurua, J.I., Catterson, V.M., Papadopoulos, Y., Chiacchio, F., Manno, G., 2017b. Improved dynamic dependability assessment through integration with prognostics. *IEEE Transactions on Reliability* 66(3), 893–913.

Aliee, H., Zarandi, H.R., 2013. A fast and accurate fault tree analysis based on stochastic logic implemented on field-programmable gate arrays. *IEEE Transactions on Reliability* 62(1), 13–22. doi:10.1109/TR.2012.2221012.

Amari, S., Dill, G., Howald, E., 2003. A new approach to solve dynamic fault trees. In: *Annual Reliability and Maintainability Symposium, 2003*, Tampa, FL, USA, IEEE, pp. 374–379.

Anand, A., Somani, A.K., 1998. Hierarchical analysis of fault trees with dependencies, using decomposition. In: *Proceedings of the Annual Reliability and Maintainability Symposium*, Anaheim, CA, USA, IEEE, pp. 69–75. doi:10.1109/RAMS.1998.653591.

Andrews, J.D., Beeson, S., 2003. Birnbaum's measure of component importance for noncoherent systems. *IEEE Transactions on Reliability* 52(2), 213–219. http://www.ncbi.nlm.nih.gov/pubmed/213219.

Arnold, F., Belinfante, A., Van der Berg, F., Guck, D., Stoelinga, M., 2013a. DFTCalc: A tool for efficient fault tree analysis. In: *International Conference on Computer Safety, Reliability, and Security*, Berlin, Springer, pp. 293–301.

Arnold, F., Belinfante, A., Van Der Berg, F., van der Berg, F.I., Guck, D., Stoelinga, M.I.A., 2013b. DFTCalc: A tool for efficient fault tree analysis (extended version). Technical Report TR-CTIT-13-13. Centre for Telematics and Information Technology (CTIT).

Askarian, M., Zarghami, R., Jalali-Farahani, F., Mostoufi, N., 2016. Fusion of micro-macro data for fault diagnosis of a sweetening unit using Bayesian network. *Chemical Engineering Research and Design* 115, 325–334.

Aslansefat, K., 2014. A novel approach for reliability and safety evaluation of control systems with dynamic fault tree. MSc. Thesis. Abbaspur Campus, Shahid Beheshti University, Tehran, Iran.

Aslansefat, K., Gogani, M.B., Kabir, S., Shoorehdeli, M.A., Yari, M., 2019. Performance evaluation and design for variable threshold alarm systems through semi-Markov process. *ISA Transactions* 97, 282–295. doi:10.1016/j.isatra.2019.08.015.

Aslansefat, K., Latif-Shabgahi, G., 2015. A systematic approach to sensitivity analysis of fault tolerant systems in NMR architecture. *Journal of Intelligent Procedures in Electrical Technology (JIPET)* 5, 3–14.

Aslansefat, K., Latif-Shabgahi, G., 2019. A hierarchical approach for dynamic fault trees solution through semi-Markov process. *IEEE Transactions on Reliability*, 1–18. doi:10.1109/TR.2019.2923893.

Auvation, 2016. OpenFTA. URL: http://www.openfta.com/.

Bahar-Gogani, M., Aslansefat, K., Shoorehdeli, M.A., 2017. A novel extended adaptive thresholding for industrial alarm systems. In: *2017 Iranian Conference on Electrical Engineering (ICEE)*, Tehran, Iran, IEEE, pp. 759–765. doi:10.1109/IranianCEE.2017.7985140.

Bavuso, S.J., Rothmann, E., Dugan, J.B., Trivedi, K.S., Mittal, N., Boyd, M.A., Geist, R.M., Smotherman, M.D., 1994. HiRel: Hybrid automated reliability predictor (HARP) integrated reliability tool system (version 7.0). NASA Technical Paper 3452, Hampton, Virginia, National Aeronautics and Space Administration, pp. 1–4.

Beeson, S., Andrews, J.D., 2003. Importance measures for noncoherent-system analysis. *IEEE Transactions on Reliability* 52(3), 301–310.

Beeson, S.C., 2002. Non-coherent fault tree analysis. Ph.D. thesis, Loughborough University. © Sally Christian Beeson, Loughborough, UK.

Bernardi, S., Campos, J., Merseguer, J., 2011. Timing-failure risk assessment of UML design using time petri net bound techniques. *IEEE Transactions on Industrial Informatics* 7(1), 90–104.

Bobbio, A., Ciancamerla, E., Franceschinis, G., Gaeta, R., Minichino, M., Portinale, L., 2003. Sequential application of heterogeneous models for the safety analysis of a control system: A case study. *Reliability Engineering and System Safety* 81(3), 269–280.

Bobbio, A., Portinale, L., Minichino, M., Ciancamerla, E., 2001. Improving the analysis of dependable systems by mapping fault trees into Bayesian networks. *Reliability Engineering and System Safety* 71(3), 249–260. doi:10. 1016/S0951-8320(00)00077-6.

Boiteau, M., Dutuit, Y., Rauzy, A., Signoret, J.P., 2006. The AltaRica dataflow language in use: Modeling of production availability of a multi-state system. *Reliability Engineering and System Safety* 91(7), 747–755.

Boudali, H., Crouzen, P., Stoelinga, M., 2007a. A compositional semantics for dynamic fault trees in terms of interactive Markov chains. In: *International Symposium on Automated Technology for Verification and Analysis*, Berlin, Springer, pp. 441–456.

Boudali, H., Crouzen, P., Stoelinga, M., 2007b. Dynamic fault tree analysis using input/output interactive Markov chains. In: *37th Annual IEEE/IFIP International Conference on Dependable Systems and Networks (DSN'07)*, Edinburgh, IEEE, pp. 708–717.

Boudali, H., Crouzen, P., Stoelinga, M., 2010. A rigorous, compositional, and extensible framework for dynamic fault tree analysis. *IEEE Transactions on Dependable and Secure Computing* 7(2), 128–143.

Boudali, H., Dugan, J.B., 2005. A discrete-time Bayesian network reliability modeling and analysis framework. *Reliability Engineering and System Safety* 87(3), 337–349.

Boudali, H., Dugan, J.B., 2006. A continuous-time Bayesian network reliability modeling, and analysis framework. *IEEE Transaction on Reliability* 55(1), 86–97.

Boudali, H., Nijmeijer, A., Stoelinga, M.I., 2009. DFTSim: A simulation tool for extended dynamic fault trees. In: *Proceedings of the 2009 Spring Simulation Multiconference, Society for Computer Simulation International*, p. 31.

Boyd, M.A., 1992. Dynamic fault tree models: Techniques for analysis of advanced fault tolerant computer systems. Ph.D. thesis. Duke University, Durham, North Carolina, USA.

Brameret, P.A., Rauzy, A., Roussel, J.M., 2015. Automated generation of partial Markov chain from high level descriptions. *Reliability Engineering and System Safety* 139, 179–187.

Chang, Y.R., Amari, S.V., Kuo, S.Y., 2004. Computing system failure frequencies and reliability importance measures using OBDD. *IEEE Transactions on Computers* 53(1), 54–68.

Chaux, P.Y., Roussel, J.M., Lesage, J.J., Deleuze, G., Bouissou, M., 2013. Towards a unified definition of minimal cut sequences. *IFAC Proceedings Volumes* 46, 1–6. doi:10.3182/20130904-3-UK-4041.00013. *4th IFAC Workshop on Dependable Control of Discrete Systems.*

Chen, G., Ge, Z., 2018. Hierarchical Bayesian network modeling framework for large-scale process monitoring and decision making. *IEEE Transactions on Control Systems Technology* 28, 671–679.

Cheng, J., Zhu, C., Fu, W., Wang, C., Sun, J., 2019. An imitation medical diagnosis method of hydro-turbine generating unit based on Bayesian network. *Transactions of the Institute of Measurement and Control* 41, 3306–3420.

Chiacchio, F., Aizpurua, J.I., Compagno, L., Khodayee, S.M., D'Urso, D., 2019. Modelling and resolution of dynamic reliability problems by the coupling of simulink and the stochastic hybrid fault tree object oriented (shyftoo). *Library and Information* 10, 283.

Chiacchio, F., Aizpurua, J.I., D'Urso, D., Compagno, L., 2018. Coherence region of the priority-and gate: Analytical and numerical examples. *Quality and Reliability Engineering International* 34(1), 107–115.

Chiacchio, F., Cacioppo, M., D'Urso, D., Manno, G., Trapani, N., Compagno, L., 2013. A Weibull-based compositional approach for hierarchical dynamic fault trees. *Reliability Engineering and System Safety* 109, 45–52.

Chiacchio, F., Compagno, L., D'Urso, D., Manno, G., Trapani, N., 2011. An open-source application to model and solve dynamic fault tree of real industrial systems. In: *2011 5th International Conference on Software, Knowledge Information, Industrial Management and Applications (SKIMA) Proceedings*, Benevento, Italy, IEEE, pp. 1–8.

Codetta-Raiteri, D., 2005. The conversion of dynamic fault trees to stochastic petri nets, as a case of graph transformation. *Electronic Notes in Theoretical Computer Science* 127(2), 45–60.

Codetta-Raiteri, D., 2015. Applying generalized continuous time Bayesian networks to a reliability case study. *IFAC-PapersOnLine* 48(21), 676–681.

Codetta-Raiteri, D., Portinale, L., 2015. Dynamic Bayesian networks for fault detection, identification, and recovery in autonomous spacecraft. *IEEE Transactions on Systems, Man, and Cybernetics: Systems* 45(1), 13–24.

Codetta-Raiteri, D., Portinale, L., 2017. Generalized continuous time Bayesian networks as a modelling and analysis formalism for dependable systems. *Reliability Engineering and System Safety* 167, 639–651.

Contini, S., Fabbri, L., Matuzas, V., 2010. A novel method to apply importance and sensitivity analysis to multiple fault-trees. *Journal of Loss Prevention in the Process Industries* 23(5), 574–584.

Contini, S., Matuzas, V., 2011. New methods to determine the importance measures of initiating and enabling events in fault tree analysis. *Reliability Engineering and System Safety* 96(7), 775–784.

Crouzen, P., 2006. Compositional analysis of dynamic fault trees using input/output interactive Markov chains. Master's thesis. University of Twente, Enschede, Netherlands.

Dominguez-Garcia, A.D., Kassakian, J.G., Schindall, J.E., Zinchuk, J.J., 2008. An integrated methodology for the dynamic performance and reliability evaluation of fault-tolerant systems. *Reliability Engineering and System Safety* 93(11), 1628–1649.

Dugan, J., 1993. Analysis of a hardware and software fault tolerant processor for critical applications. In: *9th Computing in Aerospace Conference*, San Diego,CA, IEEE, p. 4573.

Dugan, J., Sullivan, K., Coppit, D., 2000. Developing a low-cost high-quality software tool for dynamic fault-tree analysis. *IEEE Transactions on Reliability* 49(1), 49–59. doi:10.1109/24.855536.

Dugan, J.B., Bavuso, S., Boyd, M., 1992. Dynamic fault-tree models for fault- tolerant computer systems. *IEEE Transactions on Reliability* 41(3), 363–377.

Dugan, J.B., Bavuso, S.J., Boyd, M.A., 1993. Fault trees and Markov models for reliability analysis of fault-tolerant digital systems. *Reliability Engineering and System Safety* 39(3), 291–307.

Dugan, J.B., Lyu, M.R., 1995. System-level reliability and sensitivity analyses for three fault-tolerant system architectures. In: F. Cristian, G. Le Lann, T. Lunt (eds) *Dependable Computing for Critical Applications 4*, Vienna, Springer, pp. 459–477.

Dugan, J.B., Venkataraman, B., Gulati, R., 1997. DIFTree: A software package for the analysis of dynamic fault tree models. In: *Annual Reliability and Maintainability Symposium*, Philadelphia, PA, IEEE, pp. 64–70.

Dutuit, Y., Rauzy, A., 2001. Efficient algorithms to assess component and gate importance in fault tree analysis. *Reliability Engineering and System Safety* 72(2), 213–222.

Edifor, E., Walker, M., Gordon, N., 2012. Quantification of priority-or gates in temporal fault trees. In: *International Conference on Computer Safety, Reliability, and Security*, Berlin, Springer, pp. 99–110.

Ejlali, A., Miremadi, S.G., 2003. Time-to-failure tree. In: *Annual Reliability and Maintainability Symposium, 2003*, Tampa, FL, IEEE, pp. 148–152.

Ejlali, A., Miremadi, S.G., 2004. FPGA-based Monte Carlo simulation for fault tree analysis. *Microelectronics Reliability* 44(6), 1017–1028.

Elderhalli, Y., Hasan, O., Ahmad, W., Tahar, S., 2017. Dynamic fault trees analysis using an integration of theorem proving and model checking. *Clinical Orthopaedics and Related Research*. Abs/1712.02872. arXiv:1712.02872.

EPRI, 2013. EPRI CAFTA. URL: http://www.epri.com/abstracts/Pages/ProductAbstract.aspx?ProductId=000000000001015514.

Fecarotti, C., Andrews, J., Chen, R., 2016. A petri net approach for performance modelling of polymer electrolyte membrane fuel cell systems. *International Journal of Hydrogen Energy* 41(28), 12242–12260.

Fussell, J., Aber, E., Rahl, R., 1976. On the quantitative analysis of priority- and failure logic. *IEEE Transactions on Reliability* R-25, 324–326.

Garg, H., 2013. Reliability analysis of repairable systems using petri nets and vague lambda-tau methodology. *ISA Transactions* 52(1), 6–18.

Garg, H., 2014. Reliability, availability and maintainability analysis of industrial systems using pso and fuzzy methodology. *MAPAN* 29(2), 115–129.

Garg, H., 2015. An approach for analyzing the reliability of industrial system using fuzzy Kolmogorov's differential equations. *Arabian Journal for Science and Engineering* 40(3), 975–987.

Garg, H., Rani, M., Sharma, S., 2014a. An approach for analyzing the reliability of industrial systems using soft-computing based technique. *Expert Systems with Applications* 41(2), 489–501.

Garg, H., Rani, M., Sharma, S., Vishwakarma, Y., 2014b. Bi-objective optimization of the reliability-redundancy allocation problem for series-parallel system. *Journal of Manufacturing Systems* 33(3), 335–347.

Garg, H., Sharma, S., 2011. Multi-objective optimization of crystallization unit in a fertilizer plant using particle swarm optimization. *International Journal of Applied Science and Engineering* 9, 261–276.

Gascard, E., Simeu-Abazi, Z., 2018. Quantitative analysis of dynamic fault trees by means of Monte Carlo simulations: Event-driven simulation approach. *Reliability Engineering and System Safety* 180, 487–504.

Gascard, E., Simeu-Abazi, Z., Younes, J., 2011. Exploitation of built in test for diagnosis by using dynamic fault trees: Implementation in matlab simulink. In: C. Berenguer, A. Grall, C. G. Soares (eds) *Advances in Safety, Reliability and Risk Management*, Troyes, France, CRC Press, pp. 436–444.

Ge, D., Li, D., Chou, Q., Zhang, R., Yang, Y., 2016. Quantification of highly coupled dynamic fault tree using IRVPM and SBDD. *Quality and Reliability Engineering International* 32(1), 139–151.

Ge, D., Lin, M., Yang, Y., Zhang, R., Chou, Q., 2015a. Quantitative analysis of dynamic fault trees using improved sequential binary decision diagrams. *Reliability Engineering and System Safety* 142, 289–299.

Ge, D., Lin, M., Yang, Y., Zhang, R., Chou, Q., 2015b. Reliability analysis of complex dynamic fault trees based on an adapted KD Heidtmann algorithm. *Proceedings of the Institution of Mechanical Engineers, Part O: Journal of Risk and Reliability* 229, 576–586.

Ge, D., Yang, Y., 2015. Reliability analysis of non-repairable systems modeled by dynamic fault trees with priority and gates. *Applied Stochastic Models in Business and Industry* 31(6), 809–822.

Ge, D., Yang, Y., 2017. Negating a generalized cut sequence: Bridging the gap between dynamic fault trees quantification and sum of disjoint products methods. *Quality and Reliability Engineering International* 33(2), 357–367.

Getir, S., Grunske, L., van Hoorn, A., Kehrer, T., Noller, Y., Tichy, M., 2018. Supporting semi-automatic co-evolution of architecture and fault tree models. *Journal of Systems and Software* 142, 115–135.

Gheraibia, Y., Kabir, S., Aslansefat, K., Sorokos, I., Papadopoulos, Y., 2019. Safety + AI: A novel approach to update safety models using artificial intelligence. *IEEE Access* 7, 135855–135869. doi:10.1109/ACCESS.2019.2941566.

Gulati, R., Dugan, J.B., 1997. A modular approach for analyzing static and dynamic fault trees. In: *Annual Reliability and Maintainability Symposium*, Philadelphia, PA, IEEE, pp. 57–63.

Guo, W.G., Han, W., Liu, S.Y., 2011. Dynamic fault tree based on Weibull distribution. In: J. Gao (ed) *Advanced Materials Research*, Switzerland, Trans Tech Publ., pp. 1322–1327.

Hao, J., Zhang, L., Wei, L., 2014. Reliability analysis based on improved dynamic fault tree. In: J. Lee, J. Ni, J. Sarangapani, J. Mathew (eds) *Engineering Asset Management 2011*, London, Springer, pp. 283–299.

Heidtmann, K.D., 1989. Smaller sums of disjoint products by subproduct inversion. *IEEE Transactions on Reliability* 38(3), 305–311.

Hermanns, H., 2002. Interactive Markov chains. Volume 2428 of *Lecture Notes in Computer Science*, Berlin, Springer.

Herscheid, L., Tröger, P., 2014. Specification of dynamic fault tree concepts with stochastic Petri nets. In: *Eighth International Conference on Software Security and Reliability*, San Francisco, CA, IEEE, pp. 177–186.

Huang, C.Y., Chang, Y.R., 2007. An improved decomposition scheme for assessing the reliability of embedded systems by using dynamic fault trees. *Reliability Engineering and System Safety* 92(10), 1403–1412.

Huang, H., Li, Y., Sun, J., Yang, Y., Xiao, N., 2013. Fuzzy dynamic fault tree analysis for the solar array drive assembly. *Journal of Mechanical Engineering* 49(19), 70–76.

Hura, G.S., Atwood, J.W., 1988. The use of petri nets to analyze coherent fault trees. *IEEE Transactions on Reliability* 37(5), 469–474.

Iman, R.L., 1987. A matrix-based approach to uncertainty and sensitivity analysis for fault trees 1. *Risk Analysis* 7(1), 21–33.

Isograph, 2016. Isograph FaultTree+. URL: http://www.isograph.com/software/reliability-workbench/fault-tree-analysis/.

ITEM Software, 2016. ITEM ToolKit. URL: http://www.itemsoft.com/item_toolkit.html.

Jiang, G., Hongjie, Y., Peichang, L., Peng, L., 2018. A new approach to fuzzy dynamic fault tree analysis using the weakest n-dimensional t-norm arithmetic. *Chinese Journal of Aeronautics* 31(7), 1506–1514.

Junges, S., Katoen, J.P., Stoelinga, M., Volk, M., 2018. One net fits all. In: Khomenko, V., Roux, O.H. (Eds.), *Application and Theory of Petri Nets and Concurrency*, Springer International Publishing, Cham, pp. 272–293.

Kabir, S., 2017. An overview of fault tree analysis and its application in model based dependability analysis. *Expert Systems with Applications* 77, 114–135.

Kabir, S., Aslansefat, K., Sorokos, I., Papadopoulos, Y., Gheraibia, Y., 2019. A conceptual framework to incorporate complex basic events in HiP-HOPS. In: *International Symposium on Modelling-Based Safety and Assessment*, Switzerland, Springer, pp. 109–124.

Kabir, S., Edifor, E., Walker, M., Gordon, N., 2014a. Quantification of temporal fault trees based on fuzzy set theory. In: *Proceedings of the Ninth International Conference on Dependability and Complex Systems DepCoS-RELCOMEX*, Springer International Publishing, Brunów, pp. 255–264. doi:10.1007/978-3-319-07013-1_24.

Kabir, S., Walker, M., Papadopoulos, Y., 2014b. Reliability analysis of dynamic systems by translating temporal fault trees into Bayesian networks. In: Ortmeier, F., Rauzy, A. (Eds.), *Model-Based Safety and Assessment*, Springer International Publishing, Cham. Volume 8822 of *Lecture Notes in Computer Science*, pp. 96–109. doi:10.1007/978-3-319-12214-4.

Kabir, S., Geok, T.K., Kumar, M., Yazdi, M., Hossain, F., 2020. A method for temporal fault tree analysis using intuitionistic fuzzy set and expert elicitation. *IEEE Access* 8, 980–996. doi:10.1109/ACCESS.2019.2961953.

Kabir, S., Papadopoulos, Y., 2018. A review of applications of fuzzy sets to safety and reliability engineering. *International Journal of Approximate Reasoning* 100, 29–55. doi:10.1016/j.ijar.2018.05.005.

Kabir, S., Papadopoulos, Y., 2019. Applications of Bayesian networks and petri nets in safety, reliability, and risk assessments: A review. *Safety Science* 115, 154–175.

Kabir, S., Papadopoulos, Y., Walker, M., Parker, D., Aizpurua, J.I., Lampe, J., Rüde, E., 2017. A model-based extension to HiP-HOPS for dynamic fault propagation studies. In: *International Symposium on Modelling-Based Safety and Assessment*, Switzerland, Springer, pp. 163–178. doi:10.1007/978-3-319-64119-5_11.

Kabir, S., Walker, M., Papadopoulos, Y., 2015. Quantitative evaluation of pandora temporal fault trees via petri nets. *IFAC-PapersOnLine* 48(21), 458–463. doi:10.1016/j.ifacol.2015.09.569.

Kabir, S., Walker, M., Papadopoulos, Y., 2018a. Dynamic system safety analysis in HiP-HOPS with petri nets and Bayesian networks. *Safety Science* 105, 55–70.

Kabir, S., Yazdi, M., Aizpurua, J.I., Papadopoulos, Y., 2018b. Uncertainty- aware dynamic reliability analysis framework for complex systems. *IEEE Access* 6, 29499–29515. doi:10.1109/ACCESS.2018.2843166.

Kabir, S., Walker, M., Papadopoulos, Y., Rüde, E., Securius, P., 2016. Fuzzy temporal fault tree analysis of dynamic systems. *International Journal of Approximate Reasoning* 77, 20–37. doi:10.1016/j.ijar.2016.05.006.

Kaiser, B., Gramlich, C., Foürster, M., 2007. State/event fault trees - A safety analysis model for software-controlled systems. *Reliability Engineering and System Safety* 92(11), 1521–1537.

Kara-Zaitri, C., Ever, E., 2009. A hardware accelerated semi analytic approach for fault trees with repairable components. In: *2009 11th International Conference on Computer Modelling and Simulation*, Cambridge, UK, IEEE, pp. 146–151.

Knezevic, J., Odoom, E., 2001. Reliability modelling of repairable systems using petri nets and fuzzy lambda-tau methodology. *Reliability Engineering and System Safety* 73(1), 1–17.

Lampis, M., Andrews, J., 2009. Bayesian belief networks for system fault diagnostics. *Quality and Reliability Engineering International* 25(4), 409–426.

Laszlo, P., 2011. Sensitivity investigation of fault tree analysis with matrix-algebraic method. *Theory and Applications of Mathematics and Computer Science* 1, 34–44.

Le, B., Andrews, J., 2016. Petri net modelling of bridge asset management using maintenance-related state conditions. *Structure and Infrastructure Engineering* 12(6), 730–751.

Li, Y.F., Huang, H.Z., Liu, Y., Xiao, N., Li, H., 2012. A new fault tree analysis method: Fuzzy dynamic fault tree analysis. *Eksploatacja i Niezawodnosc- Maintenance and Reliability* 14, 208–214.

Li, Y.F., Mi, J., Liu, Y., Yang, Y.J., Huang, H.Z., 2015. Dynamic fault tree analysis based on continuous-time Bayesian networks under fuzzy numbers. *Proceedings of the Institution of Mechanical Engineers, Part O: Journal of Risk and Reliability* 229, 530–541. doi:10.1177/1748006X15588446.

Linard, A., Bucur, D., Stoelinga, M., 2019. Fault trees from data: Efficient learning with an evolutionary algorithm. arXiv preprint arXiv:1909.06258.

Lindhe, A., Norberg, T., Rosen, L., 2012. Approximate dynamic fault tree calculations for modelling water supply risks. *Reliability Engineering and System Safety* 106, 61–71.

Liping, H., Fuzheng, Q., 2009. Possibilistic entropy-based measure of importance in fault tree analysis. *Journal of Systems Engineering and Electronics* 20, 434–444.

Liu, D., Chen, X.J., Li, Y., Zhao, Z.W., Li, X.M., 2012. Dynamic fault trees analysis for importance measures based on cut sequence set model. In: A. Wu (ed) *Applied Mechanics and Materials*, Switzerland, Trans Tech Publ., pp. 578–582.

Liu, D., Xing, W., Zhang, C., Li, R., Li, H., 2007a. Cut sequence set generation for fault tree analysis. In: Y. H. Less, H. N. Kim, J. Kim, Y. Park, L. T. Yang, S. W. Kim (eds) *Embedded Software and Systems*, Berlin, Springer, pp. 592–603.

Liu, D., Zhang, C., Xing, W., Li, R., Li, H., 2007b. Quantification of cut sequence set for fault tree analysis. In: R. Perrott, B. M. Chapman, J. Subhlok, R. F. de Mello, L. T. Yang (eds) *High Performance Computing and Communications*, Berlin, Springer, pp. 755–765.

Liu, T., Chiou, S., 1997. The application of petri nets to failure analysis. *Reliability Engineering and System Safety* 57(2), 129–142.

Lo, H.k., Huang, C.y., Chang, Y.r., Huang, W.c., Chang, J.r., 2005. Reliability and sensitivity analysis of embedded systems with modular dynamic fault trees. In: *TENCON 2005–2005 IEEE Region 10 Conference*, Australia, IEEE, pp. 1–6.

Long, W., Sato, Y., Horigome, M., 2000. Quantification of sequential failure logic for fault tree analysis. *Reliability Engineering and System Safety* 67(3), 269–274. doi:10.1016/S0951-8320(99)00075-7.

Longo, F., Scarpa, M., Puliafito, A., 2016. WebSPN: A flexible tool for the analysis of non-Markovian stochastic petri nets. In: L. Fionella, A. Puliafito (eds) *Principles of Performance and Reliability Modeling and Evaluation*. Switzerland, Springer, pp. 255–285.

Lu, L., Jiang, J., 2007. Joint failure importance for noncoherent fault trees. *IEEE Transactions on Reliability* 56(3), 435–443.

Malhotra, M., Trivedi, K.S., 1995. Dependability modeling using petri-nets. *IEEE Transactions on Reliability* 44(3), 428–440.

Manian, R., Coppit, D.W., Sullivan, K.J., Dugan, J.B., 1999. Bridging the gap between systems and dynamic fault tree models. In: *Annual Reliability and Maintainability. Symposium. 1999 Proceedings (Cat. No. 99CH36283)*, Washington, DC, IEEE, pp. 105–111.

Manian, R., Dugan, J.B., Coppit, D., Sullivan, K.J., 1998. Combining various solution techniques for dynamic fault tree analysis of computer systems. In: *Proceedings of the Third IEEE International High-Assurance Systems Engineering Symposium*, Washington, DC, IEEE, pp. 21–28. doi:10.1109/HASE.1998.731591.

Manno, G., Chiacchio, F., Compagno, L., D'Urso, D., Trapani, N., 2012a. Matcarlore: An integrated ft and Monte Carlo simulink tool for the reliability assessment of dynamic fault tree. *Expert Systems with Applications* 39(12), 10334–10342.

Manno, G., Chiacchio, F., D'Urso, D., Trapani, N., Compagno, L., 2012b. Raatss, an extensible matlab® toolbox for the evaluation of repairable dynamic fault trees. In: *International Conference on Probabilistic Safety Assessment and Management*, Helsinki, Finland, ESREL, pp. 1–10.

Manno, G., Chiacchio, F., Compagno, L., D'Urso, D., Trapani, N., 2014. Conception of repairable dynamic fault trees and resolution by the use of raatss, a MATLAB® toolbox based on the ats formalism. *Reliability Engineering and System Safety* 121, 250–262.

Marquez, D., Neil, M., Fenton, N., 2008. Solving dynamic fault trees using a new hybrid Bayesian network inference algorithm. In: *16th Mediterranean Conference on Control and Automation*, France, IEEE, pp. 609–614.

Marquez, D., Neil, M., Fenton, N., 2010. Improved reliability modeling using Bayesian networks and dynamic discretization. *Reliability Engineering and System Safety* 95(4), 412–425.

Marseguerra, M., Zio, E., 2004. Monte Carlo estimation of the differential importance measure: Application to the protection system of a nuclear reactor. *Reliability Engineering and System Safety* 86(1), 11–24.

Marseguerra, M., Zio, E., Devooght, J., Labeau, P.E., 1998. A concept paper on dynamic reliability via Monte Carlo simulation. *Mathematics and Computers in Simulation* 47(2–5), 371–382.

Marseguerra, M., Zio, E., Podofillini, L., 2005. First-order differential sensitivity analysis of a nuclear safety system by Monte Carlo simulation. *Reliability Engineering and System Safety* 90(2–3), 162–168.

Merle, G., 2010. Algebraic modelling of dynamic fault trees, contribution to qualitative and quantitative analysis. Ph.D. thesis. ENS CACHAN, Cachan, France.

Merle, G., Roussel, J.M., Lesage, J.J., 2011. Algebraic determination of the structure function of dynamic fault trees. *Reliability Engineering and System Safety* 96(2), 267–277.

Merle, G., Roussel, J.M., Lesage, J.J., 2014. Quantitative analysis of dynamic fault trees based on the structure function. *Quality and Reliability Engineering International* 30(1), 143–156.

Merle, G., Roussel, J.M., Lesage, J.J., Bobbio, A., 2010. Probabilistic algebraic analysis of fault trees with priority dynamic gates and repeated events. *IEEE Transactions on Reliability* 59(1), 250–261.

Merle, G., Roussel, J.M., Lesage, J.J., Perchet, V., Vayatis, N., 2016. Quantitative analysis of dynamic fault trees based on the coupling of structure functions and Monte Carlo simulation. *Quality and Reliability Engineering International* 32(1), 7–18.

Mi, J., Li, Y.F., Yang, Y.J., Peng, W., Huang, H.Z., 2016. Reliability assessment of complex electromechanical systems under epistemic uncertainty. *Reliability Engineering and System Safety* 152, 1–15.

Minato, S.i., 2001. Zero-suppressed BDDs and their applications. *International Journal on Software Tools for Technology Transfer* 3(2), 156–170.

Montani, S., Portinale, L., Bobbio, A., 2005. Dynamic Bayesian networks for modeling advanced fault tree features in dependability analysis. In: *Proceedings of the Sixteenth European Conference on Safety and Reliability*, the Netherlands, AA Balkema, pp. 1415–1422.

Montani, S., Portinale, L., Bobbio, A., Codetta-Raiteri, D., 2006a. Automatically translating dynamic fault trees into dynamic Bayesian networks by means of a software tool. In: *The First International Conference on Availability, Reliability and Security*, Austria, IEEE, pp. 1–6.

Montani, S., Portinale, L., Bobbio, A., Varesio, M., Codetta-Raiteri, D., 2006b. A tool for automatically translating dynamic fault trees into dynamic Bayesian networks. In: *Annual Reliability and Maintainability Symposium*, Newport Beach, CA, IEEE, pp. 434–441.

Montani, S., Portinale, L., Bobbio, A., Codetta-Raiteri, D., 2008. Radyban: A tool for reliability analysis of dynamic fault trees through conversion into dynamic Bayesian networks. *Reliability Engineering and System Safety* 93(7), 922–932.

Ni, J., Tang, W., Xing, Y., 2013. A simple algebra for fault tree analysis of static and dynamic systems. *IEEE Transactions on Reliability* 62(4), 846–861.

Niwas, R., Garg, H., 2018. An approach for analyzing the reliability and profit of an industrial system based on the cost free warranty policy. *Journal of the Brazilian Society of Mechanical Sciences and Engineering* 40(5), 265.

Norberg, T., Rosen, L., Lindhe, A. 2009. Added value in fault tree analyses. In: *Safety, Reliability and Risk Analysis: Theory, Methods and Applications. Joint ESREL (European Safety and Reliability) and SRA-Europe (Society for Risk Analysis Europe) Conference*, Valencia, Spain, 22nd–25th September, pp. 1041–1049.

Noroozian, A., Kazemzadeh, R.B., Zio, E., Niaki, S.T.A., 2019. Importance analysis considering time-varying parameters and different perturbation occurrence times. *Quality and Reliability Engineering International*, 35, 2558–2578.

Ou, Y., 2003. Dependability and sensitivity analysis of multi-phase systems using Markov chains. PhD thesis. University of Virginia, Charlottesville, Virginia.

Ou, Y., Dugan, J.B., 2000. Sensitivity analysis of modular dynamic fault trees. In: *Proceedings of the IEEE International Computer Performance and Dependability Symposium (IPDS 2000)*, Chicago, IL, IEEE, pp. 35–43.

Ou, Y., Dugan, J.B., 2003. Approximate sensitivity analysis for acyclic Markov reliability models. *IEEE Transactions on Reliability* 52(2), 220–230.

Palshikar, G.K., 2002. Temporal fault trees. *Information and Software Technology* 44(3), 137–150.

Papadopoulos, Y., Aslansefat, K., Katsaros, P., Bozzano, M., 2019. *Model-Based Safety and Assessment*, Switzerland, Springer.

Patterson-Hine, F., Dugan, J.B., 1992. Modular techniques for dynamic fault tree-analysis. In: *Annual Reliability and Maintainability Symposium 1992 Proceedings*, Las Vegas, NV, IEEE, pp. 363–369.

Petkov, G., Pekov, M., 2009. Ageing effects sensitivity analysis by dynamic system reliability methods (GO-FLOW and ATRD). EC JRC APSA Network Meeting, pp. 1–27, Bulgaria.

Ping, Y.L., 2011. Analysis on dynamic fault tree based on fuzzy set. *Applied Mechanics and Materials*, 110–116, 2416–2420.

Piriou, P.Y., Faure, J.M., Lesage, J.J., 2017. Generalized boolean logic driven Markov processes: A powerful modeling framework for model-based safety analysis of dynamic repairable and reconfigurable systems. *Reliability Engineering and System Safety* 163, 57–68.

Piriou, P.Y., Faure, J.M., Lesage, J.J., 2019. Finding the minimal cut sequences of dynamic, repairable, and reconfigurable systems from generalized Boolean logic driven Markov process models. *Proceedings of the Institution of Mechanical Engineers, Part O: Journal of Risk and Reliability*, 1–12. doi:10.1177/1748006X19827128.

Portinale, L., Codetta-Raiteri, D., Montani, S., 2010. Supporting reliability engineers in exploiting the power of dynamic Bayesian networks. *International Journal of Approximate Reasoning* 51(2), 179–195.

Pullum, L.L., Dugan, J.B., 1996. Fault tree models for the analysis of complex computer-based systems. In: *Proceedings of the 1996 Annual Reliability and Maintainability Symposium*, Las Vegas, NV, IEEE, pp. 200–207.

Ramezani, Z., Latif-Shabgahi, G.R., Khajeie, P., Aslansefat, K., 2016. Hierarchical steady-state availability evaluation of dynamic fault trees through equal Markov model. In: *2016 24th Iranian Conference on Electrical Engineering (ICEE)*, Iran, IEEE, pp. 1848–1854.

Ranjbar, A.H., Kiani, M., Fahimi, B., 2011. Dynamic Markov model for reliability evaluation of power electronic systems. In: *2011 International Conference on Power Engineering, Energy and Electrical Drives*, Spain, IEEE, pp. 1–6.

Rao, K.D., Gopika, V., Rao, V.S., Kushwaha, H., Verma, A.K., Srividya, A., 2009. Dynamic fault tree analysis using Monte Carlo simulation in probabilistic safety assessment. *Reliability Engineering and System Safety* 94(4), 872–883.

Raptodimos, Y., Lazakis, I., 2017. Fault tree analysis and artificial neural network modelling for establishing a predictive ship machinery maintenance methodology. In: *International Conference on Smart Ship Technology*, London, UK, Royal Institution of Naval Architects, pp. 1–11.

Rauzy, A.B., 2011. Sequence algebra, sequence decision diagrams and dynamic fault trees. *Reliability Engineering and System Safety* 96(7), 785–792. doi:10.1016/j.ress.2011.02.005.

ReliaSoft, 2016. BlockSim: System reliability and maintainability analysis software tool. URL: http://www.reliasoft.com/BlockSim/index.html.

Ruijters, E., Stoelinga, M., 2015. Fault tree analysis: A survey of the state- of-the-art in modeling, analysis and tools. *Computer Science Review* 15, 29–62.

Sahner, R.A., Trivedi, K.S., 1987. Reliability modeling using Sharpe. *IEEE Transactions on Reliability* 36(2), 186–193.

Sharvia, S., Kabir, S., Walker, M., Papadopoulos, Y., 2015. Model-based dependability analysis: State-of-the-art, challenges, and future outlook. In: I. Mistrik, R. Soley, N. Ali, J. Grundy, B. Tekinerdogan (eds) *Software Quality Assurance: In Large Scale and Complex Software-Intensive Systems.* Waltham, MA, Elsevier, pp. 251–278. doi:10.1016/B978-0-12-802301-3.00012-0.

Sikorska, J., Hodkiewicz, M., Ma, L., 2011. Prognostic modelling options for remaining useful life estimation by industry. *Mechanical Systems and Signal Processing* 25(5), 1803–1836.

Simen, E., Agrell, C., Hafver, A., Børre Pedersen, F., 2018. AI + SAFETY: Safety implications for artificial intelligence. Technical Report. DNV GL, Group Technology & Research. URL: https://ai-and-safety.dnvgl.com.

Simeu-Abazi, Z., Lefebvre, A., Derain, J.P., 2011. A methodology of alarm filtering using dynamic fault tree. *Reliability Engineering and System Safety* 96(2), 257–266.

Sullivan, K.J., Dugan, J.B., Coppit, D., 1999. The galileo fault tree analysis tool. In: *Digest of Papers. Twenty-Ninth Annual International Symposium on Fault-Tolerant Computing (Cat. No. 99CB36352)*, Madison, WI, IEEE, pp. 232–235.

Taheri Kalani, J., Aslansefat, K., Latif Shabaghi, G., 2017. A systematic approach to design and analysis of univariate alarm systems using penalty approaches. *Journal of Control* 10, 1–15.

Tang, Z., Dugan, J.B., 2004. Minimal cut set/sequence generation for dynamic fault trees. In: *Annual Reliability and Maintainability Symposium*, Los Angeles, CA, IEEE, pp. 207–213. doi:10.1109/RAMS.2004.1285449.

Tannous, O., Xing, L., Dugan, J.B., 2011. Reliability analysis of warm standby systems using sequential BDD. In: *Proceedings of the Annual Reliability and Maintainability Symposium (RAMS)*, Lake Buena Vista, FL, IEEE, pp. 1–7.

Trivedi, K.S., Bobbio, A., 2017. *Reliability and Availability Engineering: Modeling, Analysis, and Applications.* Cambridge, UK, Cambridge University Press.

Vaurio, J.K., 2010. Ideas and developments in importance measures and fault-tree techniques for reliability and risk analysis. *Reliability Engineering and System Safety* 95(2), 99–107.

Verma, A., Srividya, A., Prabhudeva, S., Vinod, G., 2006. Reliability analysis of dynamic fault tree models using fuzzy sets. *Communications in Dependability and Quality Management* 9, 68–78.

Verma, A.K., Srividya, A., Karanki, D.R., 2010. System reliability modeling. *Reliability and Safety Engineering*, 71–168.

Vesely, W.E., Stamatelatos, M., Dugan, J., Fragola, J., Minarick, J., Railsback, J., 2002. *Fault Tree Handbook with Aerospace Applications*, Washington, DC, NASA.

Volk, M., Junges, S., Katoen, J.P., 2018. Fast dynamic fault tree analysis by model checking techniques. *IEEE Transactions on Industrial Informatics* 14(1), 370–379.

Volovoi, V., 2004. Modeling of system reliability petri nets with aging tokens. *Reliability Engineering and System Safety* 84(2), 149–161.

Walker, M., 2009. Pandora: A logic for the qualitative analysis of temporal fault trees. PhD thesis. University of Hull, Hull, UK.

Weber, P., Jouffe, L., 2003. Reliability modelling with dynamic Bayesian networks. *IFAC Proceedings Volumes* 36(5), 57–62.

Weber, P., Medina-Oliva, G., Simon, C., Iung, B., 2012. Overview on Bayesian networks applications for dependability, risk analysis and maintenance areas. *Engineering Applications of Artificial Intelligence* 25(4), 671–682.

Xiang, J., Machida, F., Tadano, K., Yanoo, K., Sun, W., Maeno, Y., 2012. Combinatorial analysis of dynamic fault trees with priority-and gates. In: *23rd International Symposium on Software Reliability Engineering Workshops*, Dallas, TX, IEEE, pp. 3–4. doi:10.1109/ISSREW.2012.27.

Xiang, J., Machida, F., Tadano, K., Yanoo, K., Sun, W., Maeno, Y., 2013. A static analysis of dynamic fault trees with priority-and gates. In: *Sixth Latin-American Symposium on Dependable Computing*, Brazil, IEEE, pp. 58–67. doi:10.1109/LADC.2013.14.

Xing, L., 2004. Maintenance-oriented fault tree analysis of component importance. In: *Annual Symposium Reliability and Maintainability, 2004-RAMS*, Los Angeles, CA, IEEE, pp. 534–539.

Xing, L., Dugan, J.B., 2002. Analysis of generalized phased-mission system reliability, performance, and sensitivity. *IEEE Transactions on Reliability* 51(2), 199–211.

Xing, L., Shrestha, A., Dai, Y., 2011. Exact combinatorial reliability analysis of dynamic systems with sequence-dependent failures. *Reliability Engineering and System Safety* 96(10), 1375–1385.

Xing, L., Tannous, O., Dugan, J.B., 2012. Reliability analysis of Nonrepairable cold-standby systems using sequential binary decision diagrams. *IEEE Transactions on Systems, Man, and Cybernetics - Part A: Systems and Humans* 42(3), 715–726. doi:10.1109/TSMCA.2011.2170415.

Xu, J., Wang, J., Izadi, I., Chen, T., 2011. Performance assessment and design for univariate alarm systems based on far, mar, and aad. *IEEE Transactions on Automation Science and Engineering* 9(2), 296–307.

Yassmeen Elderhalli, O.H., Tahar, S., 2019. Using machine learning to minimize user intervention in theorem proving based dynamic fault tree analysis. In: *4th Conference on Artificial Intelligence and Theorem Proving 2019*, Austria, AITP, pp. 1–2.

Yazdi, M., Kabir, S., 2017. A fuzzy Bayesian network approach for risk analysis in process industries. *Process Safety and Environmental Protection* 111, 507–519.

Yazdi, M., Kabir, S., 2018. Fuzzy evidence theory and Bayesian networks for process systems risk analysis. *Human and Ecological Risk Assessment: An International Journal* 26, 57–86.

Yevkin, O., 2010. An improved Monte Carlo method in fault tree analysis. In: *2010 Proceedings-Annual Reliability and Maintainability Symposium (RAMS)*, San Jose, CA, IEEE, pp. 1–5.

Yevkin, O., 2011. An improved modular approach for dynamic fault tree analysis. In: *2011 Proceedings-Annual Reliability and Maintainability Symposium*, Lake Buena Vista, FL, IEEE, pp. 1–5.

Yevkin, O., 2016. An efficient approximate Markov chain method in dynamic fault tree analysis. *Quality and Reliability Engineering International* 32(4), 1509–1520.

Yin, L., Smith, M.A., Trivedi, K.S., 2001. Uncertainty analysis in reliability modeling. In: *Annual Reliability and Maintainability Symposium. 2001 Proceedings. International Symposium on Product Quality and Integrity (Cat. No. 01CH37179)*, Philadelphia, PA, IEEE, pp. 229–234.

Yuge, T., Yanagi, S., 2008. Quantitative analysis of a fault tree with priority and gates. *Reliability Engineering and System Safety* 93(11), 1577–1583.

Zhang, H., Zhang, C., Liu, D., Li, R., 2011. A method of quantitative analysis for dynamic fault tree. In: *Proceedings of the Annual Reliability and Maintainability Symposium*, pp. 1–6. doi:10.1109/RAMS.2011.5754471.

Zhang, X., Miao, Q., Fan, X., Wang, D., 2009. Dynamic fault tree analysis based on petri nets. In: *8th International Conference on Reliability, Maintainability and Safety (ICRMS)*, IEEE, Chengdu, pp. 138–142.

Zhou, Z., Yan, Z., Zhou, J., Jin, G., Dong, D., Pan, Z., 2006. Design of dynamic systems based on dynamic fault trees and neural networks. In: *2006 IEEE International Conference on Automation Science and Engineering*, China, IEEE, pp. 124–128.

Zio, E., Podofillini, L., Levitin, G., 2004. Estimation of the importance measures of multi-state elements by Monte Carlo simulation. *Reliability Engineering and System Safety* 86(3), 191–204.

Zixian, L., Xin, N., Yiliu, L., Qinglu, S., Yukun, W., 2011. Gastric esophageal surgery risk analysis with a fault tree and Markov integrated model. *Reliability Engineering and System Safety* 96(12), 1591–1600.

Zonouz, S.A., Miremadi, S.G., 2006. A fuzzy-Monte Carlo simulation approach for fault tree analysis. In: *RAMS'06. Annual Reliability and Maintainability Symposium, 2006*, Newport Beach, CA, IEEE, pp. 428–433.

5 Reliability Analysis Using Condition Monitoring Approach in Thermal Power Plants

Hanumant Jagtap, Anand Bewoor, Ravinder Kumar, Mohammad H. Ahmadi, and Dipen Kumar Rajak

CONTENTS

5.1 INTRODUCTION AND BACKGROUND

With the rapid growth of modern technology, maintenance of machineries plays a critical role in many industries. The objectives of a maintenance-related decision-making process are selection and implementation of condition monitoring techniques (CMT) for industrial applications, which can be categorized and reviewed on the basis of maintenance scheduling, reliability improvement, and availability improvement. One of the major sources of electricity generation in India is the thermal power plant (TPP). Continuous electricity generation from TPPs depends on the increased availability of its major systems, subsystems, and the equipment used. High availability of TPP equipment is associated with its reliability and maintainability (Fernando Jesus Guevara Carazas & De Souza, 2009; Kuo & Ke, 2019). In order to make TPPs economical, maintenance functions need to be optimized by planning carefully and selecting suitable maintenance strategies, which will address the maintenance needs of the plant at a lower cost (Eti, Ogaji, & Probert, 2007; Niwas & Garg, 2018). The analysis based on reliability, availability, and maintainability (RAM) of critical components of TPPs is used as a standard tool for maintenance planning as well as for the function of various systems such as TPPs. Tewari and Malik (2016) reported that a failure cannot be prevented entirely but can be minimized to decrease the probability of failure. A suitable maintenance strategy is needed to retain the RAM characteristics of TPP. To achieve high performance of TPP, the equipment needs to work satisfactorily under the given working conditions. Hence, the identification of the critical equipment for TPPs is a significant and essential step in defining the maintenance strategy (Jagtap & Bewoor, 2017). In this regard, condition monitoring techniques are used to monitor a measurable parameter of the TPP equipment and help in diagnosing the fault. Also, multi-criteria decision-making methods are used to find and rank the critical equipment of the TPP (Wang, Chu, & Wu, 2007; Singh &Kulkarni, 2013).

In the case of rotating machinery, the performance parameters of machines, such as vibration, acoustic emission, wear debris in oil, thermography, and temperature, are useful indicators of the condition of the machinery (Han & Song, 2003). Successful implementation of condition monitoring programs allows the machine to operate without failures (B. R. Kumar, Ramana, & Rao, 2009). In recent years, various effective condition monitoring techniques have been developed, which includes vibration analysis, acoustic emission monitoring, wear debris analysis, thermography, temperature analysis, ultrasonic monitoring, testing, visual inspection, motor condition monitoring, and motor current signature analysis (Bagavathiappan et al., 2013). The ISO standard provides general guidelines for the selection of condition monitoring programs, appropriate measurement methods, and monitoring parameters. Generally, these describe the acceptable limits for evaluating the performance parameters of various systems, machines, or components such as rolling element bearings, shafts, pumps, fans, steam turbines, gears, centrifugal compressors, induction motors, screw compressors, large generators, and steam turbine generator sets. The selection of a maintenance strategy has a significant influence on the operational cost and the operational availability of the system. The maintenance task has been classified into three types, viz. corrective maintenance, preventive maintenance, and predictive (condition-based) maintenance. The condition-based maintenance strategy recommends that the maintenance decision should be based on the information collected through implemented condition monitoring techniques. The use of such a condition monitoring based maintenance (CMBM) strategy has not only been widely recommended by researchers but also been adopted by industries at large.

In the recent past, improvements in the reliability of TPPs have garnered considerable research interest (Zio, 2009). Therefore, focusing on RAM analysis is important for the performance improvement of TPP equipment. RAM analysis is carried out by taking into consideration the time to failure/time between failures (TTF/TBF) as well as the time to repair (TTR) record of the system. Barabady and Kumar (2008) present and discuss the characteristics of probability distributions such as the Weibull distribution, exponential distribution, and lognormal distribution for reliability as well as the availability analysis of a crushing plant. However, most of the studies report the use of RAM analysis for the thermal power plant, especially the heat exchanger of the TPP; reliability modeling and its optimization are studied by Sikos et al. (2011). The study recognized that the reliability, availability, maintenance and safety (RAMS) approach could detect the weak points in the maintenance of a heat exchanger network and highlight minute modifications, which could improve issues affecting optimality. The execution of a reliability-centered maintenance (RCM) method for steam process plants has been presented by Afefy (2010). In this study, a maintenance strategy for the thermal power plant is analyzed based on the reliability-centered maintenance model and is implemented for steam distribution dryers, process heaters, fire-tube boilers, and feed-water pumps.

F. J. G Carazas, Salazar, and Souza (2011) presented an analysis of the heat recovery steam generator for reliability and availability parameter determination, with the help of the function tree diagram and the failure mode effect analysis method. Two similar heat recovery steam generators operating from the same thermal power plant were taken up as a case study, and the reliability analysis was carried out for a five-year period. The results revealed that the reliability of both the boilers corresponded with a Weibull probability distribution representing a decrease in the failure rate. In the recent past, the RAM analysis of TPPs has been investigated by Adhikary et al. (2012), and the authors highlight that, for Unit 2, the furnace wall tube and the economizer show signs of low reliability as compared to the other subsystems. The Baffle wall tube and economizer require further improvements in maintainability, and the economizer is identified as the most critical subsystem for both the plants. R. Kumar, Sharma, and Tewari (2012) proposed the Markov approach for a simulation model based on the availability analysis of a power generation system (turbine) of TPPs under practical working conditions and investigated the effects of failures on system performance and availability. The results revealed that the turbine-governing system is most critical, and requires further improvements in maintainability in comparison with other subsystems; turbine lubrication systems are the least sensitive. The results of Bourouni's study (2013) on the availability analysis of the reverse osmosis plant using a reliability block diagram method was comparable with the results of the fault tree analysis. R. Kumar 2014) proposed and developed mathematical modeling for a boiler air circulation system of TPPs using the Markov approach. The components considered for the analysis are primary air fans, forced draft (FD) fans, induced draft (ID) fans, and air heaters. The results revealed that the breakdown of the primary air fan affects the system availability at a fast rate, while the failure of the air heater has a slight effect on the system availability.

Sabouhi et al. (2016) presented the availability and maintainability analysis of not only a gas and steam power plant but also a combined cycle power plant. Initially, reliability modes were developed from a steam turbine power plant and a gas turbine power plant. The study results revealed that the steam turbine power plant is more consistent than other power plants. Likewise, Pariaman et al. (2015) proposed a new maintenance strategy which integrates reliability-centered maintenance, reliability-based maintenance, and condition-based maintenance to enhance the overall availability of the thermal power plant. The equipment is prioritized based on a priority maintenance index as well as a failure defense task. Garg (2015) analyzed the Markov approach based on the reliability analysis of selected equipment of a thermal power plant using fuzzy Kolmogorov's differential equations. Over the years, researchers have continuously attempted to investigate reliability-based failure analysis in various process industries. Garg (2016) predicted the performance of pump systems of sumps based on a series-parallel configuration of systems using the intuitionistic fuzzy set theory approach. However, only very few cases of their applications in thermal power plants have been reported in the published literature. Furthermore, it is essential to note from the published literature that the failure data used for RAM analysis of TPPs is limited to a maximum of up to six years. It is argued that the accuracy of results can be improved by considering the maximum period for failure analysis to predict the performance of the TPP equipment. In line with the same point of view, various TPPs in India are working continuously and generating electricity at various capacities. However, case studies related to RAM analysis for prediction of the performance of TPP equipment in India are very rarely seen in the published literature. In recent studies, researchers have focused their work on system reliability optimization using various optimization techniques such as fuzzy set theory, genetic algorithm, particle swarm optimization, intuitionistic fuzzy, and artificial bee colony based lambda-tau (Garg et al., 2014; H. Garg, M. Rani,& S.P. Sharma, 2014; H. Garg & Rani, 2013; H. Garg, 2014; H. Garg & M. Rani, 2014; H. Garg, 2012; H. Garg & S.P. Sharma, 2015; Harish Garg, Rani, & Sharma, 2012; Kundu, Rossini, & Portioli-Staudacher, 2019; Qian et al., 2018).

The objective of the present work is to identify critical equipment using the reliability analysis of selected components of a thermal power plant. In addition, this research work reviews the use of different condition monitoring techniques (CMT) and their application in thermal power plants for fault detection, fault identification, and fault classification of machinery through a survey of published literature. The challenging areas in condition monitoring and fault diagnosis of the system used in the thermal power plant are highlighted. To improve the system reliability and availability of the thermal power plant, important conclusions and future research directions are provided. The details of the reliability analysis are discussed in the next section.

5.2 RELIABILITY ANALYSIS OF THE THERMAL POWER PLANT

The reliability of the plant equipment influences the effectiveness of TPPs. Reliability analysis is a useful tool for finding critical subsystems and helps to decide on a suitable maintenance strategy.

5.2.1 Basics of Reliability

Reliability of a component or system is the probability that it will perform its function for a specified period under the stated operating conditions. Mathematically, the reliability of the system $R(t)$ for the period T is taken to be the time to failure ($T \geq 0$), which can be expressed as Equation (5.1).

$$R(t) = \Pr\{T \geq t\} \tag{5.1}$$

The reliability characteristics for the Weibull distribution by neglecting the location parameter (γ) are estimated using Equation (5.2):

$$R(t) = e^{-\left(\frac{t}{\theta}\right)^{\beta}} \tag{5.2}$$

Here, β is termed as the shape parameter, θ is termed as the scale parameter, and t is the time factor.

Maintainability is defined as the probability that a failed system will be repaired or restored to a specific condition within a specific period.

Availability is the probability that a system is performing its function at a given time under the stated operating conditions.

$$\text{Availability} = \frac{\text{MTBF}}{\text{MTBF} + \text{MTTR}} \tag{5.3}$$

5.2.2 Electricity Generation Process of the Thermal Power Plant

The thermal power plant is a complex system consisting of various equipment or systems connected either in series or parallel configuration. The main objective of the plant is to produce an uninterrupted supply of electricity. The basic model of the thermal power plant is shown in Figure 5.1. The model consists of major equipment, namely (a) turbine, (b) generator, (c) induced draft fan, (d) forced draft fan, (e) primary air fan, (f) boiler feed pump, (g) cooling water pump, (h) condensate extraction pump, (i) boiler, (j) chimney, (k) coal mill, and (l) condenser.

The process of electricity generation is initiated by drawing atmospheric air by a forced draft fan through an air preheater. While passing the air through the air

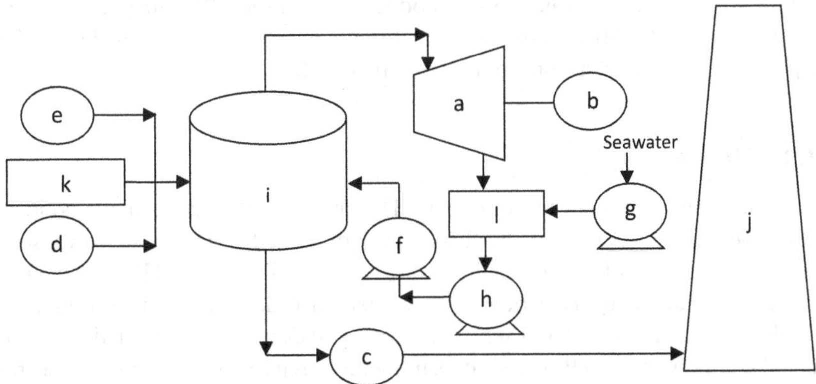

FIGURE 5.1 Model of a thermal power plant for electricity generation.

preheater, the air gets heated and is then transferred to the combustion chamber. At the same instant, the primary air fan sucks the air and supplies it to the coal mill. Inside the coal mill, the crushing of coal is done. Afterward, the primary air fan takes the pulverized coal and using a classifier supplies it to the furnace for the purpose of heating and steam is generated.

The steam formed is passed to the boiler drum and then through the superheater; the superheated steam is then sent to the turbine for electricity generation. After the steam from the turbine passes to the condenser through the condensate extraction pump, the deaerator is used as a booster and the boiler feed pump is used to increase the pressure. In this way, the cycle of the plant is completed. The scope of the present study is limited to the reliability analysis of selected equipment of the thermal power plant, namely the turbine, generator, induced draft fan, forced draft fan, primary air fan, boiler feed pump, cooling water pump, and condensate extraction pump.

In this study, the reliability of the thermal power plant is calculated based on the assumption that the selected equipment is connected in series combination. As TPP is a very complex system, it will work properly only if all the systems function appropriately. The reliability of the TPP equipment is determined using Equation (5.4) as follows:

$$Rs(t) = \prod_{i=1}^{n} Ri(t) \tag{5.4}$$

Now for the selected equipment, the overall system reliability is determined as

$$Rs = R(1) \times R(2) \times R(3) \times R(4) \times R(5) \times R(6) \times R(7) \times R(8) \tag{5.5}$$

For this study, field data related to the time between failures (TBF) are collected, sorted, and classified for the selected components of TPPs. The trend test and serial correlation test are carried out for TBF data points to confirm the independent and identical distribution. Afterward, best-fit distribution characteristics are obtained using the Kolmogorov-Smirnov (K-S) goodness test for the TBF data points of TPPs. Then, the best-fit distribution parameters of the plant equipment are evaluated. TPP equipment is prioritized according to the criticality level.

5.2.3 Data Analysis

The assumption considered for TBF and TTR data with the independent and identical distribution has been validated through the trend and serial correlation tests for selected components of the thermal power plant. The TBF and TTR failure data of the selected equipment are arranged in sequential order for the determination of the failure trend. The trend and serial correlation tests have been carried out for the selected components of TPPs. As an example, the trend plot and the serial correlation test plot of only the induced draft fan are plotted for cumulative frequency and cumulative time to failure, as shown in Figures 5.2 and 5.3, respectively.

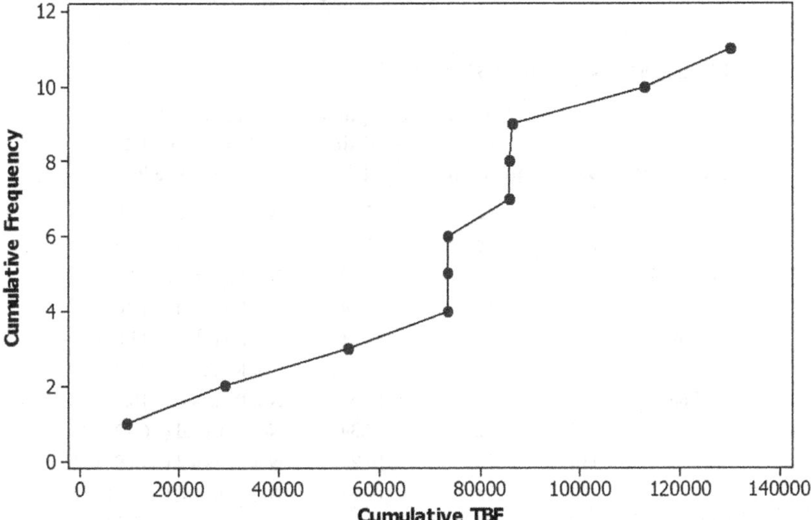

FIGURE 5.2 Trend plot for TBF of the ID fan.

It is observed from Figure 5.2 that the trend plot reveals a concave downhill, which indicates enhancement for reliability following the infant mortality region at earlier stages. Also, from the serial correlation test (Figure 5.3), it has been observed that the data points of TBF and TTR are indiscriminately scattered without any clear pattern. The calculated values of *U* test statistics for the selected equipment of the thermal power plant are tabulated in Table 5.1.

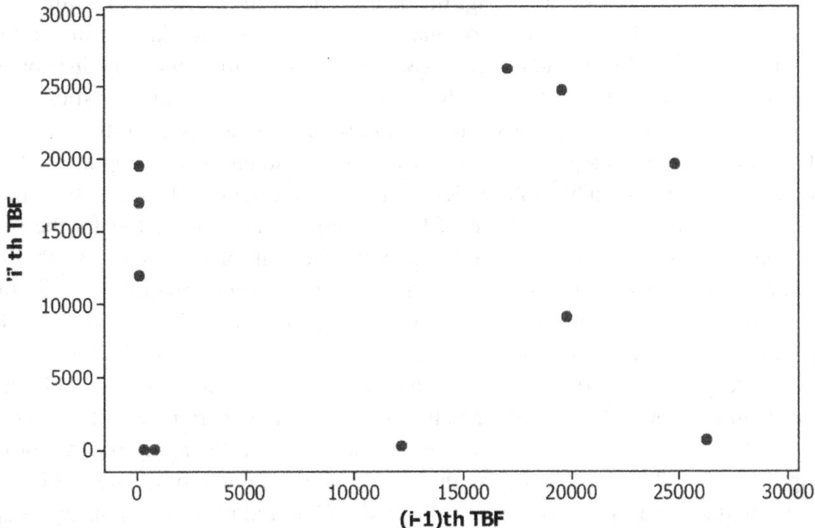

FIGURE 5.3 Serial correlation plot of the ID fan.

TABLE 5.1
***U* Test Statistics for TBF and TTR**

Equipment	Data	Degrees of freedom	Computed statistics *U*	Null hypothesis elimination at 5% significance level
Turbine	TBF	12	12.57	Not Rejected (>10.81)
	TTR	12	7.34	Not Rejected (>6.12)
Generator	TBF	12	12.57	Not Rejected (>12.44)
	TTR	12	7.34	Not Rejected (>7.16)
ID Fan	TBF	20	16.21	Not Rejected (>13.68)
	TTR	20	10.76	Not Rejected (>8.78)
FD Fan	TBF	12	12.57	Not Rejected (>10.81)
	TTR	12	7.34	Not Rejected (>6.12)
PA Fan	TBF	18	16.27	Not Rejected (>13.97)
	TTR	18	10.37	Not Rejected (>8.58)
BFP	TBF	22	16.68	Not Rejected (>15.21)
	TTR	22	11.09	Not Rejected (>9.81)
CWP	TBF	20	14.99	Not Rejected (>14.08)
	TTR	20	10.09	Not Rejected (>9.04)
CEP	TBF	12	12.57	Not Rejected (>12.29)
	TTR	12	7.34	Not Rejected (>7.06)

5.2.4 The Goodness of Fit Test

The null hypothesis of the homogeneous Poisson process having 2 (n–1) degrees of freedom is not rejected for a 5% significance level of the selected TPP equipment. The K-S test as well as the serial correlation test prove that the data points of TBF and TTR are independent, and it can be assumed that they are identically distributed. Next the trend-free data are then analyzed to find the correct characteristics of the TBF and TTR data. Subsequently, failure data points are assessed for the goodness of fit test using the chi-square and K-S test. The chi-square test is applicable for a sample size of more than 50, and the K-S test has no limitation. Hence for this study, the K-S test is conducted for TBF and TTR data points. The parameters for the best-fitted statistical distribution are calculated with the help of ReliaSoft's Weibull++ software. The software makes use of various methods to fit numerous distribution models using the given data points. The best-fit distribution results obtained for TBF data of the selected equipment of thermal power plants are tabulated in Table 5.2.

It has been observed from Table 5.2 that all the selected components of thermal power plants follow a Weibull distribution with the shape parameter $\beta > 1$ (except ID Fan). It means that an increase in failure rate is due to the aging process of the equipment. Therefore, preventive maintenance with a suitable time interval is necessary for this type of equipment. In the case of the induced draft fan, the shape parameter is $\beta < 1$ with Weibull 3P distribution. It indicates that failures occurring before its useful life can be reduced. Hence, corrective maintenance is necessary for

TABLE 5.2
Best-Fit Distribution for the TBF Data Set

Equipment	K-S goodness of fit test: Exp. 1P	Exp. 2P	Log-normal	Normal	Weibull 2P	Weibull 3P	Best-fit distribution	Parameters
Turbine	87.82	53.0954	0.4391	0.003182	0.002387	0.2149	Weibull 2 P	$\beta=3.1532$, $\theta=21{,}652.1$
Generator	87.82	53.0954	0.4391	0.003182	0.002387	0.2149	Weibull 2 P	$\beta=3.1532$, $\theta=21{,}652.1$
ID Fan	64.54	65.3528	63.9044	34.4031	42.3673	28.8333	Weibull 3 P	$\beta=0.3460$, $\theta=10{,}658.5$, $\gamma=12.40$
FD Fan	87.82	53.0954	0.4391	0.003182	0.002387	0.2149	Weibull 2 P	$\beta=3.1532$, $\theta=21{,}652.1$
PA Fan	11.85	4.2894	0.3104	0.000581	0.0008868	0.000230	Weibull 3 P	$\beta=1.5090$, $\theta=16{,}088.8$, $\gamma=596.0$
BFP	7.1094	5.4734	3.3470	8.9799	0.02990	0.5356	Weibull 2 P	$\beta=1.0798$, $\theta=12{,}557.5$
CWP	62.5631	3.2056	0.5284	7.5094	1.5049	0.02022	Weibull 3 P	$\beta=1.018$, $\theta=9849.7$, $\gamma=3314.80$
CEP	87.8240	53.0954	0.4391	0.003182	0.002387	0.2149	Weibull 2 P	$\beta=3.154$, $\theta=21{,}652.1$

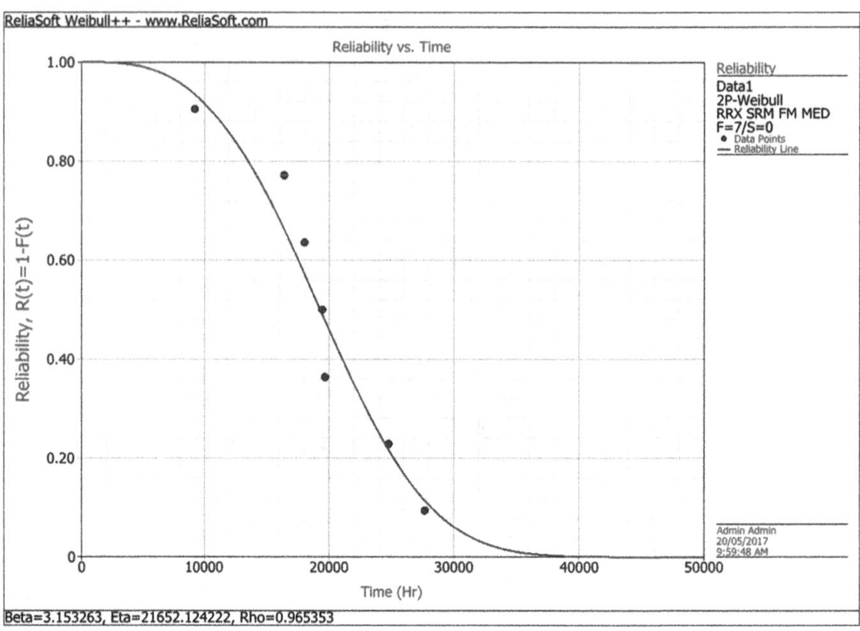

FIGURE 5.4 Weibull 2P distribution and parameters $\beta=3.1532$, $\theta=21{,}652.1242$ for forced draft fan.

the induced draft fan. As an example, the reliability plot of the forced draft fan with Weibull 2P as best-fit distribution is shown in Figure 5.4.

5.2.5 Estimation of the Time Interval for Reliability-Based Preventive Maintenance

According to the current strategy, the task of preventive maintenance of TPPs is undertaken at a fixed time interval. However, taking the cost aspect into consideration, the cost required for preventive maintenance can be reduced by setting suitable preventive maintenance intervals for critical equipment. The probabilistic approach employs statistical methods to fit a theoretical distribution of the collected failure data points of the plant equipment. This distribution is used to predict the failure pattern of the equipment, which helps in finding the preventive maintenance interval time. The reliability of the overall system is evaluated at different levels of reliability, viz. 90%, 75%, and 50% using ReliaSoft's Weibull++ software, and these are tabulated in Table 5.3.

Table 5.3 gives the time interval for the reliability of the forced draft fan of Unit 1 of TPPs. To attain the reliability level of 90% ($R=0.9$) of the forced draft fan, the maintenance task should be completed before 12,239 hr, as the completion of the maintenance task is the probability of a failure-free operation, i.e., $R=0.9$ when the forced draft fan runs for 12,239 hr. The mean time between failure (MTBF) and the mean time to repair (MTTR) for the forced draft fan are 19,370 hr and 543 hr,

TABLE 5.3
The Time Interval for Reliability-Based Preventive Maintenance for Unit 1

Component	Reliability R(1 year)	Reliable life at different reliability "t_R"(hr)		
		0.9	0.75	0.5
Turbine	0.9439	10,606	14,584	19,276
Generator	0.9439	10,606	14,584	19,276
ID Fan	0.3930	28	303	3707
FD Fan	0.9731	12,239	15,692	19,506
PA Fan	0.6432	3025	6450	12,023
BFP	0.5077	1562	3961	8943
CWP	0.5787	4394	6211	10,186
CEP	0.9439	10,606	14,584	19,276

respectively. Similarly, reliable life at 75% reliability and 50% reliability of the forced draft fan are found to be at 15,692 hr and 19,506 hr, respectively. This calculated reliability-based time interval can be considered not only to schedule servicing or repairing work and condition monitoring activities but also for replacement of the component. This analysis adds value in terms of safety and cost and for finding the type of fault in the equipment at the early stages of the operation.

Operating the equipment at the 90% reliability level may involve a high cost. Hence, the reliability of 75% level can be recommended for the initial stages, and optimum reliability levels in terms of safety, cost, and effectiveness can be recommended for more advanced stages. Similarly, MTTF and MTTR are also determined for the other selected TPP equipment, and valid recommendations for scheduling preventive maintenance can be made.

The reliability of the selected thermal power plant equipment for various intervals of time is calculated using ReliaSoft's Weibull++ software and is tabulated in Table 5.4. Also, the reliability plot of the selected TPP equipment has been plotted and shown in Figure 5.5.

TABLE 5.4
Reliability of the Thermal Power Plant Unit 1 for Various Time Intervals

Time	Turbine	Generator	ID Fan	FD Fan	PA Fan	BFP	CWP	CEP	Total
0	1	1	1	1	1	1	1	1	1
3600	0.9965	0.9965	0.50	0.9992	0.8767	0.7714	0.9731	0.9965	0.325344
7200	0.9694	0.9694	0.41	0.9877	0.7152	0.5778	0.6784	0.9694	0.103421
10,800	0.8944	0.8944	0.36	0.9384	0.5519	0.4275	0.4694	0.8944	0.026769
14,400	0.7585	0.7585	0.32	0.8160	0.4068	0.3136	0.3237	0.7585	0.004706
18,000	0.5720	0.5720	0.30	0.6060	0.2881	0.2287	0.2227	0.5720	0.000499

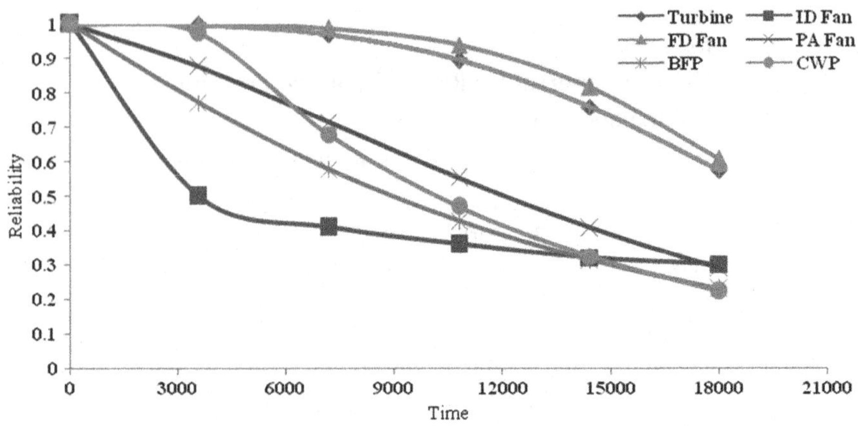

FIGURE 5.5 Reliability plot of the selected TPP equipment.

It has been observed from Table 5.4 that the probability of the TPP equipment not failing for 3600 hr of working is only 32%. It is important to note that the boiler feed pump (BFP), induced draft (ID) fan, cooling water pump (CWP), primary air (PA) fan, followed by turbine, generator, forced draft (FD) fan, and condensate extraction pump (CEP) have the most considerable effect on the overall reliability of the plant. Therefore, necessary measures must be taken to improve the reliability and availability of such critical TPP systems.

5.2.6 Comparative Analysis of Results with the Existing Approaches

Recent studies have reported the reliability and availability analyses of selected critical components of a thermal power plant using several approaches such as fault tree analysis, reliability block diagram method, Markov birth-death approach, failure mode effect and criticality analysis, Monte Carle simulation, Bayesian theory, and Poisson process-based analysis. It is quite difficult to compare the reliability results obtained from the present study directly with other techniques. The main reason is that the type of component or system is different, as well as the time considered for reliability evaluation is also different. Furthermore, the failure rate and repair rate of different systems are different. The results of previous studies are provided to understand the overall reliability evaluation of the system using the fault tree analysis approach.

Table 5.5 shows the reliability analysis of the thermal power plant using the fault tree analysis method (N. Singh Bhangu & G.L. Pahuja, 2015).

Table 5.6 shows the RAM analysis of the thermal power plant using a reliability block diagram (D.D. Adhikary, G. Bose, & S. Mitra, 2010).

Table 5.7 shows the availability analysis of the thermal power plant using the fault tree analysis method (Bhangu, Singh, & Pahuja, 2018).

It is observed from the above discussion that maintaining the life of critical components in good condition plays a major role in enhancing plant availability. Hence,

TABLE 5.5
Parameters of Eight Critical Components of Unit1

Description	No. of failures	Outage hours	MTBF	Reliability	Probability of failure
Water wall tube leakage	34	1451.85	214.95	0.0012	0.99883
Economizer leakage	0	0	∞	1.0000	0
Turbine bearing leakage	1	7.58	8752.42	0.9991	0.00086
Final/Platen Superheater (SH) leakage	0	0	∞	1.0000	0
Convection/primary SH leakage	0	0	∞	1.0000	0
Low temperature. SH leakage	0	0	∞	1.0000	0
Reheater leakage	0	0	∞	1.0000	0
Condenser tube leakage	1	31.66	8728.34	0.9964	0.00362

TABLE 5.6
The Reliability of the Power Plant at Different Time Intervals

Time (hours)	ECO	PSH	FSH	PRH	FRH	FUR	TUR	CON	TotalPlant
0	1	1	1	1	1	1	1	1	1
400	0.9191	0.9541	0.8098	0.9646	0.9898	0.9994	0.9891	0.9965	0.6679
800	0.8068	0.9013	0.7256	0.9285	0.9608	0.9961	0.9721	0.9906	0.4515
1200	0.6902	0.8475	0.6637	0.8931	0.9236	0.9901	0.9520	0.9834	0.2969
1600	0.5791	0.7945	0.6139	0.8586	0.8849	0.9817	0.9297	0.9751	0.1910
2000	0.4781	0.7431	0.5721	0.8251	0.8461	0.9713	0.9058	0.9660	0.1206
2600	0.3497	0.6696	0.5197	0.7768	0.7910	0.9535	0.8679	0.9511	0.0589
3000	0.2796	0.6234	0.4898	0.7460	0.7580	0.9419	0.8418	0.9405	0.0360
4000	0.1530	0.5184	0.4277	0.6736	0.6808	0.9082	0.7747	0.9119	0.0100
5000	0.0792	0.4281	0.3783	0.6076	0.6141	0.8729	0.7073	0.8812	0.0026
6000	0.0391	0.3515	0.3377	0.5476	0.5557	0.8389	0.6412	0.8490	0.0006
7000	0.0185	0.2872	0.3037	0.4932	0.5079	0.8051	0.5777	0.8158	0.0002
8000	0.0084	0.2337	0.2747	0.4439	0.4641	0.7734	0.5176	0.7819	0.0000
9000	0.0037	0.1894	0.2497	0.3993	0.4247	0.7422	0.4613	0.7478	0.0000
10,000	0.0016	0.1530	0.2279	0.3591	0.3936	0.7123	0.4092	0.7137	0.0000
14,000	0.0000	0.0633	0.1632	0.2339	0.2946	0.6103	0.2427	0.5809	0.0000
20,000	0.0000	0.0157	0.1055	0.1218	0.2033	0.4960	0.0997	0.4066	0.0000
26,000	0.0000	0.0037	0.0717	0.0630	0.1492	0.4091	0.0368	0.2712	0.0000
30,000	0.0000	0.0013	0.0565	0.0404	0.1230	0.3632	0.0181	0.2023	0.0000

TABLE 5.7
Availability Parameters of Unit1 for 2007–2011

Description	No. of failures	Outage hours	MTTR	MTBF	Availability	Unavailability
Water wall tube leakage	34	1451.85	42.70	987.89	0.0012	0.0414
Economizer leakage	0	0	0.00	∞	1.0000	0.0000
Turbine bearing leakage	1	7.58	7.58	35,032.42	0.9991	0.00022
Final/Platen Superheater (SH) leakage	0	0	0.00	∞	1.0000	0.0000
Convection/primary SH leakage	0	0	0.00	∞	1.0000	0.0000
Low temperature. SH leakage	0	0	0.00	∞	1.0000	0.0000
Reheater leakage	0	0	0.00	∞	1.0000	0.0000
Condenser tube leakage	1	31.66	31.66	35,008.34	0.9964	0.0009

the failure of such equipment needs to be diagnosed before any catastrophic failure. The fault of the equipment or system can be detected, identified, and diagnosed using different condition monitoring techniques. Therefore, it is important to highlight the relevant literature related to the use of condition monitoring techniques and the details are discussed in the next section.

5.3 USE OF CONDITION MONITORING TECHNIQUES IN THE THERMAL POWER PLANT

In recent years, various effective monitoring techniques have been developed for monitoring and diagnosis of machinery, which include vibration analysis, acoustic emission monitoring, wear debris analysis, temperature analysis, ultrasonic monitoring, thermography, non-destructive testing, visual inspection, motor condition monitoring, and motor current signature analysis (Parey & Singh, 2019; Hanumant P. Jagtap, 2017). The major objectives of condition monitoring techniques are (a) fault detection, (b) fault diagnosis, (c) maintenance-related decision-making, and (d) fault prediction. The vibration analysis technique is most suitable for rotating machinery. This technique is used to diagnose machinery faults such as shaft and rotor faults, gear faults, and bearing faults (Jayaswal, Wadhwani, &Mulchandani, 2008). Similarly, the other non-destructive method is acoustic analysis. This technique allows data analysis using the condition monitoring approach through crack propagation within the material (Chondros, Dimarogonas, & Yao, 2001). Furthermore, thermography is a non-contact type method that detects the fault instantly without introducing one into the object. Also, the wear debris analysis technique helps to identify the presence of both wear and contamination in the test oil sample at an early stage, which provides data on the wear trend.

Previous research in the field of condition monitoring mainly focused on a combination of two techniques such as (a) vibration and noise analysis, (b) vibration and wear analysis, and (c) vibration analysis and ultrasonic monitoring (Peng, Kessissoglou, & Cox, 2005). A high level of expertise is required to use the vibration analysis and wear debris analysis techniques, whereas a medium level of expertise is required to use infrared thermography and the oil analysis techniques. Likewise, damage in the early stages can be detected using vibration analysis, oil analysis, and the acoustic emissions techniques. Therefore, an integrated approach is needed for increasing the possibility of detecting faults before the breakdown of the system and for providing greater accuracy in diagnosing faults (Zhang et al., 2005). The integration of various condition monitoring techniques has limitations in terms of its applicability.

It is essential to discuss the reports of previous studies undertaken in the field of condition monitoring and fault diagnosis of machinery. The relevant published literature is reviewed here to address the issues related to the health condition of the machine and the associated problems. Eftekharnejad, Addali, and Mba (2012) investigated the effectiveness of a combined approach using vibration analysis and wear debris analysis techniques in an integrated machine condition monitoring program for a spur gearbox test rig. Similarly, Peng and Kessissoglou (2003) developed a correlation method with vibration analysis and wear debris analysis for a worm gearbox driven by an electric motor under three various wear conditions such as lack of lubrication, normal operation, and the presence of contaminant particles in the lubricating oil. Mohamed et al. (2011) examined the vibration characteristics of two different types of cracks in a long rotor shaft and developed a vibration detection system for fatigue crack initiation and propagation in a pre-cracked high carbon steel shaft, and it was experimentally evaluated and monitored using a vibration-based condition health monitoring method. The use of vibration monitoring techniques in intelligent manufacturing equipment was studied by Liu et al. (2013), where the performance was compared between the multibeam accelerometer and quad-beam accelerometer by performing static and dynamic experiments with two tiny sensing beams projected into a traditional quad-beam structure, which provides a solution to enhance the resonant frequency at the expense of acceptable sensitivity loss. Ziegler et al. (2013) investigated the failure analysis of a low-pressure turbine blade of a 310 MW thermal power plant. Tan et al. (2005) carried out the fault diagnosis for a marine condensate booster feed-water system using the bond graph-based approach. A model-based procedure was developed using analytical redundancy for the detection and isolation of faults on a gas turbine process, which was studied by Simani and Fantuzzi (2006). Besides, Garcia et al. (2013) analyzed the online monitoring and supervision task of a hydraulic pumping system as well as the radial shaft dynamics using shaft motion and radial vibration measurements.

5.4 CONCLUSION

In this study, reliability analysis is carried out for selected components of the thermal power plant and a critical review related to condition monitoring techniques is presented, especially with regard to the thermal power plant. For the reliability analysis, the trend and serial correlation tests are conducted for the selected equipment of TPPs with TBF and TTR failure data points. It is concluded that TBF and TTR data sets are independent and identical, which means they are free from any correlation between them. Subsequently, the K-S test is applied for the determination of the goodness of fit test for the TBF and TTR data sets of the equipment used in TPPs. It is also concluded that the selected equipment of the thermal power plant follows a Weibull distribution curve for the TBF data set. Then the reliability-based preventive maintenance interval is determined at various reliability levels (viz., 90%, 75%, and 50%), and the results reveal that CWP, PA fan, BFP, and ID fan are the most critical equipment, as their reliability becomes zero for an operation of more than 19,000 hrs. Hence, a preventive maintenance task is recommended for the reliability level of 75%.

5.5 DISCUSSION AND FUTURE RESEARCH DIRECTION

The study highlights the challenges in the area of condition monitoring and fault diagnosis of the system used in the thermal power plant. Based on the results of this research, it would be advisable to employ different condition monitoring techniques simultaneously for the fault diagnosis of the equipment of the thermal power plant. In addition, the determination of correlation between various condition monitoring parameters by adopting two or more techniques offers a greater advantage through the implementation of fault detection and diagnosis of equipment used in the thermal power plant. Therefore this research area needs to be explored in detail. Furthermore, a reliability-based maintenance model needs to be developed for enhancing the maintenance program of the thermal power plant.

NOMENCLATURE

B	Shape parameter	MTTR	Mean time to repair
θ	Scale parameter	SH	Superheater
γ	Location parameter	ECO	Economizer
CMT	Condition monitoring technique	PSH	Primary superheater
TPP	Thermal power plant	FSH	Final superheater
RAM	Reliability maintainability availability	PRH	Pendent reheater
TTF	Time to failure	FRH	Final reheater
TBF	Time between failures	FUR	Furnace
TTR	Time to repair	TUR	Turbine
MTBF	Mean time between failures	CON	Condenser

REFERENCES

Adhikary, D. D., G. K. Bose, S. Chattopadhyay, D. Bose, and S. Mitra. 2012. "RAM Investigation of Coal-Fired Thermal Power Plants: A Case Study." *International Journal of Industrial Engineering Computations* 3(3): 423–24. doi:10.5267/j.ijiec.2011.12.003.

Adhikary, D. D., G. Bose, S. Mitra, and D. Bose. 2010. "Reliability, Maintainability & Availability Analysis of a Coal Fired Power Plant in Eastern Region of India." *2nd International Conference on Production and Industrial Engineering*, (CPIE-2010) at: NIT, Jalandhar, November, 1505–1513. doi:10.1016/j.ijsu.2018.05.014.

Afefy, Islam H. 2010. "Reliability-Centered Maintenance Methodology and Application : A Case Study." *Engineering* 2(11): 863–73. doi:10.4236/eng.2010.211109.

Bagavathiappan, S., B. B. Lahiri, T. Saravanan, John Philip, and T. Jayakumar. 2013. "Infrared Thermography for Condition Monitoring - A Review." *Infrared Physics and Technology* 60: 35–55. doi:10.1016/j.infrared.2013.03.006.

Barabady, Javad, and Uday Kumar. 2008. "Reliability Analysis of Mining Equipment: A Case Study of a Crushing Plant at Jajarm Bauxite Mine in Iran." *Reliability Engineering and System Safety* 93(4): 647–53. doi:10.1016/j.ress.2007.10.006.

Bhangu, N. SIngh, G. L. Pahuja, and R. Singh. 2015. "Application of Fault Tree Analysis for Evaluating Reliability and Risk Assessment of a Thermal Power Plant." *Energy Sources, Part A: Recovery, Utilization and Enviromental Effects* 37(18): 2004–12. doi: 10.1080/15567036.2012.664608.

Bhangu, Navneet Singh, Rupinder Singh, and G. L. Pahuja. 2018. "Availability Performance Analysis of Thermal Power Plants." *Journal of the Institution of Engineers (India): Series C* 3: 439–48. doi:10.1007/s40032-018-0450-x.

Bourouni, Karim. 2013. "Availability Assessment of a Reverse Osmosis Plant: Comparison Between Reliability Block Diagram and Fault Tree Analysis Methods." *Desalination* 313: 66–76. doi:10.1016/j.desal.2012.11.025.

Carazas, F. J. G., C. H. Salazar, and G. F. M. Souza. 2011. "Availability Analysis of Heat Recovery Steam Generators Used in Thermal Power Plants." *Energy* 36(6): 3855–70. doi:10.1016/j.energy.2010.10.003.

Carazas, Fernando Jesus Guevara, and Gilberto Francisco Martha De Souza. 2009. "Availability Analysis of Gas Turbines Used in Power Plants." *International Journal of Thermodynamics* 12(1): 28–37.

Chondros, T. G., A. D. Dimarogonas, and J. Yao. 2001. "Vibration of a Beam with a Breathing Crack." *Journal of Sound and Vibration* 239(1): 57–67. doi:10.1006/jsvi.2000.3156.

Eftekharnejad, Babak, A. Addali, and D. Mba. 2012. "Shaft Crack Diagnostics in a Gearbox." *Applied Acoustics* 73(8): 723–33. doi:10.1016/j.apacoust.2012.02.004.

Eti, M. C., S. O. T. Ogaji, and S. D. Probert. 2007. "Integrating Reliability, Availability, Maintainability and Supportability with Risk Analysis for Improved Operation of the Afam Thermal Power-Station." *Applied Energy* 84(2): 202–21. doi:10.1016/j.apenergy.2006.05.001.

Garcia, Ramon Ferreiro, José Luis Calvo Rolle, Manuel Romero Gomez, and Alberto Demiguel Catoira. 2013. "Expert Condition Monitoring on Hydrostatic Self-Levitating Bearings." *Expert Systems with Applications* 40(8): 2975–84. doi:10.1016/j.eswa.2012.12.013.

Garg, H. 2014. "Reliability, Availability and Maintainability Analysis of Industrial Systems Using PSO and Fuzzy Methodology." *MAPAN-Journal of Metrology Society of India* 29(2): 115–29. doi:10.1007/s12647-013-0081-x.

Garg, H. 2016. "A Novel Approach for Analyzing the Reliability of Series Parallel System Using Credibility Theory and Different Types of Intuitionistic Fuzzy Numbers." *Journal of Brazilian Society of Mechanical Sciences and Engineering* 38(3): 1021–35. doi:10.1007/s40430-014-0284-2.

Garg, Harish. 2015. "An Approach for Analyzing the Reliability of Industrial System Using Fuzzy Kolmogorov's Differential Equations." *Arabian Journal for Science and Engineering* 40(3): 975–87. doi:10.1007/s13369-015-1584-2.

Garg, Harish, and Monica Rani. 2013. "An Approach for Reliability Analysis of Industrial Systems Using PSO and IFS Technique." *ISA Transactions* 52(6): 701–10. doi:10.1016/j.isatra.2013.06.010.

Garg, Harish, Monica Rani, and S. P. Sharma. 2012. "Fuzzy RAM Analysis of the Screening Unit in a Paper Industry by Utilizing Uncertain Data." *International Journal of Quality, Statistics, and Reliability*: 1–12. doi:10.1155/2012/203842.

Garg, H., M. Rani, and S. P. Sharma. 2014. "An Approach for Analyzing the Reliability of Industrial Systems Using Soft-Computing Based Technique." *Expert Systems with Applications* 41(2): 489–501. doi:10.1016/j.eswa.2013.07.075.

Garg, Harish, Monica Rani, S. P. Sharma, and Yashi Vishwakarma. 2014. "Bi-objective Optimization of the Reliability-Redundancy Allocation Problem for Series-Parallel System." *Journal of Manufacturing Systems* 33(3): 335–47. doi:10.1016/j.jmsy.2014.02.008.

Garg, H., M. Rani, S. P. Sharma, and Y. Vishwakarma. 2014. "Intuitionistic Fuzzy Optimization Technique for Solving Multi-Objective Reliability Optimization Problems in Interval Environment." *Expert Systems with Applications* 41(7): 3157–67. doi:10.1016/j.eswa.2013.11.014.

Garg, H., S. P. Sharma, and M. Rani. 2015. "Behavior Analysis of Pulping Unit in a Paper Mill with Weibull Fuzzy Distribution Function Using ABCBLT Technique." *International Journal of Applied Mathematics and Mechanics* 8(4): 89–96.

Garg, Harish, and S. P. Sharma. 2012. "Behavior Analysis of Synthesis Unit in Fertilizer Plant." *International Journal of Quality and Reliability Management* 29(2): 217–32. doi:10.1108/02656711211199928.

Han, Y., and Y. H. Song. 2003. "Condition Monitoring Techniques for Electrical Equipment—A Literature Survey." *IEEE Transactions on Power Delivery* 18(1): 4–13. doi:10.1109/TPWRD.2002.801425.

Jagtap, H. P., and A. K. Bewoor. 2017. "Use of Analytic Hierarchy Process Methodology for Criticality Analysis of Thermal Power Plant Equipments." *Materials Today: Proceedings* 4(2): 1927–36. doi:10.1016/j.matpr.2017.02.038.

Jagtap, Hanumant P., and Anand Bewoor. 2017. "Development of an Algorithm for Identification and Confirmation of Fault in Thermal Power Plant Using Condition Monitoring Technique." *Proceedia Engineering* 181: 690–97. doi:10.1016/j.proeng.2017.02.451.

Jayaswal, Pratesh, A. K. Wadhwani, and K. B. Mulchandani. 2008. "Machine Fault Signature Analysis." *International Journal of Rotating Machinery*: 1–10. doi:10.1155/2008/583982.

Kumar, B. Raghu, K. V. Ramana, and K. Mallikharjuna Rao. 2009. "Condition Monitoring and Fault Diagnosis of a Boiler Feed Pump Unit." *Journal of Scientific and Industrial Research* 68(9): 789–93.

Kumar, Ravinder. 2014. "Availability Analysis of Thermal Power Plant Boiler Air Circulation System Using Markov Approach." *Decision Science Letters* 3: 65–72. doi:10.5267/j.dsl.2013.08.001.

Kumar, Ravinder, Avdhesh Kr. Sharma, and P. C. Tewari. 2012. "Markov Approach to Evaluate the Availability Simulation Model for Power Generation System in a Thermal Power Plant." *International Journal of Industrial Engineering Computations* 3(5): 743–50. doi:10.5267/j.ijiec.2012.08.003.

Kundu, Kaustav, Matteo Rossini, and Alberto Portioli-Staudacher. 2019. "A Study of Kanban Assembly Line Feeding System Through Integration of Simulation and Particle Swarm Optimzation." *International Journal of Industrial Engineeirng and Computations* 10: 421–42. doi:10.5267/j.ijiec.2018.12.001.

Kuo, Ching-chang, and Jau-chuan Ke. 2019. "Availability and Comparison of Spare Systems with a Repairable Server." *International Journal of Reliability, Quality and Safety Engineering* 26(1): 1–18. doi:10.1142/S0218539319500086.

Liu, Yan, Yulong Zhao, Weizhong Wang, Lu Sun, and Zhuangde Jiang. 2013. "A High-Performance Multi-Beam Microaccelerometer for Vibration Monitoring in Intelligent Manufacturing Equipment." *Sensors and Actuators, A: Physical* 189: 8–16. doi:10.1016/j.sna.2012.08.033.

Mohamed, A. A., R. Neilson, P. MacConnell, N. C. Renton, and W. Deans. 2011. "Monitoring of Fatigue Crack Stages in a High Carbon Steel Rotating Shaft Using Vibration." *Procedia Engineering* 10: 130–35. doi:10.1016/j.proeng.2011.04.024.

Niwas, Ram, and Harish Garg. 2018. "An Approach for Analyzing the Reliability and Profit of an Industrial System Based on the Cost Free Warranty Policy." *Journal of the Brazilian Society of Mechanical Sciences and Engineering* 40(5): 1–9. doi:10.1007/s40430-018-1167-8.

Parey, Anand, and Amandeep Singh. 2019. "Gearbox Fault Diagnosis Using Acoustic Signals, Continuous Wavelet Transform and Adaptive Neuro-Fuzzy Inference System." *Applied Acoustics* 147: 133–40. doi:10.1016/j.apacoust.2018.10.013.

Pariaman, H., I. Garniwa, I. Surjandari, and B. Sugiarto. 2015. "Availability Improvement Methodology in Thermal Power Plant." *Scientific Journal of PPI-UKM* 2(1): 43–52.

Peng, Z., and N. Kessissoglou. 2003. "An Integrated Approach to Fault Diagnosis of Machinery Using Wear Debris and Vibration Analysis." *Wear* 255(7–12): 1221–32. doi:10.1016/S0043-1648(03)00098-X.

Peng, Z., N. J. Kessissoglou, and M. Cox. 2005. "A Study of the Effect of Contaminant Particles in Lubricants Using Wear Debris and Vibration Condition Monitoring Techniques." *Wear* 258(11–12): 1651–62. doi:10.1016/j.wear.2004.11.020.

Qian, Ren, Yun Wu, Xun Duan, Guangqian Kong, and Huiyun Long. 2018. "SVM Multi-Classification Optimization Research Based on Multi-Chromosome Genetic Algorithm." *International Journal of Performability Engineering* 14(4): 631–38. doi:10.23940/ijpe.18.04.p5.631638.

Sabouhi, Hamed, Ali Abbaspour, Mahmud Fotuhi-Firuzabad, and Payman Dehghanian. 2016. "Reliability Modeling and Availability Analysis of Combined Cycle Power Plants." *International Journal of Electrical Power and Energy Systems* 79: 108–19. doi:10.1016/j.ijepes.2016.01.007.

Sikos, László, Jiří Klemeš, László Sikos, and Jiří Klemeš. 2010. "Reliability, Availability and Maintenance Optimisation of Heat Exchanger Networks." *Applied Thermal Engineering* 30: 63–69. doi:10.1016/j.applthermaleng.2009.02.013.

Simani, Silvio, and Cesare Fantuzzi. 2006. "Dynamic System Identification and Model-Based Fault Diagnosis of an Industrial Gas Turbine Prototype." *Mechatronics* 16(6): 341–63. doi:10.1016/j.mechatronics.2006.01.002.

Singh, Rakesh Kumar, and Makarand S. Kulkarni. 2013. "Criticality Analysis of Power-Plant Equipments Using the Analytic Hierarchy Process." *International Journal of Industrial Engineering and Technology* 3(4): 1–14.

Tan, Mengquan, Lingen Chen, Jiashan Jin, Fengrui Sun, and Chih Wu. 2005. "Bond-Graph-Based Fault-Diagnosis for a Marine Condensate-Booster-Feedwater System." *Applied Energy* 81(4): 449–58. doi:10.1016/j.apenergy.2004.09.002.

Tewari, P. C., and Subhash Malik. 2016. "Simulation and Economic Analysis of Coal Based Thermal Power Plant : A Critical Literature Review." *IOSR Journal of Mechanical and Civil Engineering (IOSR-JMCE) Special Issue*: 36–41.

Wang, Ling, Jian Chu, and Jun Wu. 2007. "Selection of Optimum Maintenance Strategies Based on a Fuzzy Analytic Hierarchy Process." *International Journal of Production Economics* 107(1): 151–63. doi:10.1016/j.ijpe.2006.08.005.

Zhang, X., R. Xu, C. Kwan, S. Y. Liang, Q. Xie, and L. Haynes. 2005. "An Integrated Approach to Bearing Fault Diagnostics and Prognostics." In *Proceedings of the 2005, American Control Conference*, June 8–10, 2005, Portland, OR, USA, 2750–55. doi:10.1109/ACC.2005.1470385.

Ziegler, D., M. Puccinelli, B. Bergallo, and A. Picasso. 2013. "Investigation of Turbine Blade Failure in a Thermal Power Plant." *Case Studies in Engineering Failure Analysis* 1(3): 192–99. doi:10.1016/j.csefa.2013.07.002.

Zio, E. 2009. "Reliability Engineering: Old Problems and New Challenges." *Reliability Engineering and System Safety* 94(2): 125–41. doi:10.1016/j.ress.2008.06.002.

6 Reliability Evaluating of the AP1000 Passive Safety System under Intuitionistic Fuzzy Environment

Mohit Kumar and Manvi Kaushik

CONTENTS

6.1 INTRODUCTION

Safety is a major concern in nuclear engineering systems for many researchers and analysts. Safety assessment becomes even more significant for safety critical systems. Failure of these systems poses a risk to humans and the environment. While designing critical systems, the reliability and safety of their components are key factors. Problems related to safety and reliability assessment of complex systems are being solved by many researchers using different risk analysis methods. Over the last two decades, a method known as the fault tree analysis (FTA) has been widely used for risk analysis and reliability assessment. FTA is the most powerful method to assess system reliability. The fault tree of a system represents a logical relationship between the bottom events and top event. In FTA, the probability of a top event is evaluated by using the probabilities of bottom events [1]. Hence for the most appropriate results of the occurrence probability of a top event, more precise reliability data of bottom events are required. The uncertainties in reliability analysis of bottom events will result in uncertainty of the top event reliability.

In all real applications, it is not possible to get the precise reliability data of bottom events due to some environmental factors and other conditions. If there are insufficient data or any uncertainty (imprecision or vagueness) in the reliability data, then the results cannot be accurate. When the reliability data of bottom events are imprecise, insufficient or unavailable, then fuzzy set theory is used with possibility theory [2, 3] in FTA for safety and reliability assessment. In fuzzy set theory, the degree of membership of a particular object in a set is defined using the concept of the membership function which shows the belongingness of the object in the set. It has also been proved that when quantitative data are unavailable, then experts are more comfortable with expressing event failure using qualitative data described in linguistic terms. Wang et al. [4] introduced a fuzzy fault tree analysis (FFTA) for fire and explosion of crude oil tanks by using expert opinions as qualitative data. Purba et al. [5, 6] proposed a fuzzy probability based FFTA by using triangular membership functions to represent the qualitative data mathematically. Kumar [7] presented a novel weakest t-norm based FFTA approach to assess system reliability through qualitative data.

In fuzzy set theory, the uncertainty or hesitation in the membership degree cannot be assimilated. As a solution to this problem, Atanassov [8] presented the concept of intuitionistic fuzzy set (IFS) as an extension of the fuzzy set. Along with the membership function, the IFS uses the concept of the non-membership function such that their summation is less than 1 [9]. It has already been proved that in many situations, IFS is more preferable than the fuzzy set to deal with uncertainty. Related work on intuitionistic fuzzy sets can be seen in many research papers [7, 10–12]. Kumar [13] presented an area IF-defuzzification technique (AIDT) for the intuitionistic fuzzy number (IFN) to assess the reliability of nuclear bottom events. Kamyab et al. [14, 15] evaluated the effect of a passive core cooling system on the final core damage frequency of the AP1000 system. Guimaraes [2] evaluated the reliability of the AP1000 system by using a generic database to reduce a large break loss of coolant accident (LOCA). The AP1000 system is a pressurized light water reactor in which the passive concept has been implemented in its safety systems based on condensation, gravity, convection, and heat circulation [16]. The United States Nuclear Energy Commission (US-NRC) has certified the AP1000 safety system as a generation III+ reactor. Many methods have been developed to identify the sequence of components by which the undesired event occurs.

Reliability evaluation of the AP1000 passive safety system depends on the failure probabilities of its components. When quantitative failure data or probabilities of system components/events are insufficient or unavailable then qualitative data which are described as linguistic terms can be used to evaluate the system reliability. These linguistic terms can be quantified by using membership and non-membership functions of an IFN. In this chapter, an algorithm is presented to evaluate the failure probabilities of bottom events of the AP1000 passive safety system FTA by using qualitative data under intuitionistic fuzzy environment.

6.2 LITERATURE REVIEW

Tanaka et al. [17] applied the fuzzy set theory with FTA for the first time. Lin and Wang [18] combined expert elicitation with the fuzzy set theory for FTA with uncertain data. Yuhua et al. [19] proposed the FFTA to assess the failure probability of oil and gas transmission pipelines. The FFTA has also been used for the evaluation of reliabilities of nuclear power plant safety systems [5, 6, 20] and also to evaluate risk in petrochemical industries [21, 22]. A review of applications of fuzzy sets in system reliability analysis has been presented by Kabir et al. [23]. To acquire uncertain failure data, expert judgment has been used in association with the fuzzy set theory. Kumar [7] proposed the weakest triangular norm (T_w)-based FFTA to assess system reliability with qualitative data processing. Kumar [24] developed a novel T_w-based fuzzy importance measure to identify the critical basic events in FFTA.

Fuzzy set is defined by only the degree of acceptance but IFS considers the membership function (acceptance degree) and a non-membership function (rejection degree) such that the sum of both is always less than 1 [8]. Cheng et al. [25] utilized the concept of IFS theory for reliability assessment of liquefied natural gas terminal emergency shut-down system through FTA. Chang and Cheng [26] presented an approach to access the reliability of PCBA using α-cut and interval arithmetic operations on different types of vague sets. Kumar et al. [27] used triangular intuitionistic fuzzy numbers (TIFNs) and developed an approach for reliability evaluation using IFS theory. The IFS theory has been applied in various areas [28–34]. It is predetermined that the failure data of all the components of a system follow the same probability distribution, but in practical situations, it happens rarely. As a potential solution to this problem, Garg [35] proposed an alternative method to evaluate the system reliability in intuitionistic fuzzy environment. Related work on the IFS can be seen in many research studies [36–40]. Garg et al. [41] proposed Improved possibility degree method for ranking IFNs with their application in multiattribute decision-making. Kumar [42] presented an area IF-defuzzification method for IFNs to assess the failure data of nuclear events from their corresponding failure possibilities that are represented as IFNs. Kumar et al. [43] developed an approach to evaluate the importance of the attributes as an intuitionistic fuzzy correlation.

6.3 THE AP1000 PASSIVE SAFETY SYSTEM

To enhance nuclear energy resources, the execution of the safety systems of nuclear power plants is a key factor. Safety systems ensure that plants can operate normally without a high rate of risk to the staff and environment and without accidents. For the improvement of nuclear power plant safety systems, the concept of passive systems is very important. In AP1000, passive systems and equipment work in parallel configuration so that the passive system is highly reliable as compared to the active system [44]. But still there is the possibility that the passive safety system fails because of the physical circumstances on which it is defined. Therefore, the reliability study of the AP1000 passive safety system is an important issue by assuming different accident scenarios. In the present study, to evaluate the reliability of the safety systems of the nuclear power plant, fault tree analysis (FTA) has been used widely. The FTA is a top-down deductive method which helps to identify the sequence of the bottom event in a structured way, such that if all bottom events fail in the sequence, then the accident will occur. The FTA assumes both the quantitative and qualitative failure probability data of its bottom events. An AP1000 passive safety system is still in construction, and so it is not possible to get the quantitative failure probability data of all of the bottom events [45]. The reliability of the AP1000 passive safety system can be evaluated using FTA. In FTA, failure possibilities of bottom events can be taken from experts' advice and described in linguistic terms. When the boundary

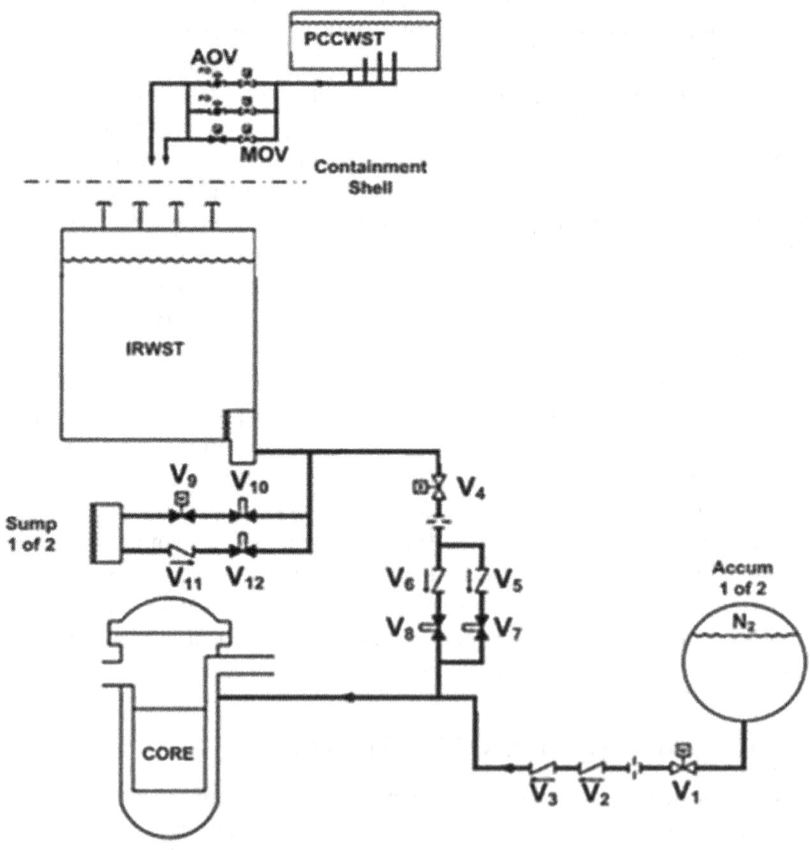

FIGURE 6.1 Mitigate a large break LOCA in the AP1000 passive safety system.

pressure of the reactor coolant breaks, then the reactor coolant flows outside; this situation is called LOCA. It happens because of a large break in the primary coolant system. The task of the primary coolant is to circulate the coolant between the reactor core and the steam generators involving two loops of heat transfer systems and one pressurizer. Each loop can be formed with one steam generator, two reactor coolant pumps, one hot leg, and two cold legs. AP1000 is designed on the basis of Westinghouse's proven pressurized water reactors. The concept of a passive system has been incorporated into its safety injection system, residual heat removal system, and containment cooling system.

The main advantage of these passive safety systems is that without any involvement of the operators and reliance on the AC power sources, both off-site or on-site, accidents can be mitigated. The three passive safety systems provided in AP1000 reactors to mitigate the large break LOCA are low pressure injection system (LIP), injection system by accumulator (AI), and long-term cooling system (LTC). The simplified working process of these safety systems is shown in Figure 6.1.

Many valves are lined up to instinctively initiate those three passive safety systems by shifting the valve positions. For high reliability, those valves are instinctively initiated to their preventive positions when they lose power or receive the actuation signal. Guimaraes et al. [2] has described the passive injection system with an accumulator, passive low pressure injection system, and passive long-term cooling system. Fault trees for these systems and probabilities of all bottom events has also been described. Probabilities of the bottom events of the safety system (Figure 6.1) are listed in Table 6.1.

TABLE 6.1
Probabilities of Bottom Events for Corresponding Component Failure

Bottom events	Component failure	Probabilities
b_1	V_1 failure	7.59 E-6
b_2	V_2 failure	1.11 E-5
b_3	V_3 failure	1.11 E-5
b_4	V_4 failure	7.59 E-6
b_5	V_5 failure	1.11 E-5
b_6	V_6 failure	1.11 E-5
b_7	V_7 failure	1.90 E-4
b_8	V_8 failure	1.90 E-4
b_9	V_9 failure	1.06 E-4
b_{10}	V_{10} failure	1.90 E-4
b_{11}	V_{11} failure	1.11 E-5
b_{12}	V_{12} failure	1.90 E-4
b_{13}	V_{13} failure	1.21 E-6
b_{14}	V_{14} failure	7.59 E-6

6.4 SOME REVIEWED DEFINITIONS

In this section, some definitions related to the present work are reviewed.

6.4.1 Intuitionistic Fuzzy Set

Atanassov [8] introduced the definition of IFS. Let X be the universe of discourse. An IFS $\tilde{A}$ defined on X is classified as:

$$\tilde{A} = \{< x, \mu_{\tilde{A}}(x), v_{\tilde{A}}(x) >: x \in X\} \tag{6.1}$$

where $\mu_{\tilde{A}} : X \to [0,1]$ and $v_{\tilde{A}} : X \to [0,1]$ are membership (degree of acceptance) and non-membership (degree of rejection) functions, respectively, such that

$$\mu_{\tilde{A}}(x) + v_{\tilde{A}}(x) \in [0,1], \quad \forall x \in X \tag{6.2}$$

In addition, the IF-index (degree of hesitation or uncertainty level) of $\tilde{A}$ is defined as:

$$\pi_{\tilde{A}}(x) = 1 - \mu_{\tilde{A}}(x) - v_{\tilde{A}}(x) \tag{6.3}$$

If $\pi_{\tilde{A}}(x) = 0, \forall x \in X$ then IFS is converted into a fuzzy set.

6.4.2 Convex Intuitionistic Fuzzy Set

An IFS $\tilde{A} = \{< x, \mu_{\tilde{A}}(x), v_{\tilde{A}}(x) >: x \in X\}$ is said to be convex IFS [46] if

- Membership function $\mu_{\tilde{A}}(x)$ is fuzzy convex

$$\mu_{\tilde{A}}\left(\lambda x_1 + (1-\lambda)x_2\right) \geq \min\left(\mu_{\tilde{A}}(x_1), \mu_{\tilde{A}}(x_2)\right) \quad \forall\, x_1, x_2 \in X, \lambda \in [0,1] \tag{6.4}$$

- Non-membership function $v_{\tilde{A}}(x)$ is fuzzy concave

$$v_{\tilde{A}}\left(\lambda x_1 + (1-\lambda)x_2\right) \leq \max\left(v_{\tilde{A}}(x_1), v_{\tilde{A}}(x_2)\right) \quad \forall\, x_1, x_2 \in X, \lambda \in [0,1] \tag{6.5}$$

6.4.3 Normal Intuitionistic Fuzzy Set

An IFS $\tilde{A} = \{< x, \mu_{\tilde{A}}(x), v_{\tilde{A}}(x) >: x \in X\}$ is said to be normal IFS [46] if $\exists$ two points x_1, x_2 s.t. $\mu_{\tilde{A}}(x_1) = 1$ and $v_{\tilde{A}}(x_2) = 1$.

6.4.4 Intuitionistic Fuzzy Number

An IFS $\tilde{A} = \{< x, \mu_{\tilde{A}}(x), v_{\tilde{A}}(x) >: x \in \mathbb{R}\}$ is said to be an IFN [46] if

(i) $\tilde{A}$ is IF-convex.
(ii) $\tilde{A}$ is IF-normal.

(iii) $\mu_{\tilde{A}}(x)$ is upper semi-continuous and $v_{\tilde{A}}(x)$ is lower semi-continuous.
(iv) $\text{Supp}(\tilde{A}) = \{x \in X : \mu_{\tilde{A}}(x) > 0\}$is bounded.

6.4.5 Triangular Intuitionistic Fuzzy Number

A triangular IFN (TIFN) [46] is denoted as *(a, b, c; a′, b, c′)* and defined in terms of the membership function $\mu_{\tilde{A}}$ and the non-membership function $v_{\tilde{A}}$ as:

$$\mu_{\tilde{A}}(x) = \begin{cases} \dfrac{x-a}{b-a}, & a \le x \le b \\ \dfrac{c-x}{c-b}, & b \le x \le c \\ 0, & \text{otherwise} \end{cases} \quad \text{and} \quad v_{\tilde{A}}(x) = \begin{cases} \dfrac{b-x}{b-a'}, & a' \le x \le b \\ \dfrac{x-b}{c'-b}, & b \le x \le c' \\ 1, & \text{otherwise} \end{cases} \tag{6.6}$$

6.4.6 Linguistic Terms

When system reliability is evaluated, then one should look for the needed probabilistic historical failure data of its components (bottom events). In case probabilistic failure data are unavailable or insufficient, then qualitative data such as analysts' or experts' opinions can be used as linguistic terms to evaluate system reliability. Seven qualitative linguistic terms are described from the analysis of a nuclear power plant's data and are listed in Table 6.2. TIFNs are used to represent the linguistic terms mathematically and are listed in Table 6.2.

6.4.7 Intuitionistic Fuzzy Failure Possibility

The intuitionistic fuzzy failure possibility (IF-failure possibility or IF-possibility) can be defined as IFN in the unit interval [0,1]. The qualitative linguistic terms are

TABLE 6.2
Failure Possibilities and Corresponding IFNs

Failure possibilities/linguistic terms	Intuitionistic fuzzy number
Very low (*VL*)	(0, 0.04, 0.08; 0, 0.04, 0.08)
Low (*L*)	(0.07, 0.13, 0.19; 0.04, 0.13, 0.22)
Reasonably low (*RL*)	(0.17, 0.27, 0.37; 0.12, 0.27, 0.42)
Moderate (*M*)	(0.35, 0.5, 0.65; 0.25, 0.5, 0.75)
Reasonably high (*RH*)	(0.63, 0.73, 0.83; 0.58, 0.73, 0.88)
High (*H*)	(0.81, 0.87, 0.93; 0.78, 0.87, 0.96)
Very high (*VH*)	(0.92, 0.96, 1; 0.92, 0.96, 1)

represented quantitatively by IF-failure possibilities. It has also been confirmed that any type of IFN could be used as IF-possibility in system reliability studies. Purba et al. [47] assigned fuzzy possibilities for seven linguistic terms based on the inductive reasoning approach and utilized these fuzzy possibilities for nuclear power plant's probabilistic safety assessment. In this study, the fuzzy possibilities are generalized as IF -possibilities to assess the reliability of the AP1000 passive system in intuitionistic fuzzy environment. A set of seven IF-possibilities is defined for a set of seven linguistic terms and listed in Table 6.2.

6.4.8 The Area IF-Defuzzification Technique (AIDT)

IF-Defuzzification is the process of converting intuitionistic fuzzy value to crisp value. Kumar [13] proposed an area IF-defuzzification technique (AIDT) to assess the reliability of bottom events of nuclear power plant. The AIDT of TIFN $\tilde{A} = (a, b, c; a', b, c')$ is defined as:

$$\text{AIDT}(\tilde{A}) = \frac{1}{36}\left[4(a + a') + 2b + (c + c')\right] \tag{6.7}$$

6.5 PROPOSED ALGORITHM

Purba [48] evaluated the failure probability of each bottom event of the AP1000 passive safety system through the fuzzy approach. In this chapter, idea of the fuzzy approach is extended to the IF-approach and a new algorithm is presented to assess the reliability of the passive safety system through qualitative data processing when the probabilistic approach cannot be used because of insufficient historical data of its components. A brief structure of the proposed algorithm is shown in Figure 6.2 and defined by the following steps:

Step 1. *Define the likelihood occurrence possibilities of bottom events*:
When historical failure data of bottom events are unavailable or insufficient, then reliability evaluation in terms of the probability data is not possible. It is already proved that credible experts are more comfortable in linguistic justification rather than numerical values. The objective of this step is to define the likelihood occurrences of the failure possibilities of bottom events. Each credible expert justifies a likelihood occurrence for each bottom event for the respective failure possibility. There are n bottom events b_i, $i = 1, 2, \ldots, n$ and m credible experts e_j, $j = 1, 2, \ldots, m$. The likelihood occurrences for the events of the passive safety system are listed in Table 6.3.

Step 2. *Evaluate IF-failure possibility of each bottom event:*
In the previous step, all bottom events are assessed by experts and linguistic judgments are given by experts to each bottom event. The objective of this step is to convert the linguistic judgments of each bottom event into an IF-failure possibility $\tilde{p}_{ij}$ where $\tilde{p}_{ij}$ is the IF-failure possibility for i^{th} bottom event justified by j^{th} expert.

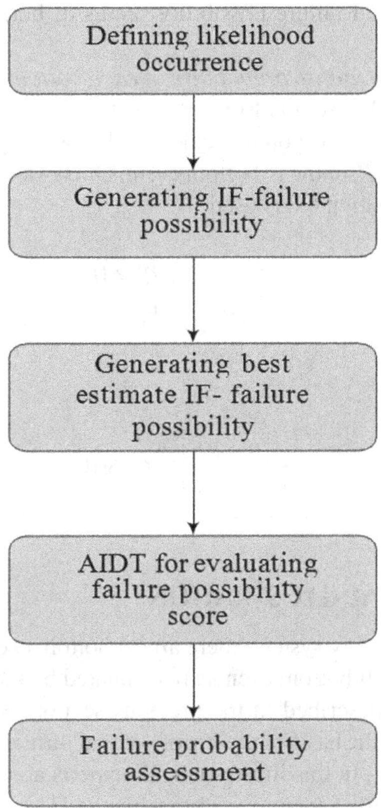

FIGURE 6.2 Structure of the proposed reliability assessment algorithm.

Step 3. *Generate the best estimate IF-failure possibilities of bottom events:* In the AP1000 passive safety system, there are seven experts and it may be possible that there is a conflict in their judgments/opinions. To avoid this problem, a weight function is given to each expert on the basis of their professional position, work experience, and educational qualification. The best estimate IF-failure possibilities of each bottom event b_i can be obtained by the IF-failure possibilities $\tilde{p}_{ij}$ as $\tilde{p}_i = \sum_{j=1}^{n} w_j * \tilde{p}_{ij}$ where w_j is the justification weight to j^{th} expert for competency such that $\sum_{j=1}^{n} w_j = 1$. For the passive safety system, we are taking 1/7 as the weight justification for each expert. The best estimate IF-failure possibilities of bottom events of the passive safety system are listed in Table 6.5.

Step 4. *Calculate bottom event failure possibility scores:* From the best-estimated failure probabilities of bottom events, failure probability scores can be generated. To generate failure possibility score R_i^s, the area IF-defuzzification technique (AIDT) (shown in Equation 6.7) is used. The generated failure possibility scores of bottom events are listed in Table 6.5.

Step 5. *Assessment of failure probability of each bottom event:*

The objective of this step is to assess the failure probability of each bottom event from the corresponding calculated possibility score. To assess the failure probability R_i of the ith bottom event, Onisawa's logarithmic function [49] shown in Equation 6.8 is used.

$$R = \begin{cases} 1/10^z; & R_i^s \neq 0 \\ 0; & R_i^s = 0 \end{cases} \tag{6.8}$$

where

$$z = \left[\frac{1 - R_i^s}{R_i^s}\right]^{1/3} * 2.301 \tag{6.9}$$

6.6 MATHEMATICAL ILLUSTRATION

In the AP1000 passive safety system, there are 14 bottom events $\{b_i: i = 1, 2, \ldots, 14\}$. Failure probabilities of all bottom events are evaluated by the IF-approach using the proposed algorithm as described in the previous section. Seven experts $\{e_j: j = 1, 2, \ldots, 7\}$ are selected on the basis of their professional position, work experience, and educational qualification. In this illustration, all experts are considered same so that the weight function for all experts are same, which is 1/7.

6.1. *Likelihood occurrence possibilities of bottom events*: All selected experts make opinions about each bottom event individually (as listed in Table 6.3) and on the basis of their judgments/opinions, likelihood occurrence of each bottom event is evaluated.

6.2. *IF-failure possibility*: In this step, all linguistic judgments for each bottom event are converted into IF-failure possibilities $\tilde{p}_{ij}$. For example, IF-failure possibilities of the event b_9 are listed in Table 6.4.

6.3. *Best estimate IF-failure possibility*: In this step, best estimate IF-failure probability $\tilde{p}_i$ for each b_i is evaluated with the help of the obtained $\tilde{p}_{ij}$ and listed in Table 6.5. For example, the evaluated IF-failure probability of b_9 is (0.444, 0.566, 0.687; 0.379, 0.566, 0.758).

6.4. *Failure possibility score*: The failure possibility scores R_i^s for all bottom events are calculated and listed in Table 6.5. For example, for the event b_9 the failure possibility score is calculated as follows:

$$\begin{aligned} R_9^s &= \frac{1}{36}\left[4 \times (0.444 + 0.379) + 2 \times 0.566 + (0.687 + 0.758)\right] \\ &= 0.162380952 \end{aligned} \tag{6.10}$$

TABLE 6.3
Bottom Event Likelihood Occurrences Assessed by Experts

Bottom event	Likelihood occurrence						
	e_1	e_2	e_3	e_4	e_5	e_6	e_7
b_1	*M*	*RL*	*RL*	*RL*	*M*	*RL*	*RL*
b_2	*M*	*L*	*RL*	*M*	*RL*	*M*	*RL*
b_3	*M*	*L*	*M*	*M*	*RL*	*M*	*L*
b_4	*M*	*RL*	*RL*	*M*	*RL*	*RL*	*RL*
b_5	*M*	*RL*	*RL*	*M*	*L*	*M*	*L*
b_6	*M*	*L*	*RL*	*M*	*RL*	*M*	*L*
b_7	*RH*	*M*	*RH*	*M*	*RH*	*M*	*H*
b_8	*RH*	*M*	*H*	*M*	*RH*	*M*	*RH*
b_9	*M*	*RH*	*M*	*M*	*RL*	*RH*	*RH*
b_{10}	*H*	*M*	*RH*	*M*	*RH*	*M*	*RH*
b_{11}	*M*	*L*	*RL*	*M*	*RL*	*M*	*RL*
b_{12}	*RH*	*M*	*H*	*M*	*RH*	*M*	*RH*
b_{13}	*RH*	*M*	*RH*	*M*	*RL*	*M*	*RH*
b_{14}	*M*	*RL*	*M*	*RL*	*RL*	*M*	*RL*

TABLE 6.4
IF-Failure Possibilities for b_9

Expert	$\tilde{p}_{ij}$
e_1	(0.35, 0.5, 0.65; 0.25, 0.5, 0.75)
e_2	(0.63, 0.73, 0.83; 0.58, 0.73, 0.88)
e_3	(0.35, 0.5, 0.65; 0.25, 0.5, 0.75)
e_4	(0.35, 0.5, 0.65; 0.25, 0.5, 0.75)
e_5	(0.35, 0.5, 0.65; 0.25, 0.5, 0.75)
e_6	(0.63, 0.73, 0.83; 0.58, 0.73, 0.88)
e_7	(0.63, 0.73, 0.83; 0.58, 0.73, 0.88)

6.5. *Failure Probability*: Using Onisawa's logarithmic function [49], failure probabilities of all bottom events are calculated and shown in Table 6.5. As an example, The evaluation of failure probability of b_9 is shown in Equation 6.12.

$$z = \left[\frac{1-0.162380952}{0.162380952}\right]^{\frac{1}{3}} \times 2.301 \tag{6.11}$$

$$= 3.975762553$$

TABLE 6.5
Evaluation of best estimate IF-failure possibilities, scores and failure probabilities of bottom events

Bottom event	Best estimate IF-failure possibility	Score	Evaluated failure probability
b_1	(2.21E-01,3.35E-01,4.5E-01;1.57E-01,3.35E-01,5.14E-01)	8.75E-02	9.39E-06
b_2	(2.33E-01,3.48E-01,4.64E-01;1.64E-01,3.48E-01,5.32E-01)	9.12E-02	1.12E-05
b_3	(2.44E-01,3.61E-01,4.78E-01;1.71E-01,3.61E-01,5.51E-01)	9.49E-02	1.32E-05
b_4	(2.21E-01,3.35E-01,4.5E-01;1.57E-01,3.35E-01,5.14E-01)	8.75E-02	9.39E-06
b_5	(2.18E-01,3.28E-01,4.38E-01;1.52E-01,3.28E-01,5.04E-01)	8.57E-02	8.61E-06
b_6	(2.18E-01,3.28E-01,4.38E-01;1.53E-01,3.28E-01,5.04E-01)	8.57E-02	8.61E-06
b_7	(5.35E-01,6.51E-01,7.67E-01;4.67E-01,6.51E-01,8.36E-01)	1.92E-01	1.93E-04
b_8	(5.36E-01,6.51E-01,7.67E-01;4.67E-01,6.51E-01,8.36E-01)	1.92E-01	1.93E-04
b_9	(4.44E-01,5.66E-01,6.87E-01;3.79E-01,5.66E-01,7.58E-01)	1.62E-01	1.06E-04
b_{10}	(5.35E-01,6.51E-01,7.67E-01;4.67E-01,6.51E-01,8.35E-01)	1.92E-01	1.93E-04
b_{11}	(2.33E-01,3.48E-01,4.64E-01;1.64E-01,3.48E-01,5.33E-01)	9.11E-02	1.12E-05
b_{12}	(5.35E-01,6.51E-01,7.67E-01;4.67E-01,6.51E-01,8.36E-01)	1.92E-01	1.93E-04
b_{13}	(4.44E-01,5.66E-01,6.87E-01;3.73E-01,5.66E-01,7.58E-01)	1.62E-01	1.06E-04
b_{14}	(2.47E-01,3.68E-01,4.9;1.75,3.68,5.61)	9.66E-02	1.422E-05

$$R = \frac{1}{10^{3.975762553}} = 0.00010574 \tag{6.12}$$

6.7 RESULTS AND DISCUSSION

Purba [5, 6] presented a fuzzy-reliability approach to assess the reliabilities of nuclear events when only qualitative data are available as experts' opinions. Kumar [7] developed a Tw-based fuzzy fault tree analysis (TBFFTA) to assess system reliability when only qualitative data such as expert decisions are available and described in linguistic terms.

- The concept of fuzzy number is extended to the concept of IFN and a novel IF-reliability approach has been presented to assess the reliability of the AP1000 passive safety system.
- Failure probabilities of the bottom events of the AP1000 passive safety system have been evaluated by the proposed approach and listed in Table 6.5. To verify the feasibility and applicability of the presented approach, the obtained results are compared with known probabilities [2]. Errors between these two results are very less as shown in Table 6.6.
- The proposed approach can evaluate the probabilities of bottom events which are involved in diminishing large break LOCA and does not have

TABLE 6.6
Comparison of Evaluated Failure Probabilities with Known Failure Probabilities of Bottom Events

Bottom event	Evaluated failure probability	Known failure probability	Relative error
b_1	9.39E-06	7.59E-06	0.230415271
b_2	1.12E-05	1.11E-05	0.00678793
b_3	1.32E-05	1.11E-05	0.187443967
b_4	9.392E-06	7.59E-06	0.230415271
b_5	8.61E-06	1.11E-05	0.22460924
b_6	8.61E-06	1.11E-05	0.22460924
b_7	1.93E-04	1.90E-04	0.122562358
b_8	1.93E-04	1.90E-04	0.017262788
b_9	1.06E-04	1.06E-04	0.0024571
b_{10}	1.93E-04	1.90E-04	0.017262788
b_{11}	1.12E-05	1.11E-05	0.00678793
b_{12}	1.93E-04	1.90E-04	0.017262788
b_{13}	1.06E-04	1.21E-04	0.126119443
b_{14}	1.42E-05	7.59E-06	0.230415271

quantitative failure data. Hence the proposed approach is applicable to evaluate the reliability of those components of the AP1000 system which are still under construction.

- Both the fuzzy-reliability approach and the IF-reliability approach can be used when historical failure data of system components are unavailable or insufficient. But the IF-reliability approach is more appropriate than the fuzzy-reliability approach because along with the degree of belongingness, it considers a degree of non-belongingness also.

6.8 CONCLUSION

To assess the reliability of the AP1000 passive safety system, an IF-reliability approach has been executed. The results of the case study show that the generated probabilities of bottom events are very close to the known failure probabilities of the AP1000 passive safety system. The obtained results confirm that the IF-reliability approach gives very realistic results when the historical probability data of bottom events are unavailable or insufficient. Meanwhile, the proposed approach is more preferable than the fuzzy-reliability approach because in many cases it may not be possible to define the membership degree with certainty, but in the IFS membership degree and non-membership degree are considered.

From a futuristic perspective, there is an engaging question to know how the IF-reliability approach will work for other components of the AP1000 passive safety system. With the help of the fault tree, the reliability of the system can be evaluated using the proposed approach. In this study, we have used TIFN. In future, other form of IFN can be used.

ACKNOWLEDGMENT

The second author would like to thank the Department of Science and Technology, Government of India, for supporting this research financially in the form of the INSPIRE fellowship.

REFERENCES

1. Verma, A.K., Ajit, S., Karanki, D.R., et al.: *Reliability and Safety Engineering* **43**, 373–392 (2010).
2. Guimaraes, A.C.F., Lapa, C.M.F., Simoes Filho, F.F.L., Cabral, D.C.: Fuzzy uncertainty modeling applied to AP1000 nuclear power plant LOCA. *Annals of Nuclear Energy* **38**(8), 1775–1786 (2011).
3. Kamyab, S., Nematollahi, M., Shafiee, G.: Sensitivity analysis on the effect of software-induced common cause failure probability in the computer-based reactor trip system unavailability. *Annals of Nuclear Energy* **57**, 294–303 (2013).
4. Wang, D., Zhang, P., Chen, L.: Fuzzy fault tree analysis for fire and explosion of crude oil tanks. *Journal of Loss Prevention in the Process Industries* **26**(6), 1390–1398 (2013).
5. Purba, J.H., Tjahyani, D.S., Ekariansyah, A.S., Tjahjono, H.: Fuzzy probability based fault tree analysis to propagate and quantify epistemic uncertainty. *Annals of Nuclear Energy* **85**, 1189–1199 (2015).
6. Purba, J.H., Tjahyani, D.S., Widodo, S., Tjahjono, H.: α-Cut method based importance measure for criticality analysis in fuzzy probability-based fault tree analysis. *Annals of Nuclear Energy* **110**, 234–243 (2017).
7. Kumar, M.: A novel weakest t-norm based fuzzy fault tree analysis through qualitative data processing and its application in system reliability evaluation. *Journal of Intelligent Systems* **29**(1), 977–993 (2019).
8. Atanassov, K.: Intuitionistic fuzzy sets. *Fuzzy Sets and Systems* **20**(1), 87–96 (1986).
9. Atanassov, K.: *Intuitionistic Fuzzy Sets*. Central Tech Library, Bulgarian Academy Science, Sofia, Bulgaria (1983).
10. Garg, H., Kumar, K.: An advanced study on the similarity measures of intuitionistic fuzzy sets based on the set pair analysis theory and their application in decision making. *Soft Computing* **22**(15), 4959–4970 (2018).
11. Garg, H., Rani, M., Sharma, S., Vishwakarma, Y.: Intuitionistic fuzzy optimization technique for solving multi-objective reliability optimization problems in interval environment. *Expert Systems with Applications* **41**(7), 3157–3167 (2014).
12. Li, D.F.: A ratio ranking method of triangular intuitionistic fuzzy numbers and its application to MADM problems. *Computers and Mathematics with Applications* **60**(6), 1557–1570 (2010).
13. Kumar, M.: An area if-defuzzification technique and intuitionistic fuzzy reliability assessment of nuclear basic events of fault tree analysis. In: Yadav, N., Yadav, A., Bansal, J.C., Deep, K., Kim, J.H. (Eds.), *Harmony Search and Nature Inspired Optimization Algorithms*, pp. 845–856. Springer, Singapore (2019).
14. Kamyab, S., Nematollahi, M., Kamyab, M., Jafari, A.: Evaluating the reliability of AP1000 passive core cooling systems with risk assessment tool. In: *World Congress on Engineering 2012*. July 4–6, 2012. London, UK, vol. 2182, pp. 1668–1673. Citeseer (2010).

15. Kamyab, S., Nematollahi, M.: Performance evaluating of the AP1000 passive safety systems for mitigation of small break loss of coolant accident using risk assessment tool-ii software. *Nuclear Engineering and Design* **253**, 32–40 (2012).
16. Cascales, M.d.S.G.: *Soft Computing Applications for Renewable Energy and Energy Efficiency*. IGI Global (2014).
17. Tanaka, H., Fan, L., Lai, F., Toguchi, K.: Fault-tree analysis by fuzzy probability. *IEEE Transactions on Reliability* **32**(5), 453–457 (1983).
18. Lin, C.T., Wang, M.J.J.: Hybrid fault tree analysis using fuzzy sets. *Reliability Engineering and System Safety* **58**(3), 205–213 (1997).
19. Yuhua, D., Datao, Y.: Estimation of failure probability of oil and gas transmission pipelines by fuzzy fault tree analysis. *Journal of Loss Prevention in the Process Industries* **18**(2), 83–88 (2005).
20. Purba, J.H.: A fuzzy-based reliability approach to evaluate basic events of fault tree analysis for nuclear power plant probabilistic safety assessment. *Annals of Nuclear Energy* **70**, 21–29 (2014).
21. Lavasani, S.M., Zendegani, A., Celik, M.: An extension to fuzzy fault tree analysis (FFTA application in petrochemical process industry. *Process Safety and Environmental Protection* **93**, 75–88 (2015).
22. Senol, Y.E., Aydogdu, Y.V., Sahin, B., Kilic, I.: Fault tree analysis of chemical cargo contamination by using fuzzy approach. *Expert Systems with Applications* **42**(12), 5232–5244 (2015).
23. Kabir, S., Papadopoulos, Y.: A review of applications of fuzzy sets to safety and reliability engineering. *International Journal of Approximate Reasoning* **100**, 29–55 (2018).
24. Kumar, M.: A novel weakest t-norm based fuzzy importance measure for fuzzy fault tree analysis of combustion engineering reactor protection system. *International Journal of Uncertainty, Fuzziness and Knowledge-Based Systems* **27**(6), 949–967 (2019).
25. Cheng, S.R., Lin, B., Hsu, B.M., Shu, M.H.: Fault-tree analysis for liquefied natural gas terminal emergency shutdown system. *Expert Systems with Applications* **36**(9), 11918–11924 (2009).
26. Chang, K.H., Cheng, C.H.: A novel general approach to evaluating the PCBA for components with different membership function. *Applied Soft Computing* **9**(3), 1044–1056 (2009).
27. Kumar, M., Prasad Yadav, S., Kumar, S.: Fuzzy system reliability evaluation using time-dependent intuitionistic fuzzy set. *International Journal of Systems Science* **44**(1), 50–66 (2013).
28. Kumar, M.: Applying weakest t-norm based approximate intuitionistic fuzzy arithmetic operations on different types of intuitionistic fuzzy numbers to evaluate reliability of pcba fault. *Applied Soft Computing* **23**, 387–406 (2014).
29. Kumar, M., Yadav, S.P.: A novel approach for analyzing fuzzy system reliability using different types of intuitionistic fuzzy failure rates of components. *ISA Transactions* **51**(2), 288–297 (2012).
30. Garg, H., Agarwal, N., Tripathi, A.: Some improved interactive aggregation operators under interval-valued intuitionistic fuzzy environment and their application to decision making process. *Scientia Iranica. Transaction E, Industrial Engineering* **24**(5), 25812604 (2017).
31. Garg, H.: Reliability, availability and maintainability analysis of industrial systems using PSO and fuzzy methodology. *MAPAN* **29**(2), 115–129 (2014).
32. Garg, H., Agarwal, N., Tripathi, A.: Choquet integral-based information aggregation operators under the interval-valued intuitionistic fuzzy set and its applications to decision-making process. *International Journal for Uncertainty Quantification* **7**(3), 249–269 (2017).

33. Garg, H., Sharma, S.: Behavior analysis of synthesis unit in fertilizer plant. *International Journal of Quality and Reliability Management* **29**(2), 217–232 (2012).
34. Kour, D., Mukherjee, S., Basu, K.: Solving intuitionistic fuzzy transportation problem using linear programming. *International Journal of System Assurance Engineering and Management* **8**(2), 1090–1101 (2017).
35. Garg, H.: A novel approach for analyzing the reliability of series-parallel system using credibility theory and different types of intuitionistic fuzzy numbers. *Journal of the Brazilian Society of Mechanical Sciences and Engineering* **38**(3), 1021–1035 (2016).
36. Garg, H.: An approach for analyzing fuzzy system reliability using particle swarm optimization and intuitionistic fuzzy set theory. *Journal of Multiple-Valued Logic and Soft Computing* **21**, 335–354 (2013).
37. Garg, H.: Reliability analysis of repairable systems using petri nets and vague lambda-tau methodology. *ISA Transactions* **52**(1), 6–18 (2013).
38. Garg, H.: An approach for analyzing the reliability of industrial system using fuzzy kolmogorov's differential equations. *Arabian Journal for Science and Engineering* **40**(3), 975–987 (2015).
39. Garg, H., Rani, M.: An approach for reliability analysis of industrial systems using PSO and IFS technique. *ISA Transactions* **52**(6), 701–710 (2013).
40. Garg, H., Rani, M., Sharma, S.: Reliability analysis of the engineering systems using intuitionistic fuzzy set theory. *Journal of Quality and Reliability Engineering* **2013**, (2013) https://doi.org/10.1155/2013/943972.
41. Garg, H., Kumar, K.: Improved possibility degree method for ranking intuitionistic fuzzy numbers and their application in multiattribute decision-making. *Granular Computing* **4**(2), 237–247 (2019).
42. Kumar, M.: Intuitionistic fuzzy measures of correlation coefficient of intuitionistic fuzzy numbers under weakest triangular norm. *International Journal of Fuzzy System Applications (IJFSA)* **8**(1), 48–64 (2019).
43. Kumar, M.: Evaluation of the intuitionistic fuzzy importance of attributes based on the correlation coefficient under weakest triangular norm and application to the hotel services. *Journal of Intelligent and Fuzzy Systems* (Preprint), **36**(4), 3211–3223 (2019).
44. Nayak, A.K., Sinha, R.K.: Role of passive systems in advanced reactors. *Progress in Nuclear Energy* **49**(6), 486–498 (2007).
45. Wang, S., Wahab, M., Fang, L.: Managing construction risks of AP1000 nuclear power plants in china. *Journal of Systems Science and Systems Engineering* **20**(1), 43–69 (2011).
46. Mahapatra, G., Roy, T.: Reliability evaluation using triangular intuitionistic fuzzy number. *International Journal of Mathematical and Statistical Sciences* **1**(1), 31–38 (2009).
47. Purba, J.H., Lu, J., Zhang, G., Pedrycz, W.: A fuzzy reliability assessment of basic events of fault trees through qualitative data processing. *Fuzzy Sets and Systems* **243**, 50–69 (2014).
48. Purba, J.H., Tjahyani, D.S.: Reliability study of the AP1000 passive safety system by fuzzy approach. *Atom Indonesia* **40**(2), 49–56 (2014).
49. Onisawa, T.: An approach to human reliability in man-machine systems using error possibility. *Fuzzy Sets and Systems* **27**(2), 87–103 (1988).

7 Fuzzy Fault Tree Analysis for Web Access Failure under Uncertainty Using a Compensatory Operator

Komal

CONTENTS

7.1 INTRODUCTION

Nowadays, engineering systems are becoming more complex due to advancements in new technologies. For the proper functioning and reliable performance of these systems, they must be maintained at regular time intervals [1–4]. In the literature, numerous qualitative and quantitative techniques exist for reliability assessment including reliability block diagram (RBD), fault tree analysis (FTA), event tree analysis (ETA), Markov models (MM), Petri-nets (PN), failure mode and effect analysis (FMEA), Bayesian approach, etc. [5] FTA is one of the most applicable reliability/failure probability assessment technique because it is simple and logical in nature [6]. Traditional FTA estimates the system's top event failure probability by using its basic events' exact failure probabilities. In certain circumstances, extraction of the system's basic events' exact failure probabilities is difficult. In such conditions, experts generally suggest the use of rough values with the help of the possibility distribution function. Using these rough values, a system's top event possibility is computed by employing Boolean algebra and this leads to the problem of uncertainty [7, 8]. To deal with uncertainty, fuzzy set, intuitionistic fuzzy set (IFS), and vague set theories are predominantly used and successfully integrated with fault tree analysis [9–14]. Specifically, the amalgam of fuzzy set theory and fault tree analysis is called as fuzzy fault tree analysis (FFTA). The development of new FFTA methods is based on the suitable choice of fuzzy sets/fuzzy numbers and appropriate fuzzy arithmetic operations. A fuzzy set is defined in terms of elements and their membership functions. In the literature, a variety of membership functions exist, and among these triangular fuzzy number (TFN) and trapezoidal fuzzy number (TPFN) are increasingly used in FFTA methods due to ease of computation [1, 7, 15–17]. Similarly, for the computation of FFTA methods, researchers either apply α-cut and interval arithmetic operations [18–20], simplified arithmetic operations [14, 17, 21–24], simplified T_ω-norm(the weakest t-norm)-based approximate fuzzy number arithmetic operations [25–27], posbist approach [28], operator theory [29], Monte Carlo simulation [30], etc. The detailed literature review is presented hereafter.

According to the reviewed literature related to different FFTA methods, it is observed that Tanaka et al. [17] used TPFN to quantify data uncertainty and employed approximate fuzzy arithmetic operations to obtain the system's top event probability. Singer [23], Cheng and Mon [18], Chen [24], Hong and Do [25], and Fuh et al. [19] analyzed the fuzzy reliability of two grinding machines, a non-repairable system, working next to each other by developing different FFTA methods. Singer [23] used the LR-type fuzzy number and extended algebraic operations; Cheng and Mon [18] applied the LR-type fuzzy number and α-cut based fuzzy arithmetic operations, while Chen [24] employed TFN with simplified fuzzy number arithmetic operations. Hong and Do [25] utilized the *LR*-type fuzzy number and T_ω (the weakest *t*-norm)-based addition and multiplication, while Fuh et al. [19] used level $(\lambda,1)$ interval-valued fuzzy numbers and fuzzy arithmetic operations. Komal [27] developed an FFTA method for analyzing patient safety risk by applying TPFNs and T_ω-based approximate fuzzy arithmetic operations. Huang et al. [28] proposed the posbist fault tree analysis method. Misra and Weber [29] adopted a possibility theory in the place of a fuzzy set theory and applied max-min operators to the find top event failure possibility. Ferdous et al. [30] presented a methodology for a fuzzy-based computer-aided FTA and applied their approach to analyze the reliability of an activated carbon filter safeguard system. Few researchers think that only specific shapes' membership functions are not adequate enough to address the uncertainty factor. So, they used different types of fuzzy membership functions such as normal, cauchy, sharp gamma, etc., for quantifying data uncertainty [31–37].

The shortcoming in fuzzy set theory is that it deals with only membership grade (acceptance degree) and does not provide any information regarding non-membership grade (rejection degree), which is also equally important in some situations. So, few researchers applied the IFS or vague set theory for incorporating membership and non-membership degrees of any basic event of the system [14, 20–22, 26, 38, 39]. Based on the above literature survey, it is inferred that:

- Most of the researchers used TFN or TPFN to quantify uncertainty and applied different types of fuzzy arithmetic operations in FFTA methods to get the fuzzy possibility of a top event.
- Some researchers used different types of fuzzy numbers such as normal, cauchy, sharp gamma, etc., for quantifying data uncertainty and applied different kinds of operations.
- Using membership and non-membership grades, few researchers applied IFS or vague set theory and different arithmetic operations to get the system's top event failure possibility.

It has been observed from the above findings that most of the existing FFTA methods assumed the system's basic events, either fuzzy sets, IFS, or vague sets, and connected these through different operators without any compensation in the aggregated values [40]. To aggregate the system's basic events researchers generally apply pairwise *t*-norm and *t*-conorm related operators [41]. In general, a *t*-norm provides no compensation while *t*-conorm provides full compensation between small and large degrees of membership of an event [42]. Compensation with the help of a parameter provides good and a wide range of solutions which is very useful in decision making [42, 43]. It is observed from the study that the solutions obtained by using compensatory operators are always efficient and more credible due to the fact that these operators have more interactions with human decision making [43].

Considering the advantage of compensation, the main objective of this chapter is to develop a flexible and more realistic FFTA method to evaluate the fault interval of a system by considering compensation in aggregated fuzzy values. This chapter proposes a novel FFTA method in which TFNs are used to represent basic events, a *compensatory* operator is applied to connect the basic events, while α-cut based fuzzy arithmetic operations are employed to obtain fault and reliability intervals of the system which is the main contribution of this study [21, 40, 42]. The proposed method used the compensatory *and* [40, 42] and *or* 21] operators to connect basic events. The main benefit of the compensatory operator in the proposed method is that the decision maker has a range of solutions and this may be very useful for developing more efficient maintenance policies to enhance system performance. The proposed approach has been applied to analyze the fuzzy fault tree of a web access failure model [44–47].

7.2 PRELIMINARIES

In this section, some preliminary concepts and definitions are briefly described that will be used in the proposed FFTA method.

7.2.1 Fuzzy Sets

According to Zadeh [9], a fuzzy set $\tilde{A}$ in the universe of discourse U is mathematically expressed as,

$$\tilde{A} = \{(x, \mu_{\tilde{A}}(x)) : x \in U\} \tag{7.1}$$

where $\mu_{\tilde{A}}(x) \in [0,1]$ is the membership degree of an element x in the fuzzy set $\tilde{A}$.

7.2.2 Triangular Fuzzy Number (TFN)

A TFN $\tilde{A} = (a_1, a_2, a_3)$ is a triplet which is used to represent an imprecise real number $a_2 \in R$ in the range $a_1 \le a_2 \le a_3$ with the help of a membership function defined as [1, 7]:

$$\mu_{\tilde{A}}(x) = \begin{cases} \dfrac{x - a_1}{a_2 - a_1}, & a_1 \le x \le a_2 \\ \dfrac{a_3 - x}{a_3 - a_2}, & a_2 \le x \le a_3. \end{cases} \tag{7.2}$$

7.2.3 α-cut set

The α-cut of a fuzzy set $\tilde{A}$ is a crisp set $A^{(\alpha)}$ which contains only those elements of $\tilde{A}$ whose membership degrees are greater than or equal to α. Mathematically, α-cut of a fuzzy set $\tilde{A}$ is defined as

$$A^{(\alpha)} = \{x \in U : \mu_{\tilde{A}}(x) \ge \alpha\} \tag{7.3}$$

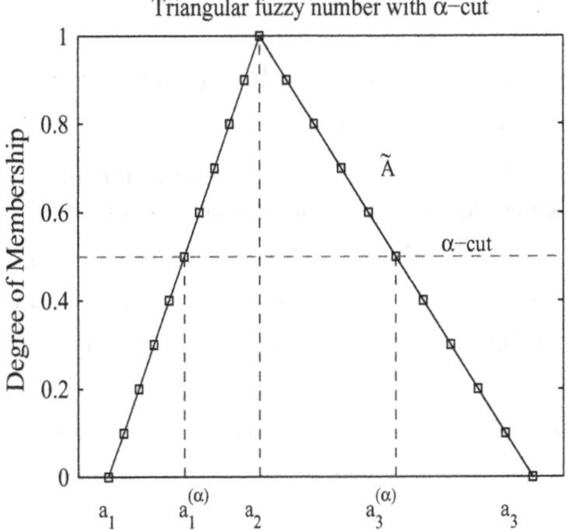

FIGURE 7.1 A triangular fuzzy number $\tilde{A}$ with an α-cut at $\alpha = 0.5$.

where α is a parameter in the range $0 \le \alpha \le 1$ and U is the universe of discourse. If $\tilde{A} = (a_1, a_2, a_3)$ is a TFN, then its α-cut is given by

$$A^{(\alpha)} = \left[a_1^{(\alpha)}, a_3^{(\alpha)}\right] \tag{7.4}$$

and shown in Figure 7.1 for $\alpha = 0.5$.

7.2.4 Different Types of Fuzzy Arithmetic Operations

Let $\tilde{A} = (a_1, a_2, a_3)$ and $\tilde{B} = (b_1, b_2, b_3)$ be two TFNs representing two vague numbers "a_2" and "b_2" in the range $a_1 \le a_2 \le a_3$ and $b_1 \le b_2 \le b_3$, respectively. Three different types of fuzzy arithmetic operations based on simple TFN arithmetic, α-cut coupled interval arithmetic operations, and T_ω (the weakest t-norm) definition are given in Tables 7.1 through 7.3, respectively, with α-cut definition of TFNs $\tilde{A}$ and $\tilde{B}$ as given below [18, 24, 25, 31].

$A^{(\alpha)} = [a_1^{(\alpha)}, a_3^{(\alpha)}]$, $B^{(\alpha)} = [b_1^{(\alpha)}, b_3^{(\alpha)}]$, $\alpha \in [0,1]$

7.2.5 Fuzzy Reliability

As far as the standard definition of reliability is concerned, "Reliability is the probability that a component or system will perform its required function for the given period of time when used under the stated operating conditions" [5]. It is an important factor in equipment maintenance because lower equipment reliability means higher maintenance, higher operational cost, and hence lower profit. To compute system reliability, precise information about system failure/repair is required which is sometimes not available in the exact form. In that case researchers preferably use the fuzzy set theory and compute system fuzzy reliability [4, 10–12]. The fuzzy

TABLE 7.1
Simplified Fuzzy Arithmetic Operations Defined on TFNs [24]

Operation	Fuzzy expression for triangular fuzzy numbers
Addition	$\tilde{A} \oplus \tilde{B} = (a_1 + b_1, a_2 + b_2, a_3 + b_3)$
Subtraction	$\tilde{A} \ominus \tilde{B} = (a_1 - b_3, a_2 - b_2, a_3 - b_1)$
Multiplication	$\tilde{A} \otimes \tilde{B} = (a_1 b_1, a_2 b_2, a_3 b_3)$
Division	$\tilde{A} \oslash \tilde{B} = (a_1 / b_3, a_2 / b_2, a_3 / b_1)$
Compliment	$\tilde{1} \ominus \tilde{A} = (1 - a_3, 1 - a_2, 1 - a_1)$

TABLE 7.2
α-Cut Coupled Fuzzy Arithmetic Operations Defined for Any Fuzzy Number [18]

Operation	Fuzzy expression using α-cut and interval arithmetic
Addition	$\tilde{A} \oplus^{\alpha} \tilde{B} = [a_1^{(\alpha)} + b_1^{(\alpha)}, a_3^{(\alpha)} + b_3^{(\alpha)}]$
Subtraction	$\tilde{A} \ominus^{\alpha} \tilde{B} = [a_1^{(\alpha)} - b_3^{(\alpha)}, a_3^{(\alpha)} - b_1^{(\alpha)}]$
Multiplication	$\tilde{A} \otimes^{\alpha} \tilde{B} = \left[\min\left(a_1^{(\alpha)} b_1^{(\alpha)}, a_1^{(\alpha)} b_3^{(\alpha)}, a_3^{(\alpha)} b_1^{(\alpha)}, a_3^{(\alpha)} b_3^{(\alpha)}\right), \max\left(a_1^{(\alpha)} b_1^{(\alpha)}, a_1^{(\alpha)} b_3^{(\alpha)}, a_3^{(\alpha)} b_1^{(\alpha)}, a_3^{(\alpha)} b_3^{(\alpha)}\right) \right]$
Division	$\tilde{A} \oslash^{\alpha} \tilde{B} = \left[\min\left(\frac{a_1^{(\alpha)}}{b_1^{(\alpha)}}, \frac{a_1^{(\alpha)}}{b_3^{(\alpha)}}, \frac{a_3^{(\alpha)}}{b_1^{(\alpha)}}, \frac{a_3^{(\alpha)}}{b_3^{(\alpha)}} \right), \max\left(\frac{a_1^{(\alpha)}}{b_1^{(\alpha)}}, \frac{a_1^{(\alpha)}}{b_3^{(\alpha)}}, \frac{a_3^{(\alpha)}}{b_1^{(\alpha)}}, \frac{a_3^{(\alpha)}}{b_3^{(\alpha)}} \right) \right]$
Compliment	$\tilde{1} \ominus^{\alpha} \tilde{A} = [1 - a_3^{(\alpha)}, 1 - a_1^{(\alpha)}]$

TABLE 7.3
T_ω Based Fuzzy Arithmetic Operations Defined on TFNs [25]

Operation	Fuzzy expression using T_ω-norm on triangular fuzzy number arithmetic
Addition	$\tilde{A} \oplus_{T_\omega} \tilde{B} = (a_2 + b_2 - \max((a_2 - a_1), (b_2 - b_1)), a_2 + b_2, a_2 + b_2 + \max((a_3 - a_2), (b_3 - b_2)))$
Subtraction	$\tilde{A} \ominus_{T_\omega} \tilde{B} = (a_2 - b_2 - \max((a_2 - a_1), (b_3 - b_2)), a_2 - b_2, a_2 - b_2 + \max((a_3 - a_2), (b_2 - b_1)))$
Multiplication	$\tilde{A} \otimes_{T_\omega} \tilde{B} = (a_2 b_2 - \max((a_2 - a_1) b_2, (b_2 - b_1) a_2), a_2 b_2, a_2 b_2 + \max((a_3 - a_2) b_2, (b_3 - b_2) a_2))$
Division	$\tilde{A} \oslash_{T_\omega} \tilde{B} = (a_2 / b_2 - \max((a_2 - a_1) / b_2, (1/b_2 - 1/b_3) a_2), a_2 / b_2, a_2 / b_2 + \max((a_3 - a_2) / b_2, (1/b_1 - 1/b_2) a_2))$
Compliment	$\tilde{1} \ominus_{T_\omega} \tilde{A} = (1 - a_3, 1 - a_2, 1 - a_1)$

reliability of a system is the fuzzy probability that the system will perform its intended task successfully without failing when the system is transferring from one state to the other state.

7.3 TRADITIONAL FTA, POSBIST FTA, AND FFTA METHODS

This section gives a brief idea about different fault tree methods used for a comparative study.

7.3.1 Traditional FTA [6, 48] Method

A fault tree provides the graphical interaction between the top, intermediate, and basic events of any system through various nodes and logic-gates (OR, AND) in a top down manner. In conventional FTA, a system's top event failure possibility function can be easily formulated in terms of its basic events' failure possibilities [6, 30]. The mathematical equations to find the system's top event failure possibility due to its n basic events with failure possibilities q_i and connected either in series (OR) or parallel (AND) configuration are given in Table 7.4 (first row) [48].

7.3.2 Posbist FTA [28] Method

Huang et al. [28] introduced a posbist FTA approach by using newly defined AND and OR operator definitions based on system minimal cut sets. The mathematical operations applied by Huang et al. [28] in their posbist FTA approach are given in Table 7.4 (second row) for n basic events connected either in OR AND configurations with failure possibilities $q_i = P_{oss}(Xi), i = 1,2,...,n$.

7.3.3 Different FFTA Methods

Traditional FTA methods assume the basic events' failure possibilities as crisp values. However, in real-life situations, the failure possibilities of the system's basic events are ambiguous, qualitatively incomplete, ill-defined, and inaccurate [16, 49]. To overcome these limitations, traditional FTA has been integrated with the fuzzy set theory in various forms [18, 24, 25, 33, 34] (Table 7.4).

7.3.3.1 Cheng and Mon [18] FFTA Method

Cheng and Mon [18] extended traditional FTA by replacing the crisp failure probabilities of the basic events with TFNs and applied α-cut-based interval arithmetic operations (Table 7.2). The mathematical equations used by Cheng and Mon [18] are given in Table 7.4 (third row) for OR and AND gates. In the TFN form, the fuzzy failure possibility of the ith basic event is represented by $\tilde{q}_i = (q_{i1}, q_{i2}, q_{i3}), i = 1,2,...,n$.

TABLE 7.4
Equations used in traditional FTA, posbist FTA, existing FFTA, and proposed FFTA for number of basic events

Row no.	Method	Gate	Equation	Data type (q_i)	Approach
1.	Traditional [6]	OR	$P_{\text{OR}} = 1-[(1-q_1)\times(1-q_2)\times\cdots\times(1-q_n)]$	Crisp	Ordinary arithmetic
	FTA	AND	$P_{\text{AND}} = q_1 \times q_2 \times \cdots \times q_n$		
2.	Huang et al. [28]	OR	$P_{\text{oss}}^{\text{OR}} = \text{Max}(P_{\text{oss}}(X1), P_{\text{oss}}(X2),\ldots, P_{\text{oss}}(Xn))$	Crisp	Possibility theory
	Posbist FTA	AND	$P_{\text{oss}}^{\text{AND}} = \text{Min}(P_{\text{oss}}(X1), P_{\text{oss}}(X2),\ldots, P_{\text{oss}}(Xn))$		
3.	Cheng and Mon [18]	OR	$\tilde{P}_{\text{OR}}^{(\alpha)} = \tilde{1} \ominus^{\alpha} [(\tilde{1} \ominus^{\alpha} \tilde{q}_1) \otimes^{\alpha} (\tilde{1} \ominus^{\alpha} \tilde{q}_2) \otimes^{\alpha} \ldots \otimes^{\alpha} (\tilde{1} \ominus^{\alpha} \tilde{q}_n)]$	Triangular fuzzy numbers	α-cut and interval arithmetic
	FFTA	AND	$\tilde{P}_{\text{AND}}^{(\alpha)} = \tilde{q}_1 \otimes^{\alpha} \tilde{q}_2 \otimes^{\alpha} \ldots \otimes^{\alpha} \tilde{q}_n$		
4.	Chen [24]	OR	$\tilde{P}_{\text{OR}} = \tilde{1} \ominus [(\tilde{1} \ominus \tilde{q}_1) \otimes (\tilde{1} \ominus \tilde{q}_2) \otimes \ldots \otimes (\tilde{1} \ominus \tilde{q}_n)]$	Triangular fuzzy numbers	Fuzzy number arithmetic
	FFTA	AND	$\tilde{P}_{\text{AND}} = \tilde{q}_1 \otimes \tilde{q}_2 \otimes \ldots \otimes \tilde{q}_n$		
5.	Hong and Do [25]	OR	$\tilde{P}_{\text{OR}} = \tilde{1} \ominus_{T_\omega} \left[\left(\tilde{1} \ominus_{T_\omega} \tilde{q}_1\right) \otimes_{T_\omega} \left(\tilde{1} \ominus_{T_\omega} \tilde{q}_2\right) \otimes_{T_\omega} \ldots \otimes_{T_\omega} \left(\tilde{1} \ominus_{T_\omega} \tilde{q}_n\right) \right]$	Triangular fuzzy numbers	T_ω based fuzzy number arithmetic
	FFTA Method	AND	$\tilde{P}_{\text{AND}} = \tilde{q}_1 \otimes_{T_\omega} \tilde{q}_2 \otimes_{T_\omega} \ldots \otimes_{T_\omega} \tilde{q}_n$		
6.	Shu et al. [21]	OR	$\tilde{P}_{\text{OR}} = \tilde{1} \ominus [(\tilde{1} \ominus \tilde{q}_1) \otimes (\tilde{1} \ominus \tilde{q}_2) \otimes \ldots \otimes (\tilde{1} \ominus \tilde{q}_n)]$	Triangular fuzzy numbers	Fuzzy number arithmetic, and-by-min and or operations
	FFTA	AND	$\tilde{P}_{\text{AND}} = (\min(q_{11}, q_{21}, \ldots, q_{n1}), \min(q_{12}, q_{22}, \ldots, q_{n2}), \min(q_{13}, q_{23}, \ldots, q_{n3}))$		
7.	Proposed FFTA	OR	$\tilde{P}_{\text{OR}} = \tilde{1} \ominus [(\tilde{1} \ominus \tilde{q}_1) \otimes (\tilde{1} \ominus \tilde{q}_2) \otimes \ldots \otimes (\tilde{1} \ominus \tilde{q}_n)]$	Triangular fuzzy numbers	Fuzzy number arithmetic
	Method	AND	$\tilde{P}_{\text{AND}} = (\tilde{q}_1 \otimes \tilde{q}_2 \otimes \ldots \otimes \tilde{q}_n)^{(1-\gamma)} \otimes (\tilde{1} \ominus [(\tilde{1} \ominus \tilde{q}_1) \otimes (\tilde{1} \ominus \tilde{q}_2) \otimes \ldots \otimes (\tilde{1} \ominus \tilde{q}_n)])^{\gamma}$		

7.3.3.2 Chen [24] FFTA Method

Chen [24] developed a new FFTA method in which the basic events' uncertain failure probabilities are quantified through TFNs, and simplified fuzzy number arithmetic operations as given in Table 7.1 are applied to compute the top event fuzzy failure possibility. The mathematical equations used by Chen [24] are given in Table 7.4 (fourth row) for OR and AND gates.

7.3.3.3 Hong and Do [25] FFTA Method

To reduce the accumulating phenomenon of fuzziness in fuzzy number arithmetic operations, Hong and Do [25] replaced the basic events' uncertain failure probabilities with *LR*-type fuzzy numbers and applied T_ω based fuzzy number arithmetic operations as given in Table 7.3 [50]. The mathematical equations used by them are given in Table 7.4 (fifth row) in the context of TFNs. TFNs are a special case of *LR*-type fuzzy numbers where $L(x) = R(x) = x$ [50, 51].

7.3.3.4 Shu et al. [21] FFTA Method

Shu et al. [21] incorporate the membership and non-membership values of basic events' uncertain failure probabilities through IFS and apply two newly defined "and-by-min" and "and-by-product" operations along with "or" operation defined on IFS. This study used only "and-by-min" and "or" operations because "and-by-product" operations coincide with Chen's [24] AND operation in the context of TFNs. The mathematical equations used by Shu et al. [21] are given in Table 7.4 (sixth row) in the context of TFNs.

7.4 PROPOSED FFTA USING COMPENSATORY OPERATORS

First, the *compensatory "and"* and *"or"* operations are defined in definitions 7.1 and 7.2, respectively.

Definition 7.1. Let $\tilde{A} = (a_1, a_2, a_3)$ and $\tilde{B} = (b_1, b_2, b_3)$ be two TFNs, then the failure possibility $F(\tilde{A} \cap \tilde{B})$ of an AND event for $\tilde{A} > 0$ and $\tilde{B} > 0$ can be defined using a compensatory operator [40, 42, 43] for the parameter $\gamma \in [0,1)$ as

$$\begin{aligned}
F(\tilde{A} \cap \tilde{B}) &= (F(\tilde{A}) \otimes F(\tilde{B}))^{(1-\gamma)}(1 \ominus ((1 \ominus F(\tilde{A})) \otimes (1 \ominus F(\tilde{B}))))^{\gamma} \\
&= ((a_1, a_2, a_3) \otimes (b_1, b_2, b_3))^{(1-\gamma)}(1 \ominus ((1 \ominus (a_1, a_2, a_3)) \otimes (1 \ominus (b_1, b_2, b_3))))^{\gamma} \\
&= (a_1 b_1, a_2 b_2, a_3 b_3)^{(1-\gamma)}(1 \ominus ((1 - a_3, 1 - a_2, 1 - a_1) \otimes (1 - b_3, 1 - b_2, 1 - b_1)))^{\gamma} \\
&= (a_1 b_1, a_2 b_2, a_3 b_3)^{(1-\gamma)}(1 \ominus ((1 - a_3)(1 - b_3), (1 - a_2)(1 - b_2), (1 - a_1)(1 - b_1)))^{\gamma} \\
&= (a_1 b_1, a_2 b_2, a_3 b_3)^{(1-\gamma)}(1 - (1 - a_1)(1 - b_1), 1 - (1 - a_2)(1 - b_2), \ldots \\
&\quad 1 - (1 - a_3)(1 - b_3))^{\gamma}
\end{aligned} \tag{7.5}$$

Definition 7.2. Let $\tilde{A} = (a_1, a_2, a_3)$ and $\tilde{B} = (b_1, b_2, b_3)$ be two TFNs, then the failure possibility $F(\tilde{A} \cup \tilde{B})$ for $\tilde{A} > 0$ and $\tilde{B} > 0$ can be defined using "*or*" operator [21] as

$$\begin{aligned} F(\tilde{A} \cup \tilde{B}) &= 1 \ominus ((1 \ominus F(\tilde{A})) \otimes (1 \ominus F(\tilde{B}))) \\ &= 1 \ominus ((1 \ominus (a_1, a_2, a_3)) \otimes (1 \ominus (b_1, b_2, b_3))) \\ &= 1 \ominus ((1 - a_3, 1 - a_2, 1 - a_1) \otimes (1 - b_3, 1 - b_2, 1 - b_1)) \\ &= 1 \ominus ((1 - a_3)(1 - b_3), (1 - a_2)(1 - b_2), (1 - a_1)(1 - b_1)) \\ &= (1 - (1 - a_1)(1 - b_1), 1 - (1 - a_2)(1 - b_2), \ldots \\ &\quad 1 - (1 - a_3)(1 - b_3)) \end{aligned} \tag{7.6}$$

To rank the critical basic events of the fault tree, this chapter modifies Tanaka et al.'s [17] *V*-index definition as following:

Definition 7.3. Suppose $\tilde{q}_T$ denotes the possible failure interval of the system's top event while $\tilde{q}_{Ti}$ represents the possible failure interval of system's top event when its *i*th component failure is not considered. Then the index, *V*, measures the difference between $\tilde{q}_T$ and $\tilde{q}_{Ti}$ which is defined as

$$V(\tilde{q}_T, \tilde{q}_{Ti}) = (q_{T1} - q_{Ti1}) + (q_{T2} - q_{Ti2}) + (q_{T3} - q_{Ti3}) > 0 \tag{7.7}$$

where $\tilde{q}_T = (q_{T1}, q_{T2}, q_{T3})$ and $\tilde{q}_{Ti} = (q_{Ti1}, q_{Ti2}, q_{Ti3})$ are in the form of TFNs. $V(\tilde{q}_T, \tilde{q}_{Ti})$ indicates the extent of improvement in eliminating the failure of the *i*th component. If $V(\tilde{q}_T, \tilde{q}_{Ti}) \geq V(\tilde{q}_T, \tilde{q}_{Tj})$, then preventing the failure of the *i*th component is more effective than the preventing the failure of the *j*th component.

The following steps are performed for developing the proposed FFTA with a flowchart as shown in Figure 7.2.

Step 1. Construct the fault tree logic diagram.
The fault tree is constructed based on interactions (series/parallel) among the components using logical gates OR/AND and the entire process is traced back from the top to basic events for the considered problem.

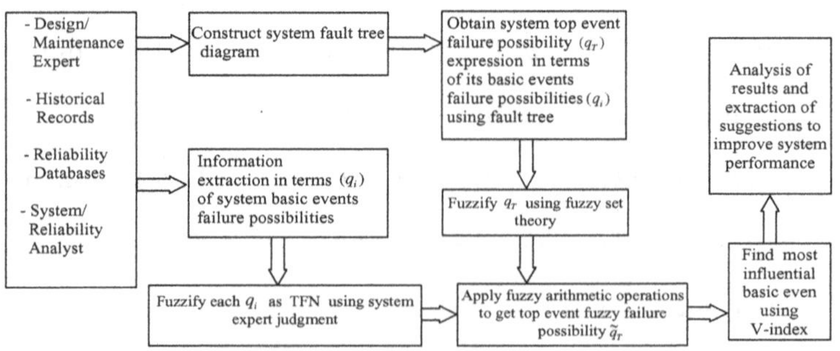

FIGURE 7.2 Flowchart of the proposed FFTA method.

Step 2. *Obtain the possible failure of bottom event in terms of TFN.*
The possible failure distribution for each basic event is acquired from different sources and then integrated with system experts. Experts' judgments are aggregated and represented in terms of TFNs.

Step 3. *Calculate the possible failure of systems using fuzzy arithmetic operations on TFNs.*
Using the fault tree diagram and the possible failure of bottom events represented in the form of TFNs, the system's top event possible failure can be calculated utilizing fuzzy arithmetic operations defined in TFNs as given in Table 7.2 and Equations 5 and 6.

Step 4. *Compute the fuzzy reliability of top event.*
The fuzzy reliability of the top event is equal to one minus the fuzzy failure of the top event.

Step 5. *Find the most influential bottom event of system reliability.*
Using the definition of index V as defined in Equation 7.7, calculate $V(\tilde{q}_T, \tilde{q}_{Ti}), \forall i$ by deleting the ith bottom event in the fault tree diagram and find the most influential power (i.e. $\max_i V(\tilde{q}_T, \tilde{q}_{Ti})$) for the whole system.

Step 6. *Analyze the results and suggestions.*

7.5 CASE STUDY

In this chapter, web access has been considered as the main system. Many studies have been carried out to analyze the web server LOG [44–46]. Several works reported that the LOG file illustrates some statistical patterns of metrics such as file transfer, documents popularity, and the files size of the requests. The web server reliability analysis frequently depends on incomplete and vague data such as the LOG data or web tools' result.

7.5.1 Fault Tree and Top Event Failure Possibility Equation

LOG files normally contain information regarding the services requested from a system, the responses provided, and the origin of the requests. The details of the server activities are logged into text files. The LOG file or HTTP failures are due to the user/client and the server. The error status code for the client's error is indicated by 4xx while 5xx for the server error [46]. The common errors experienced by the user are code 400 (bad request), code 401 (unauthorized), code 403 (forbidden), and code 404 (not found), while for the server code 500 (internal server error) and code 503 (service unavailable) are frequently occurred errors [46]. The aim of this chapter is to illustrate the inter-relationship among these HTTP basic events symbolically and to indicate the physical connections in highly time-varying web processes. The fault tree is constructed to pictorially represent the propagation of all the combinations of basic events leading to web service failure and is illustrated in Figure 7.3 [46].

For connecting the fault tree diagram of web access failure, this research uses a logical node to describe the "AND" gate with the sign of $\cap$ and "OR" gate with the

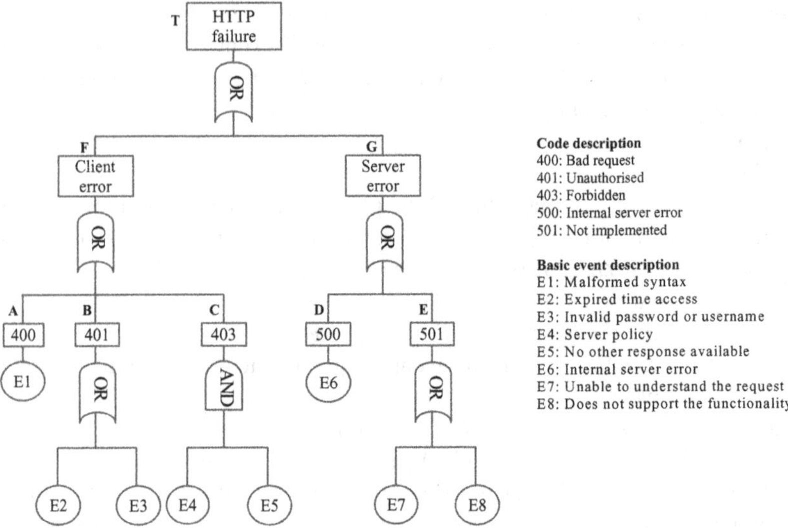

FIGURE 7.3 Web access fault tree.

sign of $\cup$. Their relationship can be represented in parallel and series. So, the top event *T*, i.e., "HTTP failure" can be expressed as:

$$\begin{aligned} T &= F \cup G \\ &= [A \cup B \cup C] \cup [D \cup E] \\ &= [E1 \cup (E2 \cup E3) \cup (E4 \cap E5)] \cup [E6 \cup (E7 \cup E8)] \end{aligned} \tag{7.8}$$

Let q_j represent the failure possibility of the bottom event *j*, then the failure possibility of events B, C, and E can be easily calculated using the mathematical expressions given in Table 7.4 (first row) as follows:

$$q_B = 1 - (1 - q_{E2})(1 - q_{E3})$$

$$q_C = q_{E4} q_{E5}$$

$$q_E = 1 - (1 - q_{E7})(1 - q_{E8})$$

Then, the failure possibility of the top event *T*, i.e., "HTTP failure" can be described as:

$$\begin{aligned} q_T &= 1 - (1 - q_F)(1 - q_G) \\ &= 1 - (1 - q_A)(1 - q_B)(1 - q_C)(1 - q_D)(1 - q_E) \\ &= 1 - (1 - q_{E1})(1 - q_{E2})(1 - q_{E3})(1 - q_{E4} q_{E5})(1 - q_{E6})(1 - q_{E7})(1 - q_{E8}) \end{aligned} \tag{7.9}$$

7.5.2 Fuzzification of Input Data

Sometimes, the extracted data are imprecisely known or insufficient for probabilistically estimating the basic event failure probabilities, and so experts suggest a rough estimate of failure probabilities which may have some sort of uncertainties [24, 38]. In our study, the LOG file is able to provide the frequency of the access failure within a certain time period. However, the information given is incomplete and vague in general. Before using these data to estimate the top event failure possibility, uncertainties contained in the limited data should be quantified. So, using experts' knowledge and opinion, the available uncertain failure intervals of basic events are transformed in TFNs as given in Table 7.5 [46].

7.5.3 Experimental Results and Discussion

The proposed approach for analyzing web access failure has been implemented along with six existing fault tree methods, as discussed in Section 3, i.e. traditional [6], Huang et al. [28], Cheng and Mon [18], Chen [24], Hong and Do [25] and Shu et al. [21] (in context of fuzzy sets) methods, and the computed results are compared.

7.5.3.1 Traditional Method [6]

In the traditional fault tree method, the possibility theory and ordinary arithmetic operations are used to compute the top event failure possibility [5, 6, 48]. To compute the failure possibility of the top event "HTTP failure", Equation 7.9 and the data given in Table 7.5 (column 3) are used, and the calculations are as given below.

$$\begin{aligned} q_T &= 1-(1-0.4064)(1-0.0273)(1-0.0273)(1-0.1118\times 0.1118) \\ &\quad \times(1-0.3016)(1-0.0056)(1-0.0056) \\ &= 0.616985 \end{aligned} \tag{7.10}$$

TABLE 7.5
The Possible Range of Bottom Event Failure [46]

Basic event	Cause description	Failure possibility	Crisp q_{Ei}	TFN representations $\tilde{q}_{Ei} = (q_{Ei1}, q_{Ei2}, q_{Ei3})$
E1	400-Malformed syntax	$\tilde{q}_{E1}$	0.4064	(0.0809, 0.4064, 0.8128)
E2	Expired time access	$\tilde{q}_{E2}$	0.0273	(0.0055, 0.0273, 0.0546)
E3	Invalid password or username	$\tilde{q}_{E3}$	0.0273	(0.0055, 0.0273, 0.0546)
E4	Server policy	$\tilde{q}_{E4}$	0.1118	(0.0224, 0.1118, 0.2236)
E5	No other response available	$\tilde{q}_{E5}$	0.1118	(0.0224, 0.1118, 0.2236)
E6	500-Internal server error	$\tilde{q}_{E6}$	0.3016	(0.0603, 0.3016, 0.6032)
E7	Unable to understand the request	$\tilde{q}_{E7}$	0.0056	(0.0019, 0.0056, 0.0168)
E8	Does not support the functionality	$\tilde{q}_{E8}$	0.0056	(0.0019, 0.0056, 0.0168)

From the above calculation, it is inferred that the failure possibility of the top event "HTTP failure" is 0.616985. The reliability of "HTTP failure free working" is 0.383015. The computed failure possibility of top event "HTTP failure" is listed in Table 7.6.

7.5.3.2 Huang et al.'s [28] Method

Huang et al. [28] developed a posbist FTA method which works well when the basic events' failure probabilities are extremely small or essential data are scarce. Using Equation 7.8, mathematical equations from Table 7.4 (second row), and data given in Table 7.5 (column 3), calculations have been done to analyze the failure possibility of top event "HTTP failure" which are as follows:

$$\begin{aligned} P_{oss}(B) &= \max(P_{oss}(E2), P_{oss}(E3)) \\ &= \max(0.0273, 0.0273) \\ &= 0.0273 \end{aligned} \tag{7.11}$$

$$\begin{aligned} P_{oss}(C) &= \min(P_{oss}(E4), P_{oss}(E5)) \\ &= \min(0.1118, 0.1118) \\ &= 0.1118 \end{aligned} \tag{7.12}$$

$$\begin{aligned} P_{oss}(E) &= \max(P_{oss}(E7), P_{oss}(E8)) \\ &= \max(0.0056, 0.0056) \\ &= 0.0056 \end{aligned} \tag{7.13}$$

$$\begin{aligned} P_{oss}(F) &= \max(P_{oss}(A), P_{oss}(B), P_{oss}(C)) \\ &= \max(P_{oss}(E1), P_{oss}(B), P_{oss}(C)) \\ &= \max(0.4064, 0.0273, 0.1118) \\ &= 0.4064 \end{aligned} \tag{7.14}$$

$$\begin{aligned} P_{oss}(G) &= \max(P_{oss}(D), P_{oss}(E)) \\ &= \max(P_{oss}(E6), P_{oss}(E)) \\ &= \max(0.3016, 0.0056) \\ &= 0.3016 \end{aligned} \tag{7.15}$$

Then, the top event failure possibility can be calculated as:

$$\begin{aligned} P_{oss}(T) &= \max(P_{oss}(F), P_{oss}(G)) \\ &= \max(0.4064, 0.3016) \\ &= 0.4064 \end{aligned} \tag{7.16}$$

TABLE 7.6
Comparison of Results Obtained from Existing FTA and FFTA Methods with Proposed FFTA Method

∝	Traditional method [6]	Huang et al. [28]	Chen and Mon [18]	Chen [24]	Hong and Do [25]	Shu et al. [21]
1.0	0.616985	0.406400	[0.616985, 0.616985]	[0.616985, 0.616985]	[0.616985, 0.616985]	[0.655500, 0.655500]
.9	0.616985	0.406400	[0.579019, 0.662194]	[0.570233, 0.649190]	[0.595983, 0.643208]	[0.606760, 0.684967]
.8	0.616985	0.406400	[0.539135, 0.704362]	[0.523481, 0.681394]	[0.574980, 0.669430]	[0.558020, 0.714434]
.7	0.616985	0.406400	[0.497318, 0.743541]	[0.476729, 0.713599]	[0.553977, 0.695653]	[0.509280, 0.743902]
.6	0.616985	0.406400	[0.453554, 0.779786]	[0.429977, 0.745803]	[0.532975, 0.721875]	[0.460540, 0.773369]
.5	0.616985	0.406400	[0.407829, 0.813154]	[0.383226, 0.778007]	[0.511972, 0.748098]	[0.411800, 0.802836]
.4	0.616985	0.406400	[0.360133, 0.843705]	[0.336474, 0.810212]	[0.490970, 0.774321]	[0.363060, 0.832303]
.3	0.616985	0.406400	[0.310455, 0.871502]	[0.289722, 0.842416]	[0.469967, 0.800543]	[0.314320, 0.861770]
.2	0.616985	0.406400	[0.258789, 0.896609]	[0.242970, 0.874621]	[0.448964, 0.826766]	[0.265580, 0.891237]
.1	0.616985	0.406400	[0.205127, 0.919095]	[0.196218, 0.906825]	[0.427962, 0.852988]	[0.216840, 0.920704]
.0	0.616985	0.406400	[0.149466, 0.939030]	[0.149466, 0.939030]	[0.406959, 0.879211]	[0.168100, 0.950171]

Proposed Method Results with Different Values of γ

α	$\gamma = 0$	$\gamma = 0.2$	$\gamma = 0.4$	$\gamma = 06$	$\gamma = 08$	$\gamma \rightarrow 1.0$
.0	[0.616985, 0.616985]	[0.620670, 0.620670]	[0.627155, 0.627155]	[0.638569, 0.638569]	[0.658658, 0.658658]	[0.694015, 0.694015]
.9	[0.570233, 0.649190]	[0.573611, 0.652671]	[0.579600, 0.658757]	[0.590244, 0.669407]	[0.609235, 0.688058]	[0.643287, 0.720745]
.8	[0.523481, 0.681394]	[0.526553, 0.684671]	[0.532044, 0.690358]	[0.541919, 0.700245]	[0.559811, 0.717459]	[0.592559, 0.747475]
.7	[0.476729, 0.713599]	[0.479494, 0.716672]	[0.484489, 0.721960]	[0.493594, 0.731083]	[0.510388, 0.746859]	[0.541831, 0.774205]
.6	[0.429977, 0.745803]	[0.432436, 0.748673]	[0.436934, 0.753562]	[0.445269, 0.761921]	[0.460965, 0.776260]	[0.491103, 0.800934]
.5	[0.383226, 0.778007]	[0.385377, 0.780674]	[0.389378, 0.785164]	[0.396944, 0.792758]	[0.411541, 0.805660]	[0.440375, 0.827664]
.4	[0.336474, 0.810212]	[0.338319, 0.812675]	[0.341823, 0.816765]	[0.348619, 0.823596]	[0.362118, 0.835061]	[0.389647, 0.854394]
.3	[0.289722, 0.842416]	[0.291260, 0.844675]	[0.294268, 0.848367]	[0.300294, 0.854434]	[0.312695, 0.864461]	[0.338919, 0.881124]
.2	[0.242970, 0.874621]	[0.244202, 0.876676]	[0.246713, 0.879969]	[0.251969, 0.885272]	[0.263271, 0.893862]	[0.288191, 0.907854]
.1	[0.196218, 0.906825]	[0.197144, 0.908677]	[0.199157, 0.911571]	[0.203644, 0.916110]	[0.213848, 0.923262]	[0.237463, 0.934583]
.0	[0.149466, 0.939030]	[0.150085, 0.940678]	[0.151602, 0.943172]	[0.155319, 0.946948]	[0.164425, 0.952663]	[0.186735, 0.961313]

Thus, the failure possibility of the top event "HTTP failure" is 0.4064 and the reliability of "HTTP failure free working" is 0.5936. The computed failure possibility of the top event "HTTP failure" from this method is listed in Table 7.6.

7.5.3.3 Cheng and Mon's [18] Method

Cheng and Mon [18] used TFNs to quantify the uncertainty present in failure data and applied α-cuts coupled with fuzzy arithmetic operations to obtain the fuzzy possibility of the top event. To implement Cheng and Mon's [18] method, fuzzified Equation 7.9, fuzzy arithmetic operations (Table 7.2), the definitions of OR and AND gates given in Table 7.4 (third row), and fuzzified data in TFN form given in Table 7.5 (fourth column) have been used. The mathematical calculations are given below.

$$\begin{aligned}\tilde{q}_T &= \tilde{1} \ominus^{\alpha} [(\tilde{1} \ominus^{\alpha} \tilde{q}_{E1}) \otimes^{\alpha} (\tilde{1} \ominus^{\alpha} \tilde{q}_{E2}) \otimes^{\alpha} (\tilde{1} \ominus^{\alpha} \tilde{q}_{E3}) \otimes^{\alpha} (\tilde{1} \ominus^{\alpha} (\tilde{q}_{E4} \otimes^{\alpha} \tilde{q}_{E5})) \otimes^{\alpha} \ldots \\ &\quad (\tilde{1} \ominus^{\alpha} \tilde{q}_{E6}) \otimes^{\alpha} (\tilde{1} \ominus^{\alpha} \tilde{q}_{E7}) \otimes^{\alpha} (\tilde{1} \ominus^{\alpha} \tilde{q}_{E8})] \\ &= (0.149466, 0.616985, 0.939030)\end{aligned}$$

The computed fuzzy failure possibility of the top event "HTTP failure" for 11 α-cuts is listed in Table 7.6, while the fuzzy reliability of "HTTP failure free working" can be easily computed by applying the α-cut based complement definition (Table 7.2). The defuzzified values of fuzzy failure possibility and fuzzy reliability are computed by the center of gravity (COG) method [27, 52] and is given as 0.588435 and 0.411565, respectively.

7.5.3.4 Chen's [24] Method

Chen [24] extended Cheng and Mon's [18] FFTA method by replacing α-cut coupled fuzzy arithmetic operations with simplified fuzzy arithmetic operations defined on TFNs as given in Table 7.1. Now, to apply Chen's [24] method, fuzzified Equation 7.9, the definitions of OR and AND gates given in Table 7.4 (fourth row), simplified fuzzy arithmetic operations of fuzzy numbers as given in Table 7.1, and the fuzzy failure probabilities of the fundamental events in TFNs as given in Table 7.5 (column 4) have been used.

$$\begin{aligned}\tilde{q}_T &= \tilde{1} \ominus [(\tilde{1} \ominus \tilde{q}_{E1}) \otimes (\tilde{1} \ominus \tilde{q}_{E2}) \otimes (\tilde{1} \ominus \tilde{q}_{E3}) \otimes (\tilde{1} \ominus (\tilde{q}_{E4} \otimes \tilde{q}_{E5})) \otimes \ldots \\ &\quad (\tilde{1} \ominus \tilde{q}_{E6}) \otimes (\tilde{1} \ominus \tilde{q}_{E7}) \otimes (\tilde{1} \ominus \tilde{q}_{E8})] \\ &= (0.149466, 0.616985, 0.939030)\end{aligned}$$

The computed failure possibility of the top event "HTTP failure" in the form of TFN is 0.149466, 0.616985, and 0.939030. The fuzzy reliability of "HTTP failure free working" in TFN form is 0.060970, 0.383015, and 0.850534. The failure possibility of the top event computed by this method in terms of 11 α-cuts for $\alpha = 0, 0.1, 0.2, \ldots, 1$ is listed in Table 7.6. The defuzzified values of fuzzy failure possibility and fuzzy reliability obtained from the COG method are 0.568494 and 0.431506, respectively.

7.5.3.5 Hong and Do's [25] Method

To reduce the accumulating phenomenon of fuzziness, Hong and Do [25] developed a novel FFTA method in which TFNs are used to quantify uncertainty in data, and T_ω based fuzzy arithmetic operations defined on TFNs are used to obtain the fuzzy possibility of the top event. To implement Hong and Do's [25] method in the present study, fuzzified Equation 7.9, T_ω based fuzzy arithmetic operations (Table 7.3), the definitions of OR and AND gates given in Table 7.4 (fifth row), and fuzzified data Table 7.5 (column 4) have been used. The mathematical calculations are given below.

$$\tilde{q}_T = \tilde{1} \ominus_{T_\omega} [(\tilde{1} \ominus_{T_\omega} \tilde{q}_{E1}) \otimes_{T_\omega} (\tilde{1} \ominus_{T_\omega} \tilde{q}_{E2}) \otimes_{T_\omega} (\tilde{1} \ominus_{T_\omega} \tilde{q}_{E3}) \otimes_{T_\omega} (\tilde{1} \ominus_{T_\omega} (\tilde{q}_{E4} \otimes_{T_\omega} \tilde{q}_{E5})) \otimes_{T_\omega} \ldots$$
$$(\tilde{1} \ominus_{T_\omega} \tilde{q}_{E6}) \otimes_{T_\omega} (\tilde{1} \ominus_{T_\omega} \tilde{q}_{E7}) \otimes_{T_\omega} (\tilde{1} \ominus \tilde{q}_{E8})]$$
$$= (0.406959, 0.616985, 0.879211)$$

The computed failure possibility of the top event "HTTP failure" in the form of TFN is 0.406959, 0.616985, and 0.879211. The fuzzy reliability of "HTTP failure free working" in the TFN form is 0.120789, 0.383015, and 0.593041. The computed fuzzy failure possibility of top event "HTTP failure" from this method for 11 α-cuts is listed in Table 7.6, while the fuzzy reliability of "HTTP failure free working" can be easily computed by applying α-cut based complement definition (Table 7.3). The defuzzified values of the fuzzy failure possibility and fuzzy reliability are 0.634385 and 0.365615, respectively.

7.5.3.6 Shu et al.'s [21] Method in the Context of Fuzzy Sets

Shu et al. [21] used a triangle vague set to quantify uncertainty in data, and applied arithmetic operations defined on vague sets to obtain the fuzzy possibility of the top event. Since this study uses TFNs, to implement Shu et al.'s [21] method in the context of fuzzy sets, fuzzified Equation 7.9, simplified fuzzy arithmetic operations (Table 7.1), the definitions of OR and AND gates given in Table 7.4 (sixth row), and fuzzified data (Table 7.5) (column 4) have been used. The mathematical calculations are given below.

$$\tilde{q}_T = \tilde{1} \ominus [(\tilde{1} \ominus \tilde{q}_{E1}) \otimes (\tilde{1} \ominus \tilde{q}_{E2}) \otimes (\tilde{1} \ominus \tilde{q}_{E3}) \otimes (\tilde{1} \ominus (\tilde{q}_{E4} \otimes \tilde{q}_{E5})) \otimes \ldots$$
$$(\tilde{1} \ominus \tilde{q}_{E6}) \otimes (\tilde{1} \ominus \tilde{q}_{E7}) \otimes (\tilde{1} \ominus \tilde{q}_{E8})]$$
$$= (0.1681000.6555000.950171)$$

The computed failure possibility of the top event "HTTP failure" in the form of TFN is 0.168100, 0.655500, and 0.950171. The fuzzy reliability of "HTTP failure free working" in TFN form is 0.049829, 0.344500, and 0.831900. The computed fuzzy failure possibility of the top event "HTTP failure" for only 11 α-cuts is listed in Table 7.6, while the fuzzy reliability of "HTTP failure free working" can be easily computed by applying the α-cut based complement definition (Table 7.2). The defuzzified values of fuzzy failure possibility and fuzzy reliability are 0.591257 and 0.408743, respectively.

7.5.3.7 Proposed Method

To compute the fuzzy failure possibility of the top event using developed FFTA, fuzzified Equation 7.9, simplified fuzzy arithmetic operations given in Table 7.1, and the definitions of OR and AND gates given in Table 7.4 (seventh row) have been used. The mathematical equation to compute the top event "HTTP failure" possibility is established as follows.

$$\begin{aligned}\tilde{q}_T &= \tilde{1}\ominus[(\tilde{1}\ominus\tilde{q}_{E1})\otimes(\tilde{1}\ominus\tilde{q}_{E2})\otimes(\tilde{1}\ominus\tilde{q}_{E3})\otimes\ldots\\ &(\tilde{1}\ominus(\tilde{q}_{E4}\otimes\tilde{q}_{E5})^{(1-\gamma)}(\tilde{1}\ominus((\tilde{1}\ominus q_{E4})\otimes(\tilde{1}\ominus q_{E5})))^{\gamma})\otimes\ldots\\ &(\tilde{1}\ominus\tilde{q}_{E6})\otimes(\tilde{1}\ominus\tilde{q}_{E7})\otimes(\tilde{1}\ominus\tilde{q}_{E8})]\end{aligned} \tag{7.17}$$

Now using Equation 7.17, fuzzified data (Table 7.5) (column 4), and the simplified fuzzy arithmetic operations given in Table 7.1, the top event's fuzzy failure possibility have been computed for different values of the compensatory parameter, i.e. $\gamma = 0, 0.2, 0.4, 0.6, 0.8$ and $\gamma \to 1$, and tabulated in Table 7.7 along with fuzzy reliability in the TFN form. The computed results for fuzzy failure possibility are also tabulated α-cut wise in Table 7.6 and plotted in Figure 7.4 for different values of the compensatory parameter γ along with the results obtained in the above discussed methods. The defuzzified values of the fuzzy failure possibility of the top event "HTTP failure" are 0.568494, 0.570478, 0.573976, 0.580279 ,0.591915, and 0.614021, while the defuzzified values of fuzzy reliability of "HTTP failure free working" are 0.431506, 0.429522, 0.426024, 0.419721, 0.408085, and 0.385979 obtained for $\gamma = 0, 0.2, 0.4, 0.6, 0.8$ and $\gamma \to 1$, respectively. Comparing these results, it is easily seen that the developed FFTA method provides a series of solutions based on the compensatory parameter, γ, values. The benefit of the proposed approach is that it incorporates not only fuzzy factors in the decision-making process but also the designers' experience and knowledge.

$[\gamma = 0]$ $[\gamma = 0.2]$
$[\gamma = 0.4]$ $[\gamma = 0.6]$
$[\gamma = 0.8]$ $[\gamma \to 1]$

TABLE 7.7
Top Event Failure Possibility Using Proposed Method with Different Values of γ

γ	Top event failure possibility $\tilde{q}_T = (q_{TL}, q_{TM}, q_{TR})$	Reliability interval $\tilde{R}_s = (R_L, R_M, R_R)$
.0	(0.1495, 0.6170, 0.9390)	(0.0610, 0.3830, 0.8505)
.2	(0.1501, 0.6207, 0.9407)	(0.0593, 0.3793, 0.8499)
.4	(0.1516, 0.6272, 0.9432)	(0.0568, 0.3728, 0.8484)
.6	(0.1553, 0.6386, 0.9469)	(0.0531, 0.3614, 0.8447)
.8	(0.1644, 0.6587, 0.9527)	(0.0473, 0.3413, 0.8356)
→1.0	(0.1867, 0.6940, 0.9613)	(0.0387, 0.3060, 0.8133)

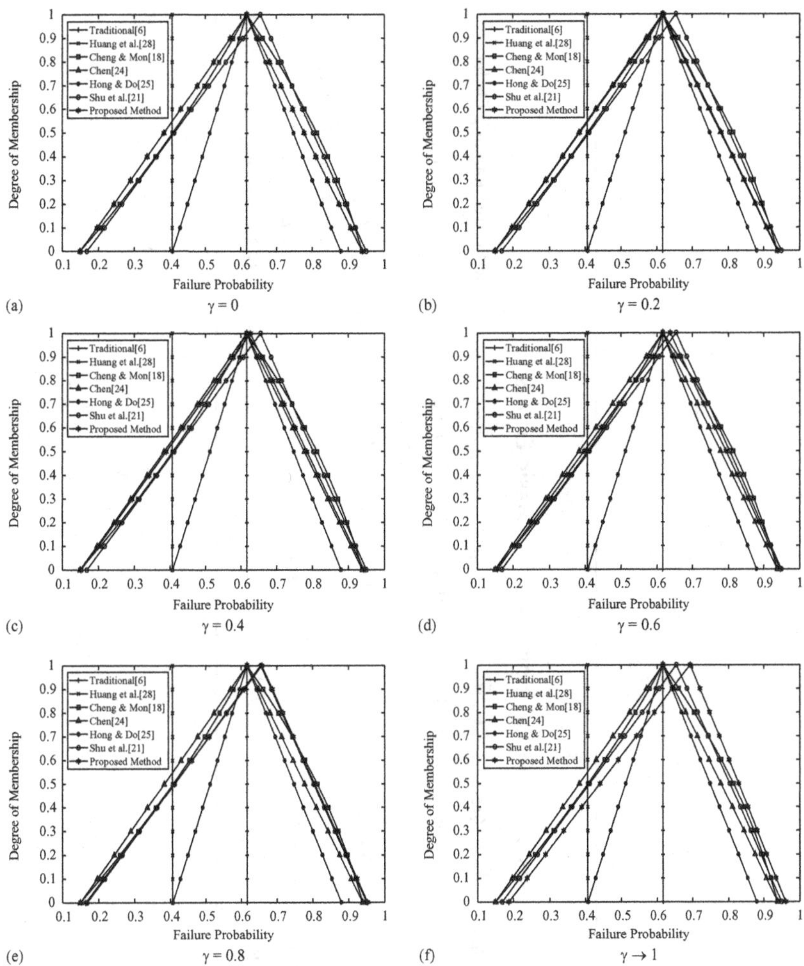

FIGURE 7.4 Membership function for top event at different values of compensatory parameter γ.

7.5.4 Ranking of Critical Components

To rank the critical events of "HTTP failure" in preferential order, Tanaka et al.'s [17] V-index definition given in Equation 7.7 in the context of TFNs has been used. The calculated values of $V(\tilde{q}_T, \tilde{q}_{Ti})$ for each basic event and different values of γ are given in Table 7.8. Based on the results in Table 7.8 for each value of γ, it is analyzed that the most critical basic event is E1 (400-Malformed syntax), which has the strongest contribution for improving the failure access possibility of the top event and validates the conclusion made by Cheong and Lan's [46] study. The order of all the basic events are also matched with Cheong and Lan's [46] and given below.

$$E1 \succ E6 \succ (E2, E3) \succ (E4, E5) \succ (E7, E8)$$

TABLE 7.8
The Failure Difference $V(\tilde{q}_T, \tilde{q}_{Ei})$ at Different Values of γ and Ranking of Critical Basic Events

γ	$V(\tilde{q}_T, \tilde{q}_{E1})$	$V(\tilde{q}_T, \tilde{q}_{E2})$	$V(\tilde{q}_T, \tilde{q}_{E3})$	$V(\tilde{q}_T, \tilde{q}_{E4})$	$V(\tilde{q}_T, \tilde{q}_{E5})$	$V(\tilde{q}_T, \tilde{q}_{E6})$	$V(\tilde{q}_T, \tilde{q}_{E7})$	$V(\tilde{q}_T, \tilde{q}_{E8})$	Basic Events' Ranking Order
.0	0.6018	0.0190	0.0190	0.0085	0.0085	0.3127	0.0048	0.0048	$E1 \succ E6 \succ (E2, E3) \succ (E4, E5) \succ (E7, E8)$
.2	0.5921	0.0188	0.0188	0.0144	0.0144	0.3085	0.0048	0.0048	$E1 \succ E6 \succ (E2, E3) \succ (E4, E5) \succ (E7, E8)$
.4	0.5767	0.0184	0.0184	0.0249	0.0249	0.3018	0.0047	0.0047	$E1 \succ E6 \succ (E2, E3) \succ (E4, E5) \succ (E7, E8)$
.6	0.5521	0.0179	0.0179	0.0438	0.0438	0.2909	0.0045	0.0045	$E1 \succ E6 \succ (E2, E3) \succ (E4, E5) \succ (E7, E8)$
.8	0.5128	0.0169	0.0169	0.0787	0.0787	0.2730	0.0043	0.0043	$E1 \succ E6 \succ (E2, E3) \succ (E4, E5) \succ (E7, E8)$
	0.4490	0.0153	0.0153	0.0683	0.0683	0.2431	0.0039	0.0039	$E1 \succ E6 \succ (E2, E3) \succ (E4, E5) \succ (E7, E8)$

7.6 CONCLUSIONS AND FUTURE RESEARCH DIRECTIONS

In this chapter, a systematic procedure for implementing fuzzy fault tree analysis has been proposed. The proposed approach uses *compensatory "and"* operator to provide the range of fault intervals at different values of the compensatory parameter γ and the triangular fuzzy number for representing possibilities of failure bottom events of the system. The benefit of the proposed approach is that it incorporates not only uncertainty factors in the decision-making process but also the designer's experience and knowledge. The analysis is based on experts' knowledge, and so it plays a critical role. If at any stage, expert judgment is not up to the mark, then the computed results and consequently suggestions extracted for improvement in system performance may not be appropriated and useful. The proposed approach has been implemented on web access failure. The proposed method is also compared with the existing techniques of fault tree methods. Moreover, this chapter also modifies Tanaka et al.'s fuzzy fault tree definition to find the critical bottom events of the system. For different values of parameter γ, modified Tanaka et al.'s technique has been implemented and the computed results are tabulated. It is concluded that the most critical basic event is E1 (400-Malformed syntax), which provides the largest contribution to improve the failure access possibility of the top event. Based on the plotted and tabulated results, a system analyst/decision maker may take appropriate action for better managerial decision making and could formulate future system maintenance strategies for failure-free web access. The results show that the proposed method could estimate failure intervals more flexibly than previous methods and the decision makers find the most critical system component to improve the reliability.

Future work can be oriented in the following directions:

- The developed FFTA method may be applied to analyze some other problems related to healthcare, clean energy, waste management, etc.
- The proposed FFTA method may be further improved by developing some other operations and operators.
- The discussed FFTA method can also be tested by adopting different types of fuzzy membership functions with the help of experts.
- The proposed FFTA method can be further extended for IFS or vague sets.

ACKNOWLEDGMENT

The author would like to thank the editor and anonymous referees for their useful suggestions for improving the quality of the chapter.

NOMENCLATURE

$\tilde{A}$	Fuzzy set $\tilde{A}$	α-cut	Alpha cut of a fuzzy set $\tilde{A}$
U	Universal set	R	Set of real numbers
T_ω	Weakest t-norm	$(\lambda,1)$	Interval-valued fuzzy number of level $(\lambda,1)$
$\mu_{\tilde{A}}(x)$	Membership value of x in fuzzy set $\tilde{A}$	$\tilde{A}=(a_1,a_2,a_3)$	Triangular fuzzy number with left end a_1, middle value a_2, and right end a_3
$\tilde{q}_T$	Fuzzy failure possibility of system top event	$\tilde{q}_i$	Fuzzy failure possibility of system ith component
$\tilde{q}_{Ti}$	Fuzzy failure probability of system top event when $\tilde{q}_i=\tilde{0}$	$q_i=(q_{i1},q_{i2},q_{i3})$	Fuzzy failure possibility of system ith component in TFN form
$A^{(\alpha)}$	Alpha cut of a fuzzy set $\tilde{A}$	V	Index V measures the difference between $\tilde{q}_T$ and $\tilde{q}_{Ti}$
$a_1^{(\alpha)}$	Lower bound of an alpha cut of a TFN $\tilde{A}$	$a_3^{(\alpha)}$	Upper bound of an alpha cut of a TFN $\tilde{A}$
$L(x),R(x)$	Left and right reference functions of LR-type fuzzy number	$P_{oss}(Xi)$	Failure possibilities of ith event X_i
$\cup$	Union used to denote system components connected by OR gate, i.e., in series configuration	$\cap$	Intersection used to denote system components connected by AND gate, i.e., in parallel configuration
$P_{\text{OR}},P_{\text{AND}}$	Probability of top event due to events connected by OR and AND gates	$P_{\text{oss}}^{\text{OR}},P_{\text{oss}}^{\text{AND}}$	Possibility of top event due to events connected by OR and AND gates
$\tilde{P}_{\text{OR}},\tilde{P}_{\text{AND}}$	Fuzzy possibility of top event due to events connected by OR and AND gates	$\tilde{P}_{\text{OR}}^{(\alpha)},\tilde{P}_{\text{AND}}^{(\alpha)}$	Possibility of top event at any cut level α due to events connected by OR and AND gates
$F(\tilde{A}\cup\tilde{B})$	Fuzzy failure possibility for events connected by OR gate	$F(\tilde{A}\cap\tilde{B})$	Fuzzy failure possibility for events connected by AND gate
$\oplus,\ominus,\otimes,\oslash$	Simplified addition, subtraction, multiplication, and division between two TFNs	$\oplus^\alpha,\ominus^\alpha,\otimes^\alpha,\oslash^\alpha$	Simplified addition, subtraction, multiplication, and division between any two fuzzy numbers at cut level α
$\oplus_{T_\omega},\ominus_{T_\omega},\otimes_{T_\omega},\oslash_{T_\omega}$	T_ω based addition, subtraction, multiplication, and division between two TFNs	γ	Compensatory parameter

REFERENCES

1. J.P. Sawyer and S.S. Rao. Fault tree analysis of fuzzy mechanical systems. *Microelectronics and Reliability*, 34(4):653–667, 1994.
2. R.S. Chanda and P.K. Bhattacharjee. A reliability approach to transmission expansion planning using fuzzy fault-tree model. *Electric Power Systems Research*, 45(2):101–108, 1998.
3. Komal and S.P. Sharma. Fuzzy reliability analysis of repairable industrial systems using soft-computing based hybridized techniques. *Applied Soft Computing*, 24:264–276, 2014.
4. K.Y. Cai, C.Y. Wen and M.L. Zhang. Fuzzy variables as a basis for a theory of fuzzy reliability in the possibility context. *Fuzzy Sets and Systems*, 42(2):145–172, 1991.
5. C. Ebeling. *An Introduction to Reliability and Maintainability Engineering*. Tata McGraw-Hill Company Ltd., New York, 2001.
6. P. Kales. *Reliability: For Technology, Engineering, and Management*. Prentice-Hall, Upper Saddle River, NJ, 1998.
7. P.V. Suresh, A.K. Babar and V.V. Raj. Uncertainty in fault tree analysis: A fuzzy approach. *Fuzzy Sets and Systems*, 83(2):135–141, 1996.
8. K.D. Rao, H.S. Kushwaha, A.K. Verma and A. Srividya. Quantification of epistemic and aleatory uncertainties in level-1 probabilistic safety assessment studies. *Reliability Engineering and System Safety*, 92(7):947–956, 2007.
9. L.A. Zadeh. Fuzzy sets. *Information and Control*, 8(3):338–353, 1965.
10. K.Y. Cai and C.Y. Wen. Street-lighting lamps replacement: A fuzzy viewpoint. *Fuzzy Sets and Systems*, 37(2):161–172, 1990.
11. K.Y. Cai, C.Y. Wen and M.L. Zhang. Posbist reliability behavior of typical systems with two types of failures. *Fuzzy Sets and Systems*, 43(1):17–32, 1991.
12. K.Y. Cai, C.Y. Wen and M.L. Zhang. Fuzzy states as a basis for a theory of fuzzy reliability. *Microelectronic Reliability*, 33(15):2253–2263, 1993.
13. K.T. Atanassov. Intuitionistic fuzzy sets. *Fuzzy Sets and Systems*, 20(1):87–96, 1986.
14. J.R. Chang, K.H. Chang, S.H. Liao and C.H. Cheng. The reliability of general vague fault tree analysis on weapon systems fault diagnosis. *Soft Computing*, 10(7):531–542, 2006.
15. W. Pedrycz. Why triangular membership functions? *Fuzzy Sets and Systems*, 64(1):21–30, 1994.
16. Y.A. Mahmood, A. Ahmadi, A.K. Verma, A. Srividya and U. Kumar. Fuzzy fault tree analysis: A review of concept and application. *International Journal of System Assurance Engineering and Management*, 4(1):19–32, 2013.
17. H. Tanaka, L.T. Fan, F.S. Lai and K. Toguchi. Fault-tree analysis by fuzzy possibility. *IEEE Transactions on Reliability*, R-32(5):453–457, 1983.
18. C.H. Cheng and D.L. Mon. Fuzzy system reliability analysis by interval of confidence. *Fuzzy Sets and Systems*, 56(1):29–35, 1993.
19. C.F. Fuh, R. Jea and J.S. Su. Fuzzy system reliability analysis based on level (λ, 1) interval-valued fuzzy numbers. *Information Sciences*, 272:185–197, 2014.
20. S.R. Cheng, B. Lin, B.M. Hsu and M.H. Shu. Fault-tree analysis for liquefied natural gas terminal emergency shutdown system. *Expert Systems with Applications*, 36(9):11918–11924, 2009.
21. M.H. Shu, C.H. Cheng and J.R. Chang. Using intuitionistic fuzzy sets for fault-tree analysis on printed circuit board assembly. *Microelectronics Reliability*, 46(12):2139–2148, 2006.
22. K.H. Chang, C.H. Cheng and Y.C. Chang. Reliability assessment of an aircraft propulsion system using IFS and OWA tree. *Engineering Optimization*, 40(10):907–921, 2008.

23. D. Singer. A fuzzy set approach to fault tree and reliability analysis. *Fuzzy Sets and Systems*, 34(2):145–155, 1990.
24. S.M. Chen. Fuzzy system reliability analysis using fuzzy number arithmetic operations. *Fuzzy Sets and Systems*, 64(1):31–38, 1994.
25. D.H. Hong and H.Y. Do. Fuzzy system reliability analysis by the use of T_ω (the weakest t-norm) on fuzzy number arithmetic operations. *Fuzzy Sets and Systems*, 90(3):307–316, 1997.
26. M. Kumar and S.P. Yadav. The weakest f-norm based intuitionistic fuzzy fault-tree analysis to evaluate system reliability. *ISA Transactions*, 51(4):531–538, 2012.
27. Komal. Fuzzy fault tree analysis for patient safety risk modeling in healthcare under uncertainty. *Applied Soft Computing*, 37:942–951, 2009.
28. H.Z. Huang, X. Tong and M.J. Zuo. Posbist fault tree analysis of coherent systems. *Reliability Engineering and System Safety*, 84(2):141–148, 2004.
29. K.B. Misra and G.G. Weber. A new method for fuzzy fault tree analysis. *Microelectronics Reliability*, 29(2):195–216, 1989.
30. R. Ferdous, F. Khan, B. Veitch and P.R. Amyotte. Methodology for computer aided fuzzy fault tree analysis. *Process Safety and Environmental Protection*, 87(4):217–226, 2009.
31. Komal. A novel FFTA for handwashing process to maintain hygiene with events following different membership function. *Arabian Journal of Science and Engineering*, 42(7):3007–3019, 2017.
32. S.G. Chowdhury and K.B. Misra. Evaluation of fuzzy reliability of a non-series parallel network. *Microelectronics Reliability*, 32(1–2):1–4, 1992.
33. D.L. Mon and C.H. Cheng. Fuzzy system reliability analysis for components with different membership functions. *Fuzzy Sets and Systems*, 64(2):145–157, 1994.
34. S.M. Chen. New method for fuzzy system reliability analysis. *Cybernetics and Systems: An International Journal*, 27(4):385–401, 1996.
35. H. Garg. Analysis of an industrial system under uncertain environment by using different types of fuzzy numbers. *International Journal of System Assurance Engineering and Management*, 9(2):525–538, 2018.
36. Komal. Fuzzy reliability analysis of a phaser measurement unit using generalized fuzzy lambda-tau(GFLT) technique. *ISA Transactions*, 76:31–42, 2018a.
37. Komal. Fuzzy reliability analysis of DFSMC system in LNG carriers for components with different membership function. *Ocean Engineering*, 155:278–294, 2018b.
38. K.H. Chang and C.H. Cheng. A novel general approach to evaluating the PCBA for components with different membership function. *Applied Soft Computing*, 9(3):1044–1056, 2009.
39. M. Kumar. Applying weakest f-norm based approximate fuzzy arithmetic operations on different types of intuitionistic fuzzy numbers to evaluate reliability of PCBA fault. *Applied Soft Computing*, 23:387–406, 2014.
40. M.K. Luhandjula. Compensatory operators in fuzzy linear programming with multiple objectives. *Fuzzy Sets and Systems*, 8(3):245–252, 1982.
41. W. Wang and X. Liu. Intuitionistic fuzzy information aggregation using Einstein operations. *IEEE Transactions on Fuzzy Systems*, 20(5):923–938, 2012.
42. E.P. Klemenet, R. Mesiar and E. Pap. On the relationship of compensatory operators to triangular norms and conorms. *International Journal of Uncertainty, Fuzziness and Knowledge-Based Systems*, 4(2):129–144, 1996.
43. H.J. Zimmermann and P. Zysno. Latent connectives in human decision making. *Fuzzy Sets and Systems*, 4(1):37–51, 1980.
44. M. Crovella and A. Bestavros. *Explaining World Wide Web Traffic Self-Similarity*. Tech. Rep. BUCS-TR-95F-015. Boston University, CD Dept, Boston, MA (1995).

45. P.S. Louis. A model of web server performance. In *6th International World Wide Web Conference*, 7–11 April, Santa Clara, CA, 1997.
46. C.W. Cheong and A.L.H. Lan. Web access failure analysis-fuzzy reliability approach. *International Journal of the Computer, the Internet and Management*, 12(1):65–73, 2004.
47. Komal. Fuzzy fault tree analysis of web access failure using compensatory operator. In *Presented in 18th Online World Conference on Soft-Computing in Industrial Applications (WSC18)*, 1–12 December, Paper ID: WSC18–2014–0012, 2014.
48. R. Ferdous, F. Khan, R. Sadiq, P. Amyotte and B. Veitch. Analyzing system safety and risks under uncertainty using a bow-tie diagram: An innovative approach. *Process Safety and Environmental Protection*, 91(1–2):1–18, 2013.
49. Mentes and I.H. Helvacioglu. An application of fuzzy fault tree analysis for spread mooring systems. *Ocean Engineering*, 38(2–3):285–294, 2011.
50. K.P. Lin, M.J. Wu, K.C. Hung and Y. Kuo. Developing a T_{ω} (the weakest ŕ-norm) fuzzy GERT for evaluating uncertain process reliability in semiconductor manufacturing. *Applied Soft Computing*, 11(8):5165–5180, 2011a.
51. K.P. Lin, W. Wen, C.C. Chou, C.H. Jen and K.C. Hung. Applying fuzzy GERT with approximate fuzzy arithmetic based on the weakest ŕ-norm operations to evaluate repairable reliability. *Applied Mathematical Modelling*, 35(11):5314–5325, 2011b.
52. T.J. Ross. *Fuzzy Logic with Engineering Applications*. 2nd edn. Wiley, New York, 2004.

8 Entropy-Based Fuzzy Reliability–Redundancy Allocation Model

A New Interactive Approach

Sahidul Islam

CONTENTS

8.1 INTRODUCTION

Reliability engineering appeared in the early 1950s and was first applied to communication and transportation systems. Most of the early reliability work was confined to the analysis of the performance aspects of systems. In the first half the century, numerous well-written books on reliability were made available. The primary goal of the reliability engineer has always been to find the best way to increase system reliability. A balance between system reliability and resource consumption is essential. The diversity of system resources, resource constraints, and options for reliability improvement have led to the construction and analysis of several optimization models.

Here we have considered an entropy-based fuzzy reliability optimization model to produce a highly reliable system and also to maximize the entropy amount of the system subject to the available cost, weight, and volume of each component. The goal of the present study is to apply an efficient and modified optimization technique to find the optimum number of redundant components of the proposed model. So an interactive fuzzy multi-objective decision-making (IFMODM) method is used here to design a high productivity system. In other existing methods like the fuzzy multi-objective nonlinear programming (FMONLP) method and the fuzzy multi-objective goal programming (FMOGP) method, the objective value and the maximum tolerances for resources are given initially. In a real world situation, it is unrealistic to initially ask the decision maker (DM) to give goals and tolerances without providing any information about them. Therefore, the obtained solution may not be satisfactory for the DM. However, the IFMODM technique considers an expansive assortment of circumstances that the DM might meet when solving a nonlinear programming problem. In this method, the DM may be satisfied with the initially obtained solution; if not, he may alter the original model to get an agreeable solution. This procedure will then proceed until the point when the DM will be satisfied with the outcome. So clearly this method gives a more reliable system than other existing methods. An illustrative example is provided to show the utility of IFMODM on the proposed reliability model, and the result of the proposed approach is compared with FMONLP and FMOGP approach at the end of the chapter.

8.2 LITERATURE REVIEW

Presently, our society is for the most part subject to current mechanical systems and it is almost certain that these innovative systems have enhanced the productivity, health, and prosperity of our society. For each of the complex systems, the system reliability assumes a vital job. The reliability of any system is essential to producers, architects, and furthermore to the clients. During the design period of an item, reliability engineers are called upon to quantify the reliability of that item. They want higher reliability of their items which raise the production cost of the things. In such a case, the question arises how to meet the objective of system reliability. Thus, increase in the production cost affects the client's financial plan. In this way, the design reliability optimization problem is stated as reliability improvement at the least cost possible. In this regard, a well-known strategy for maximizing the reliability of a system is to present a few redundant components. For better planning, system components with known cost, reliability, weight, and different qualities are utilized, and the comparing problem can be figured as a combinatorial optimization problem, where either system reliability is maximized or system cost is minimized. Subsequently both the plans for the most part include limitations on permissible weight as well as cost and are least focused on the system reliability level. The comparing problem is known as the reliability redundancy allocation problem. The essential target of the reliability redundancy allocation problem is to choose the best mix of components and levels of redundancy either to maximize the system reliability as well as to minimize the system cost subject to a few constraints.

Several researchers have considered reliability redundancy problem, such as Ghare and Taylor, (1969) who described the optimal redundancy for reliability in a series system, Misra (1975), who considered the reliability design of a system containing mixed redundancies, Tillman et al. (1977), who described the optimization technique for system reliability with redundancy, and Kim and Yum (1993), who used a heuristic method for solving redundancy optimization problems in complex systems. Researchers like Allella et al. (2005), Boland and EL-Neweihi (1995), Bulfin and Liu (1985), Chern (1992), Coit and Smith (1998), and Mettas (2000) have solved the reliability redundancy allocation problem. Coit and Konak (2006) applied the multiple weighted objectives heuristic for the redundancy allocation problem. Kim et al. (2006) considered simulated annealing algorithms to solve the reliability–redundancy optimization problem, and Liang and Chen (2007) solved a redundancy allocation of series-parallel systems using a variable neighborhood search algorithm. Liang and Smith (2004) presented an ant colony optimization algorithm for the redundancy allocation problem. Tillman et al. (1977) discussed the determination of component reliability and redundancy for optimum system reliability. Onishi et al. (2007) presented an improved surrogate constraint method for solving a redundancy allocation problem with a mix of components. Ramirez-Marquez and Coit (2004) applied a heuristic method for solving the redundancy allocation problem.

To solve the reliability optimization problems, a few scientists have developed different optimization methods which include approximate methods, exact methods, heuristics, multi-objective optimization methods, and so on. Goal programming, dynamic programming, branch, and bound give the exact solution for reliability optimization problems. A multi-objective formulation of the reliability allocation problem to maximize system reliability and minimize system cost has been described by Sakawa (1978) using surrogate worth trade-off methods. Chern and Jan (1985) provided a parametric programming approach; also Kuo and Prasad (2000, 2007) and Kuo et al. (2001) presented some suitable methods for the reliability optimization of systems.

In the reliability redundancy allocation problem, a reliability optimization problem based on entropy that is fuzzy in nature is rarely discussed. Wang (2009) presented a report to estimate the uncertainty of possible failure events of a redundancy system based on the cross-entropy method. Caserta and Noder (2009) discussed a cross-entropy based algorithm for maximizing the reliability of a complex system through the optimal redundancy allocation. Kang and Kwak (2009) investigated the application of a maximum entropy principle for reliability-based design optimization. Ridder (2005) applied cross entropy on Markovian reliability systems. Also Mahapatra (2009) used the global criterion method to an entropy-based reliability optimization problem.

This was the first time when Park (1987) applied fuzzy optimization techniques to the problem of reliability apportionment for a series system. In recent times Gong et al. (2012) discussed the fuzzy entropy clustering approach to reliability, and several authors like Garg (2013, 2015, 2016, 2017), Garg et al. (2013, 2014, 2018), Mahapatra (2011), and Dancese et al. (2014) discussed various fuzzy optimization techniques and soft computing based techniques to solve the reliability optimization problem. In other existing methods like the fuzzy multi-objective nonlinear programming method described by Lai and Hwang (1994) and also Zangiabadi and Maleki (2007), Hwang and Lee (2004) presented a fuzzy multi-objective goal programming method to solve the reliability optimization problem. Islam and Roy (2010) presented an entropy-based multi-objective problem in the fuzzy environment.

8.3 RELIABILITY THEORY: BASIC CONCEPTS AND TERMINOLOGIES

Improvements in science and innovation and the requirements of present-day society are competing against one another. In the industrial process, industries are eager to promote more automation to fulfill the ever-increasing demands of society. The industrial system together with their products become complex day by day. Therefore, the effectiveness of such complex systems needs improvement. The system effectiveness is studied to understand the suitability and efficiency of the system for satisfying the intended tasks. The skill of doing the intended task is basically judged by the reliability and quality of the system.

Nowadays we are totally dependent on the significant performance of machinery equipment, appliances, entertainment centers, robots, and so on, both within and outside the home. But if we do not pay regular attention, the appliances may fail and hence the reliability and maintainability decrease. Reliability engineering is required: to plan, exhibit, produce, and deliver these products to the user such that all of these components and products are efficient and highly reliable during their life span at optimum costs.

Since the late 1940s and the early 1950s, the subject of reliability optimization appeared was considered prospective research and was initially applied to communication and transportation systems. In the early days, majority of the reliability work was restricted to the exploration of the performance aspects of an operating system. The fundamental objective of the decision maker has always been to find the ideal method to increase the reliability of the system. As the systems become more complex, it results in the unreliable behavior of the systems with respect to cost, effort, and so on, and this can be done by considering the following principles:

(i) Maintaining the simplicity of the system so that it can be compatible with the performance requirements.
(ii) To increase reliability for the less reliable components and to use parallel redundancy.
(iii) For certain components when failure occurs, use standby redundancy to activate them.
(iv) For failed components which are replaced but not automatically switched in, use repair maintenance.
(v) For exchangeable components, use better arrangement.

To enhance system reliability, following the above conditions will result in the regular utilization of resources. Consequently, harmony between system reliability and resource utilization is very essential.

8.3.1 Definition of Reliability (Kuo et al., 2001)

Reliability is the best accentual measure of the integrity of a designed part, equipment, product, or system. The definition of reliability is as follows:

> "Reliability is the conditional probability that parts, equipment, products or systems will perform its intended functions for a desired length of time or function period at a given confidence level".

So, reliability may be considered as the probability which is given by the ratio of the number of successful missions undertaken to the total number of missions undertaken.

8.3.2 System Reliability (Kuo et al., 2001)

"System reliability is a proportion of how well a system meets its outline goal and it is normally expressed as far as the reliabilities of the subsystems of parts".

For the most part, to decide on the reliability factor of a system, the system is blown up into broken down to subsystems and components whose individual reliability variables can be evaluated or decided. Contingent upon the way in which these subsystems and components are associated will comprise the given system. The combinatorial rule is applied to achieve system reliability.

8.3.3 Fundamental System Configurations (Kuo et al., 2001)

A system, by and large, isn't made of a single segment. We generally need to assess the reliability of a simple and, in addition, complex/muddled system. Let us consider the reliability of a system comprising a number of parts. These parts can be equipment or human or even programming. If a portion of the parts is software products, then at that point the demonstration requires exceptional considerations. We will here discuss a few essential reliability configurations. Let $R_S(t)$ denote the system reliability for time t and when t is fixed, the system reliability is simply denoted by R_S.

8.3.3.1 Series Configurations (Kuo et al., 2001)

The series configuration is the easiest and among the most widely recognized structures in the reliability system. In this configuration, every one of the parts must work together to guarantee the system task. As such, the system falls short when any of the components falls short.

8.3.3.2 Parallel Configuration (Kuo et al., 2001)

A parallel system is a system that is not considered failed except if all the parts have failed. This is in some cases called redundant configuration. The word "redundant" is utilized just when the system configuration is purposely changed to deliver extra parallel paths keeping in mind the end goal to enhance the system reliability. In a parallel configuration comprising various components, the system works if any of those parts is working.

8.3.3.3 Series-Parallel Configuration (Kuo et al., 2001)

In this configuration, consider a system which consists of k subsystems connected in parallel and each i-th subsystem consisting of n_i components in series for $i = 1, 2, \ldots, k$ (Figure 8.1).

8.3.3.4 Parallel-Series Configuration (Kuo et al., 2001)

Let us suppose a configuration consisting of k subsystems in a series with i-th subsystem (for $i = 1, 2, \ldots, k$) consisting of n_i components in parallel (Figure 8.2).

8.3.3.5 Hierarchical Series-Parallel Systems (Kuo et al., 2001)

A system is known as a hierarchical series-parallel system if the system can be seen as an arrangement of subsystems masterminded in a series-parallel; every subsystem has a comparable arrangement; subsystems of every subsystem have a comparable design and so on. This system has a non-direct and non-divisible structure and comprises a settled parallel and series system. For example, we consider the following system as follows (Figure 8.3):

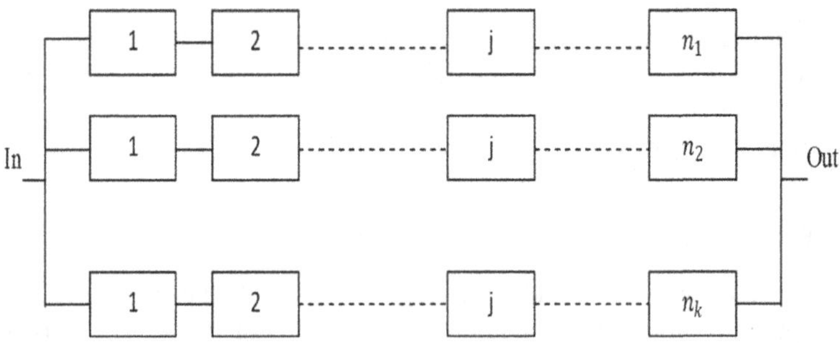

FIGURE 8.1 A series-parallel system.

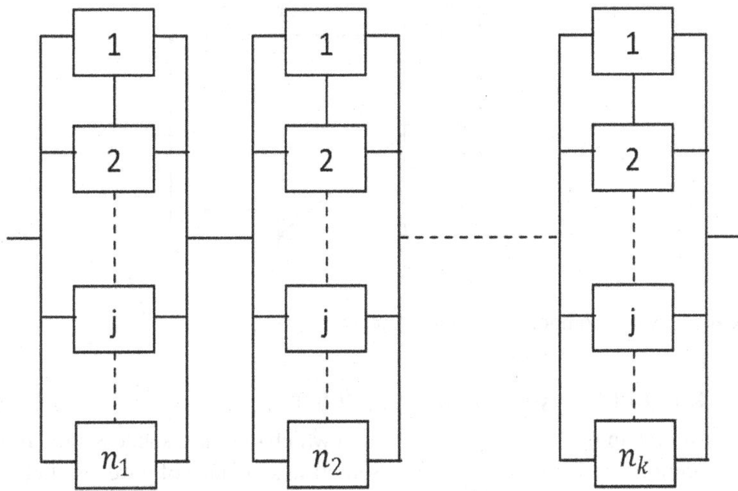

FIGURE 8.2 A parallel-series system.

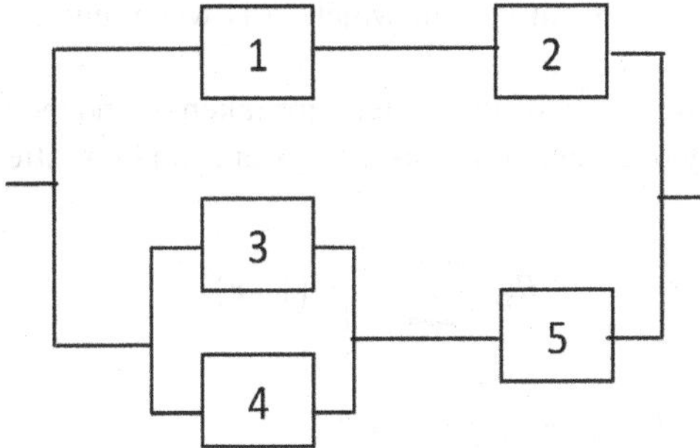

FIGURE 8.3 Five-component hierarchical series-parallel system.

The above system with five components 1, 2, 3, 4, and 5 consists of two subsystems {1,2} and {3,4,5} in parallel. The first subsystem has components 1 and 2 in series, whereas the second one has a series arrangement of component 5 and a subsystem which has components 3 and 4 in parallel.

8.3.3.6 Complex Configuration (Kuo et al., 2001)

Sometimes a reliability system can't be diminished to series and parallel arrangements. In light of the fact that there exist blends of segments which are associated neither in a series nor in parallel, that system is called a complex or non-parallel-series system. A five-component complex system can be considered as follows (Figure 8.4):

The reliability of the above complex system at time t is given by

$$R_S = 1-(1-r_2)(1-r_5)r_3-(1-r_1r_2)(1-r_4r_5)(1-r_3) \tag{8.1}$$

where $r_i\left(i=1,2,..5\right)$ is the reliability of each component.

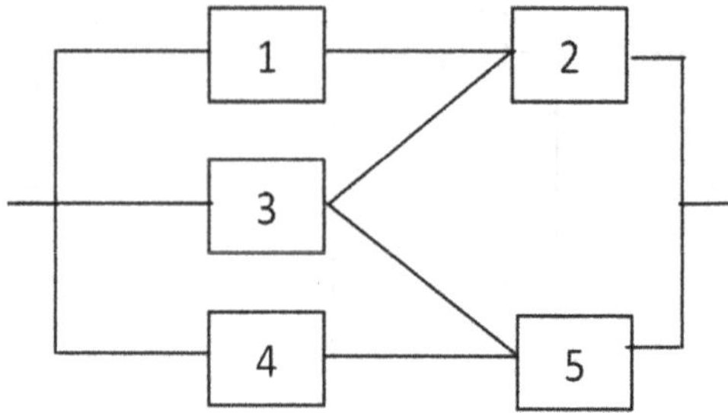

FIGURE 8.4 A five-component complex system.

8.3.3.7 K-out-of-N System (Kuo et al., 2001)

A k-out-of-n system is an n component system which operates successfully, if at least k of its n components work satisfactorily. Sometimes in place of pure parallel system, we can use this redundant system. It is also referred to as k-out-of-n: G system. For example, we can say an n component series system as n-out-of-n: G system and an n component parallel system a 1-out-of-n: G system. Also a k-out-of-n: F system is referred to as an n component system which fails when any k-out-of-n component fails.

When all the components of the system are taken as independent and identical, the system reliability of a general k-out-of-n system can be written as

$$R_S = \sum_{i=k}^{n} \binom{n}{i} r^i (1-r)^{n-i} \tag{8.2}$$

where r is the reliability of each component.

8.3.4 Some Reliability Optimization Models

At the point when redundancy is utilized to enhance the system reliability, the comparing issue is known as the redundancy allocation problem. The goal of this problem is to locate the quantity of redundant parts that expands the system reliability under a few resource limitations. This problem is a standout among the most famous ones in reliability optimization since the 1950s due to its probability for wide applications. When it is hard to enhance the reliability of unreliable parts, system reliability can without much of a stretch be upgraded by including the redundancies of those segments. Be that as it may, for configuration engineers, enhancing the component reliability has been by and large favored over including redundancy, in light of the fact that, in numerous cases, redundancy is hard to add to genuine systems because of specialized restrictions; what's more, generally extensive amounts of assets, for example, weight, volume, and cost, are required.

To solve reliability optimization problems, several researchers have considered various methods as a solution procedure which can be classified into several classes:

(i) heuristics for redundancy allocation;
(ii) exact methods for redundancy allocation;
(iii) heuristics for reliability redundancy allocation;
(iv) multiple objective optimization in reliability system, etc.

8.3.4.1 Model I: Redundancy allocation model (Kuo et al., 2001)

The problem of finding optimal redundancy levels $x_1, x_2, \ldots, x_n$ for maximizing system reliability subject to resource constraints is called the redundancy allocation problem and can be described so as to

maximize $R_S = f(x_1, x_2, \ldots, x_n)$,
subject to,

$$g_i(x_1, x_2, \ldots, x_n) \le b_i, \quad \text{for } i = 1, 2, \ldots, m.$$

$$l_j \le x_j \le u_j, \quad \text{for } j = 1, 2, \ldots, n. \tag{8.3}$$

x_j being the integer.

8.3.4.2 Model II: Reliability–redundancy allocation model (Kuo et al., 2001)

The problem of finding optimal redundancy levels $x_1, x_2, \ldots, x_n$ and optimal component reliabilities $r_1, r_2, \ldots, r_n$ that maximize system reliability subject to resource constraints is called the reliability–redundancy allocation problem and can be expressed as to

maximize $R_S = f(x_1, x_2, \ldots, x_n;\ r_1, r_2, \ldots, r_n)$,
subject to, $g_i(x_1, x_2, \ldots, x_n;\ r_1, r_2, \ldots, r_n) \le b_i, \quad \text{for } i = 1, 2, \ldots, m$.

$$l_j \le x_j \le u_j, \quad \text{for } j = 1, 2, \ldots, n.$$

$$r_j^l \ge r_j \ge r_j^u, \quad \text{for } j = 1, 2, \ldots, n. \tag{8.4}$$

x_j being the integer.

8.3.5 Prerequisite Mathematics

Fuzzy Set: (Zadeh (1965)) A fuzzy set $\tilde{A}$ in U is a set of ordered pairs

$$\tilde{A} = \{(x, \mu_{\tilde{A}}(x)) \mid x \in U\},$$

where U is a collection of objects denoted generically by x and $\mu_{\tilde{A}}(x): U \to [0,1]$ is called the membership function or grade of membership of x in $\tilde{A}$.

Fuzzy Number: A fuzzy number is a quantity which is imprecise, as opposed to the case with a single valued number. A fuzzy number measurement not alluded to one single value is an associated set of conceivable qualities, where every conceivable quality has a weight somewhere in the range of 0 and 1. This weight is called the membership function, which has the following form:

$$\mu_{\tilde{B}}(x) : R \rightarrow [0,1].$$

where $\mu_{\tilde{B}}(x)$ is a membership function of the fuzzy set $\tilde{B}$.

Generalized Fuzzy Number (GFN): S. H. Chen (1985, 1999) represents a generalized trapezoidal fuzzy number (GTrFN) $\tilde{A}$ as $\tilde{A} = (a_1, a_2, a_3, a_4; w)$, where $0 < w \leq 1$, and a_1, a_2, a_3 and a_4 are real numbers. The generalized fuzzy number (GFN) $\tilde{A}$ is a fuzzy subset of real line *R*, whose membership function $\mu_{\tilde{A}}(x)$ satisfies the following properties:

(a) $\mu_{\tilde{A}}(x)$ is a continuous mapping from *R* to the closed interval $[0,1]$;
(b) $\mu_{\tilde{A}}(x) = 0$, for all $x \in (-\infty, a_1]$;
(c) $\mu_{\tilde{A}}(x)$ is strictly increasing with constant rate on $[a_1, a_2]$;
(d) $\mu_{\tilde{A}}(x) = w$ for all $x \in [a_2, a_3]$;
(e) $\mu_{\tilde{A}}(x)$ is strictly decreasing with constant rate on $[a_3, a_4]$;
(f) $\mu_{\tilde{A}}(x) = 0$ where $x \in [a_4, \infty)$;

Note: Ã is a normalized fuzzy number when $w = 1$, and it is non-normalized for $w \neq 1$.

Generalized Trapezoidal Fuzzy Number (GTrFN): A generalized trapezoidal fuzzy number (GTrFN) $(a_1, a_2, a_3, a_4; w)$ is a fuzzy set of the real line *R* whose membership function: $\mu_{\tilde{A}}(x) : R \rightarrow [0, w]$ is defined as (Figure 8.5)

$$\mu_{\tilde{A}}^{w}(x) = \begin{cases} \mu_{L\tilde{A}}^{w}(x) = w\left(\dfrac{x - a_1}{a_2 - a_1}\right), & a_1 \leq x \leq a_2; \\ w, & a_2 \leq x \leq a_3; \\ \mu_{R\tilde{A}}^{w}(x) = w\left(\dfrac{x - a_4}{a_3 - a_4}\right), & a_3 \leq x \leq a_4; \\ 0, & \text{otherwise;} \end{cases} \tag{8.5}$$

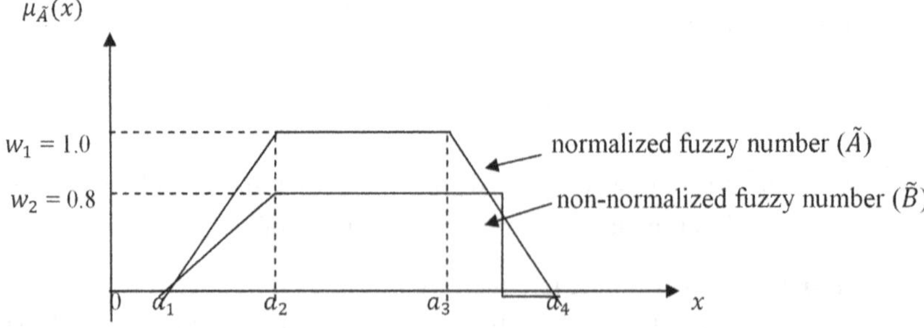

FIGURE 8.5 Two generalized trapezoidal fuzzy numbers $\tilde{A}$ and $\tilde{B}$.

where $\mu_{L\tilde{A}}^{w}$ and $\mu_{R\tilde{A}}^{w}$ are the left and right membership functions of $\tilde{A}$, respectively, and the inverse functions $h_{L\tilde{A}}^{w}:[0,w]\rightarrow[a,b]$ and $h_{R\tilde{A}}^{w}:[0,w]\rightarrow[\text{c},\text{d}]$ are defined as

$$h_{L\tilde{A}}^{w}(y)=+\frac{(b-a)}{w}y \tag{8.6}$$

$$h_{R\tilde{A}}^{w}(y)=d+\frac{(c-d)}{w}y;\quad y\in[0,w].$$

Now according to T.S.Liou and Wang's (1992) integral value method, we have for a non-normal fuzzy number $\tilde{A}$, the corresponding membership function $\mu_{\tilde{A}}(x)$ which can be normalized by dividing the maximal value of $\mu_{\tilde{A}}(x)$, i.e. w. Let $\bar{\tilde{A}}$ and $\mu_{\bar{\tilde{A}}}$ be the normalized fuzzy number and the corresponding membership function.

Let $k\in[0,1]$ be the index of optimism which represents the degree of optimism of a decision maker (DM). Then the total k-integral value is defined as

$$I_k^w(\tilde{A})=\left[kI_R^w(\tilde{A})+(1-k)I_L^w(\tilde{A})\right]$$

where $I_{wR}(\tilde{A})$ and $I_{wL}(\tilde{A})$ represents the right and left integral values of $\tilde{A}$, respectively.

Now, when $\tilde{A}$ is being ranked, we have

$$I_L^w(\tilde{A})=\int_0^1 h_{L\tilde{A}}^{w}(y)dy=\frac{1}{2}(a+b)$$

and (8.7)

$$I_R^w(\tilde{A})=\int_0^1 h_{R\tilde{A}}^{w}(y)dy=\frac{1}{2}(c+d)$$

Thus, $I_k^w(\tilde{A})=\frac{1}{2}\left[k(c+d)+(1-k)(a+b)\right]$ which does not depend on the value of w, i.e. whether $\tilde{A}$ is normal or not. A larger value of k indicates a higher degree of optimism.

Now for k = 0, the total k-integral value is $I_0^w(\tilde{A})=\frac{1}{2}(a+b)=I_w^L(\tilde{A})$, which represents a pessimistic viewpoint of a DM and for optimistic DM's viewpoint, i.e. for k = 1, $I_1^w(\tilde{A})=\frac{1}{2}(c+d)=I_w^R(\tilde{A})$. When k = 0.5, the total k-integral value $I_{0.5}^w(\tilde{A})=\frac{1}{2}\left[I_w^R(\tilde{A})+I_w^L(\tilde{A})\right]$ which reflects a moderately optimistic DM's viewpoint.

Property 1.1:

(a) If $\tilde{B}=(b_1,b_2,b_3,b_4;\,w)$ and $y = qb$, $q > 0$ then $y = q\tilde{b}$ is a fuzzy number $(qb_1,qb_2,qb_3,qb_4;\,w)$.

(b) If $y = qb$, $q < 0$ then $y = q\tilde{b}$ is a fuzzy number $(qb_4,qb_3,qb_2,qb_1;\,w)$.

Proof: Islam and Roy (2006).

Property 1.2:

If $\tilde{A}_1=(a_1,b_1,c_1,d_1;\,w_1)$ and $\tilde{A}_2=(a_2,b_2,c_2,d_2;\,w_2)$. Then $\tilde{A}_1 \oplus \tilde{A}_2$ is a fuzzy number $(a_1 + a_2, b_1 + b_2, c_1 + c_2, d_1 + d_2; \min(w_1, w_2))$.

Proof: Islam and Roy (2006).

8.4 MATHEMATICAL MODEL

A reliability model is developed under the following notations (Figure 8.6):

R_s	System reliability
H_N	Entropy
C_{lim}	Cost limitations/Available cost
Q_{lim}	Weight limitations
V_{lim}	Volume limitations
q_j	Weight of each component of stage j
v_j	Volume of each component of stage j
R_j	Reliability of each component of stage j
x_j	Number of redundancy component in stage j
t	Mission time
α_j, β_j	Constants representing the physical characteristics of each component at stage j
$R_{j,\text{min}}$	Lower bound on the reliability of the component j

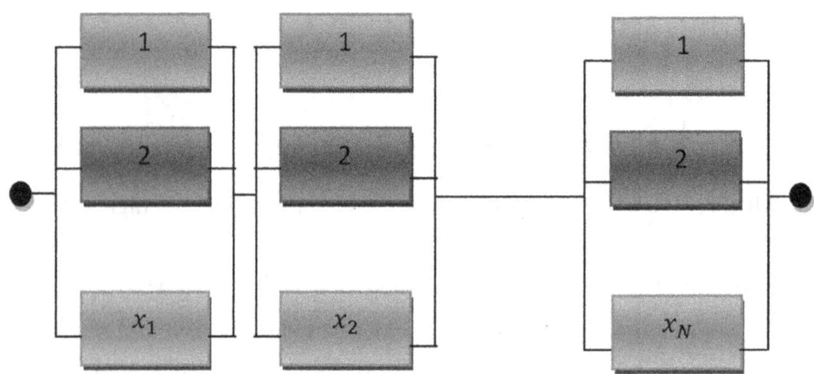

FIGURE 8.6 A mixed system having N-stage in series and components in parallel at each stage.

8.4.1 Optimal Redundancy Allocation in Single Objective Problem

Here considering an n-stage series system our object is to maximize the system reliability through the selection of component reliabilities as well as redundancy levels at the stages subjected to some resource constraints such as volume, weight, and cost.

Therefore the problem becomes,

$$\text{Maximize } R_s(R) = \prod_{j=1}^{N} \left\{1-(1-R_j)^{x_j}\right\}$$

Subject to,

$$C_S(R) = \sum_{j=1}^{N} \alpha_j \left[\frac{-t}{(m+1)\ln(R_j)}\right]^{\beta_j} \left[x_j + \exp\left(\frac{x_j}{4}\right)\right] \le C_{\text{lim}}$$

$$V_S(R) = \sum_{j=1}^{N} v_j x_j^{2} \le V_{\text{lim}} \tag{8.8}$$

$$Q_s(R) = \sum_{j=1}^{N} q_j x_j \exp\left(\frac{x_j}{4}\right) \le Q_{\text{lim}}$$

$$0.5 \le R_{j,\min} \le R_j \le 1, \quad 0 \le R_s \le 1, \quad \text{for } j = 1,2,\ldots,N.$$

$$x_j \ge 1, \; j = 1,2,\ldots,N, \; x_j \text{ and } m \text{ being integer } \& \; m > -1.$$

Here the component failure rate is nonlinear in behavior and in the form of Kt^m, $m > -1$. If the two parameters K and m are chosen appropriately, then the component failure rate is constant for $m = 0$, and linearly increasing for $m = 1$.

8.4.2 Entropy in Redundancy Allocation Problem

Entropy is the measure of information associated with each conceivable data value and is the negative logarithm of the probability mass function for the value. In this way, when the data source has a lower probability value, the event carries more "information" than when the source data has a higher probability value. The amount of information passed on by each occasion characterized along these lines turns into a random variable whose normal value is the information entropy. Generally, entropy alludes to turmoil or uncertainty, and the meaning of entropy utilized in information theory is specifically analogous to the definition utilized in statistical thermodynamics. The idea of entropy was presented by Claude Shannon in 1948.

The basic model of a data communication framework is made up of three components, a wellspring of data, a communication channel, and a recipient, and, as communicated by Shannon (1948), the "fundamental problem of communication" is for the collector to have the capacity to distinguish what data was generated by the source, based on the signal it gets through the channel. The entropy gives an absolute limit on the most limited conceivable average length of a lossless pressure encoding of the data created by a source, and if the entropy of the source is not as much as the channel capacity of the communication channel, the data generated by the source can be reliably communicated to the receiver.

As indicated by the maximization-entropy principle, given some fractional data about a random variate, scalar or vector, we ought to pick that probability distribution for it, which is reliable with the given data, yet has generally maximum uncertainty related with it.

The concept of "entropy" is introduced to provide a quantitative measure of uncertainty and this is usually used in physics. In mathematics, a more abstract definition is used. The Shannon (1948) entropy of a random variable X is defined as

$$H(X) = -\sum_{x} p(x) \ln p(x) \tag{8.9}$$

where $p(x)$ is the probability that X is in the state x, and we assume 0 ln 0 by 0.

In this redundancy allocation problem, a n-stage series-parallel system with $x_j\left(j = 1,2,\ldots,N\right)$ number of redundant components is considered and the total number of components is given by Σx_j. Now normalizing the redundant component x_j by dividing them by the total number of redundant components Σx_j, a probability distribution $p_j = \frac{x_j}{\Sigma x_j}$ is found (Mahapatra, 2009).

So (8.9) becomes

$$H_N(R) = -\sum \left(\frac{x_j}{\Sigma x_j}\right) \ln \left(\frac{x_j}{\Sigma x_j}\right) \tag{8.10}$$

with each $p_j \geq 0$ (j = 1, 2, ..., N) & $\Sigma p_j = 1$ and $H_N(R)$ is maximum when $p_j = \frac{1}{N}$.

8.4.3 Multi-Objective Reliability Optimization Problem

Consider entropy as an additional objective function, the problem (8.8) becomes

$$\text{Maximize } R_s\left(R\right) = \prod_{j=1}^{N} \left\{1 - \left(1 - R_j\right)^{x_j}\right\}$$

$$\text{Maximize } H_N\left(R\right) = -\sum \left(\frac{x_j}{\Sigma x_j}\right) \ln \left(\frac{x_j}{\Sigma x_j}\right)$$

subject to the same constraints defined in Equations 8.8.

8.4.4 Multi-objective Fuzzy Reliability Optimization Problem

In daily life, due to some uncertainty in judgments and lack of evidence, it is not always possible to get consequential data for the reliability system and also it cannot be represented by random variables selected from a probability distribution. But using a fuzzy number, it can be possible for the decision maker to specify the membership functions of the constraints and the goal of the optimization model. To make the reliability optimization model more flexible and acceptable to the decision maker in practical life, here the resource constraints (such as weight, volume, and cost) are considered as a generalized trapezoidal fuzzy number.

Therefore, the above reliability optimization model (Equation 8.8) in a fuzzy environment becomes

$$\text{Maximize } R_s(R)=\prod_{j=1}^{N}\left\{1-\left(1-R_j\right)^{x_j}\right\}$$

$$\text{Maximize } H_N(R)=-\sum\left(\frac{x_j}{\sum x_j}\right)\ln\left(\frac{x_j}{\sum x_j}\right)$$

subject to,

$$C_S(R)=\sum_{j=1}^{N}\alpha_j\left[\frac{-t}{(m+1)\ln(R_j)}\right]^{\beta_j}\left[x_j+\exp\left(\frac{x_j}{4}\right)\right]\le\tilde{C}_{\text{lim}}$$

$$V_S(R)=\sum_{j=1}^{N}v_jx_j^{\,2}\le\tilde{V}_{\text{lim}}$$

$$Q_s(R)=\sum_{j=1}^{N}q_jx_j\exp\left(\frac{x_j}{4}\right)\le\tilde{Q}_{\text{lim}}$$

$$0.5\le R_{j,\min}\le R_j\le 1,\quad 0\le R_s\le 1,\quad \text{for } j=1,2,\ldots,N.$$

$$x_j\ge 1,\ j=1,2,\ldots,N,\ x_j \text{ and } m \text{ being integer \& } m>-1. \tag{8.11}$$

8.5 MATHEMATICAL ANALYSIS

8.5.1 Interactive Fuzzy Multiple Objective Decision-Making Method

An interactive fuzzy multiple objective decision-making proposal here provides an efficient and systematic approach of Multiple Objective Decision-Making (MODM) techniques. The decision maker (DM) may be satisfied with the solutions obtained in the MODM problem or may want to change the original model when the DM is not satisfied with the particular solutions or resources used. In that case the DM proceeds with the interactive process to design a high productivity system.

Definition 8.1: A solution $x \in X$ is called *fuzzy efficient* if there is no $x^* \in X$ such that:

$$Z_p(x^*) \geq Z_p(x) \quad \text{for } p = 1,2,\ldots,n. \text{ and}$$

$$f_i(x^*) \geq f_i(x) \quad \text{for } i = 1,2,\ldots,m.$$

as well as

$$Z_p(x^*) > Z_p(x) \quad \text{for at least one } p \in \{1,2,\ldots,n\} \text{ and}$$

$$f_i(x^*) > f_i(x) \quad \text{for at least one } i \in \{1,2,\ldots,m\}$$

The solution procedure using IFMODM approaches can be summarized in the following steps:

Step 1: Develop a fuzzy MODM problem as

$$\begin{aligned} &\text{Maximize} \quad (Z_1(x), Z_2(x), \ldots, Z_n(x)) \\ &\text{Subject to,} \quad g_i(x) \leq \tilde{b}_i \\ &\qquad i = 1,2,\ldots,m, \; x \geq 0. \end{aligned} \tag{8.12}$$

Step 2: Determine the best upper bound and worst lower bound for constructing the membership functions as follows:

$$Z_p^- = Z_p(x_p^-) = \max_{x \in X^-} Z_p(x) \; \forall p$$

$$\text{where } X^- = \{x : g_i(x) \leq \tilde{b}_i, \forall i, \; x \geq o\};$$

and

$$Z_p^+ = Z_p(x_p^+) = \max_{x \in X^+} Z_p(x) \; \forall p$$

$$\text{where } X^+ = \{x : g_i(x) \leq \tilde{b}_i + \delta_i, \forall i, \; x \geq o\};$$

where δ_i is the maximum tolerance corresponding to the fuzzy resource $\tilde{b}_i$.

Step 3: Construct the membership functions of the objective functions and constraints as follows:

$$f_p^{w_p}(Z_p(x)) = \begin{cases} 0 & \text{if } Z_p(x) < Z_{p,\min}; \\ w_p\left(\dfrac{Z_p(x) - Z_{p,\min}}{Z_p^+ - Z_{p,\min}}\right) & \text{if } Z_{p,\min} \le Z_p(x) \le Z_p^+; \quad \forall\, p = 1,2,\ldots,n. \\ w_p & \text{if } Z_p^+ < Z_p(x); \end{cases}$$

(8.13)

and

$$f_i\left(g_i(x)\right) = \begin{cases} 0 & \text{if } b_i + \delta_i < g_i(x); \\ 1 - \left(\dfrac{g_i(x) - b_i}{\delta_i}\right) & \text{if } b_i \le g_i(x) \le b_i + \delta_i; \quad \forall\, i = 1,2,\ldots,m. \\ 1 & \text{if } g_i(x) < b_i; \end{cases} \tag{8.14}$$

where $Z_{p,\min} = \min_p\left[Z_p\left(x_p^+\right), Z_p\left(x_p^-\right)\right]$ (p = 1,2,..., n) gives the pessimistic values, and the optimistic values are given by the diagonal in the upper half of Table 8.1, which is the maximum achievable value of the corresponding objectives.

Step 4: Now based on the min-operator introduced by Bellman and Zadeh (1970), the fuzzy decision is defined as

fuzzy decisions (D) = fuzzy objective goals (G) $\cap$ fuzzy constraints (C)
Thus, the membership function is characterized by

$$f_D(x) = \min\left(f_G(x), f_C(x)\right)$$

So, introduce the variable λ, where

TABLE 8.1
Pay-off Matrix of Efficient Extreme Solutions

	Z_1	Z_2		Z_n
x_1^+	$Z_1^*\left(x_1^+\right)$	$Z_2\left(x_1^+\right)$	$\cdots$	$Z_n\left(x_1^+\right)$
x_2^+	$Z_1\left(x_2^+\right)$	$Z_2^*\left(x_2^+\right)$	$\cdots$	$Z_n\left(x_2^+\right)$
$\vdots$	$\vdots$	$\vdots$		$\vdots$
x_n^+	$Z_1\left(x_n^+\right)$	$Z_2\left(x_n^+\right)$	$\cdots$	$Z_n^*\left(x_n^+\right)$
x_1^-	$Z_1\left(x_1^-\right)$	$Z_2\left(x_1^-\right)$	$\cdots$	$Z_n\left(x_1^-\right)$
x_2^-	$Z_1^-\left(x_2^-\right)$	$Z_2^-\left(x_2^-\right)$		$Z_n\left(x_2^-\right)$
$\vdots$	$\vdots$	$\vdots$		$\vdots$
x_n^-	$Z_n\left(x_n^-\right)$	$Z_2\left(x_n^-\right)$	$\cdots$	$Z_n^-\left(x_n^-\right)$

$$\lambda = \min\left[f_1^{w_1}\left(Z_1\left(x\right)\right),\ldots,f_n^{w_n}\left(Z_n\left(x\right)\right)\right]$$

$$= \min\left[f_1^{w}\left(Z_1\left(x\right)\right),\ldots,f_n^{w}\left(Z_n\left(x\right)\right)\right]$$

$$= \min\left[wf_1\left(Z_1\left(x\right)\right),\ldots,wf_n\left(Z_n\left(x\right)\right)\right]$$

where

$$w = \min\left(w_1, w_2,\ldots, w_n\right)$$

$$= \min_p(w_p), \text{ for } p = 1,2,\ldots,n.$$

Step 5: Obtain a preferred solution by solving the following λ-Fuzzy minimization problem:

$$\text{Maximize } \lambda$$

$$\text{Subject to, } wf_p(Z_p(x)) \geq \lambda; \quad \forall\, p = 1,2,\ldots,n;$$

$$f_i\left(g_i\left(x\right)\right) \geq \lambda; \quad \forall i = 1,2,\ldots,m;$$

$$\lambda \in \left[0, w\right] \text{ and } x \geq 0,\ w \in \left(0,1\right]$$

Now using positive weights $W_p(p = 1,2, \ldots, n)$ for the objectives $Z_p(x)$, we have

$$\text{Maximize } \lambda$$

$$\text{Subject to, } W_p\ wf_p(Z_p(x)) \geq \lambda; \quad \forall\, p = 1,2,\ldots,n;$$

$$f_i\left(g_i\left(x\right)\right) \geq \lambda; \quad \forall i = 1,2,\ldots,m;$$

$$\lambda \in \left[0, w\right] \text{ and } x \geq 0,\ w \in \left(0,1\right],\ \sum_{p=1}^{n} W_p = 1.$$

where positive weights $W_p(p = 1,2, \ldots, n)$ reflect the decision maker's preferences regarding the relative importance of each objective goal.

Step 6: If one of the fuzzy efficient extreme solutions in Table 8.1 or the preferred solution obtained from above equation is satisfactory for the DM, then the process is successfully concluded and stops. Otherwise go to the next step.

Step 7: Now to get a suitable result to the DM, modify membership functions of the objectives and constraints and also assuming the linear membership functions.

Step 8: If the preferred optimal fuzzy solution of the modified model is acceptable to the DM, then stop. Otherwise go to step 1.

There are some restrictions on modifying the membership functions of objectives and fuzzy constraints. Only the following variations are acceptable for modification:

(1) Increase of $Z_{p,\min}$: Increase of $Z_{p,\min}$ leads to the rise of requirement on the p-th objective. All feasible solutions x with $Z_p(x) < Z_{p,\min}$ (new) are eliminated from the new feasible solution set. Now, we should increase as few requirements as possible in each iteration to avoid the possibility of getting into an empty feasible solution set because of excess increases of $Z_{p,\min}$. We must be very careful to modify Z_p^+ when the decision maker insists on changing Z_p^+, because reduction of the upper bound Z_p^+ can lead to an inefficient solution (Figure 8.7).

(2) For the constraints $\leq, \geq, =$, the decrease of δ_i is an acceptable modification which can guarantee an efficient solution in the recalculated compromise solution step. The consequence of an increase of δ_i with $\leq, \geq, =$ constraints might be, for example, that the feasible solution set increases and new possible solutions are included in the investigation (Figure 8.8).

8.5.2 Application of the Interactive Method

8.5.2.1 Interactive method to Solve the Multi-objective Fuzzy Reliability Optimization Problem (MOFROP)

To solve the above defined model (Equation 8.11), using Section (8.5.1), formulate the table of extreme solutions as follows:

	$R_S(R)$	$H_N(R)$
R_1^+	$R_S^*\left(R_1^+\right)$	$H_N\left(R_1^+\right)$
R_2^+	$R_S\left(R_2^+\right)$	$H_N^*\left(R_2^+\right)$
R_1^-	$R_S\left(R_1^-\right)$	$H_N\left(R_1^-\right)$
R_2^-	$R_S\left(R_2^-\right)$	$H_N\left(R_2^-\right)$

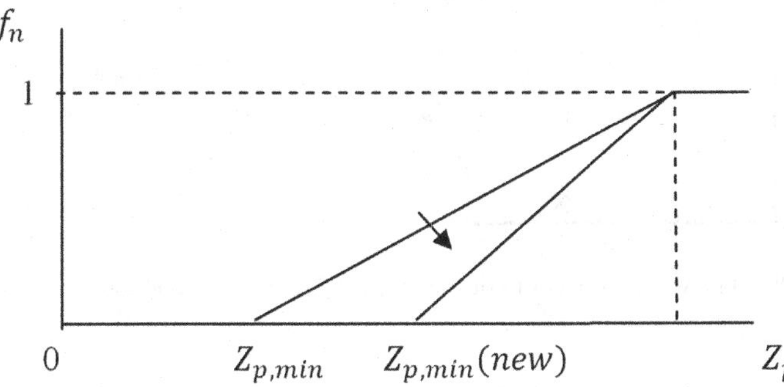

FIGURE 8.7 The effects of changing $Z_{p,\min}$.

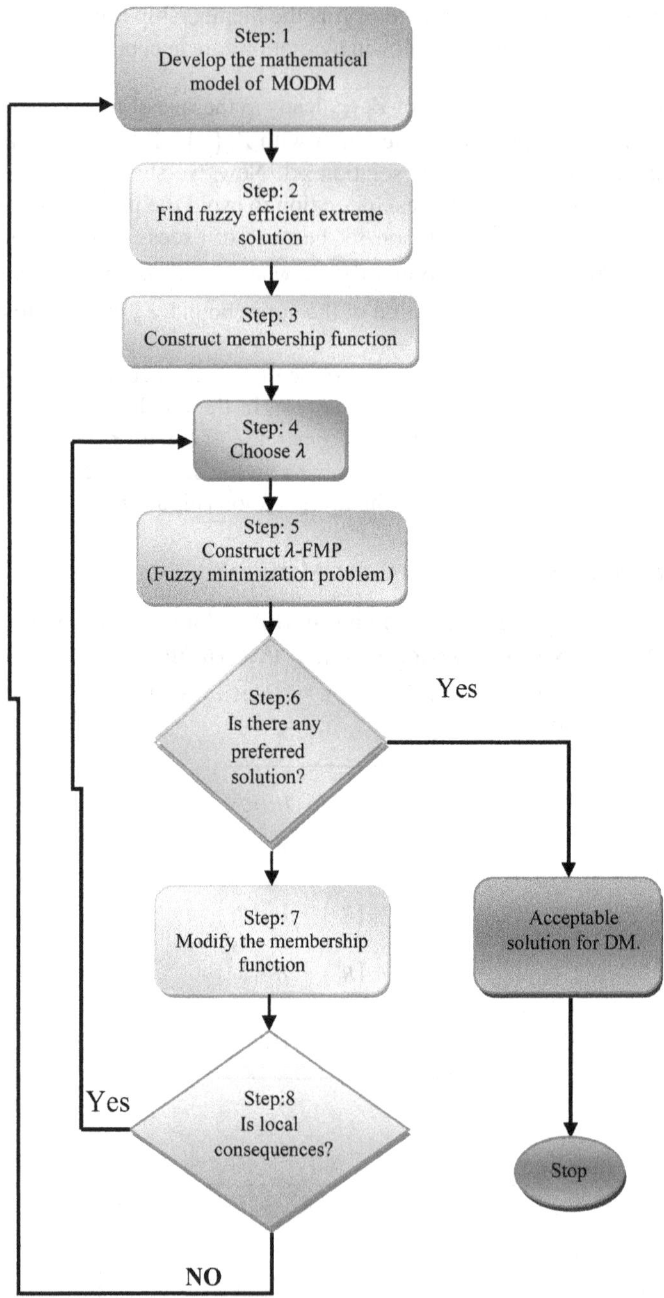

FIGURE 8.8 The whole procedure of the proposed interactive method.

Now the optimistic and pessimistic values are identified, which are given by $R_S^*\left(R_1^+\right)$, $H_N^*\left(R_2^+\right)$ and $R_S^{\min}$, $H_N^{\min}$ respectively.

where $R_S^{\min} = \min\left[R_S\left(R_1^+\right), R_S\left(R_2^+\right), R_S\left(R_1^-\right), R_S\left(R_1^-\right)\right]$

and $H_N^{\min} = \min\left[H_N\left(R_1^+\right), H_N\left(R_2^+\right), H_N\left(R_1^-\right), H_N\left(R_2^-\right)\right]$

Here, the linear membership functions for the objectives $R_S(R)$, $H_N(R)$ and constraints $C_S(R)$, $V_S(R)$, $W_S(R)$ are defined as follows:

$$f_{R_S}^{w_1}(R_S(R)) = \begin{cases} 0 & \text{if } R_S(R) < R_S^{\min}; \\ w_1\left(\dfrac{R_S(R) - R_S^{\min}}{R_S^*\left(R_1^+\right) - R_S^{\min}}\right) & \text{if } R_S^{\min} \le R_S(R) \le R_S^*\left(R_1^+\right); \\ w_1 & \text{if } R_S^*\left(R_1^+\right) < R_S(R); \end{cases} \tag{8.15}$$

$$f_{H_N}^{w_2}(H_N(R)) = \begin{cases} 0 & \text{if } H_N(R) < H_N^{\min}; \\ w_2\left(\dfrac{H_N(R) - H_N^{\min}}{H_N^*\left(R_2^+\right) - H_N^{\min}}\right) & \text{if } H_N^{\min} \le H_N(R) \le H_N^*\left(R_2^+\right); \\ w_2 & \text{if } H_N^*\left(R_2^+\right) < H_N(R); \end{cases} \tag{8.16}$$

$$f_{C_S}(C_S(R)) = \begin{cases} 0 & \text{if } C_{\lim} + \delta_{C_S} < C_S(R); \\ 1 - \left(\dfrac{C_S(R) - C_{\lim}}{\delta_{C_S}}\right) & \text{if } C_{\lim} \le C_S(R) \le C_{\lim} + \delta_{C_S}; \\ 1 & \text{if } C_S(R) < C_{\lim}; \end{cases} \tag{8.17}$$

$$f_{V_S}(V_S(R)) = \begin{cases} 0 & \text{if } V_{\lim} + \delta_{V_S} < V_S(R); \\ 1 - \left(\dfrac{V_S(R) - V_{\lim}}{\delta_{V_S}}\right) & \text{if } V_{\lim} \le V_S(R) \le V_{\lim} + \delta_{V_S}; \\ 1 & \text{if } V_S(R) < V_{\lim}; \end{cases} \tag{8.18}$$

$$f_{Q_S}(Q_S(R)) = \begin{cases} 0 & \text{if } Q_{\lim} + \delta_{Q_S} < Q_S(R); \\ 1 - \left(\dfrac{Q_S(R) - Q_{\lim}}{\delta_{Q_S}}\right) & \text{if } Q_{\lim} \le Q_S(R) \le Q_{\lim} + \delta_{Q_S}; \\ 1 & \text{if } Q_S(R) < Q_{\lim}; \end{cases} \tag{8.19}$$

Now after electing the membership functions, according to step 5 in Section (8.5.1), the crisp model is formulated as follows

$$\text{Maximize } \lambda$$

$$\text{subject to, } W_1\, w = \left(\frac{R_S(R) - R_S^{\min}}{R_S^*(R_1^+) - R_S^{\min}} \right) \geq \lambda$$

$$W_2\, w \left(\frac{H_N(R) - H_N^{\min}}{H_N^*(R_2^+) - H_N^{\min}} \right) \geq \lambda$$

$$\delta_{C_S}(1-\lambda) - \left(\sum_{j=1}^{N} \alpha_j \left[\frac{-t}{(m+1)\ln(R_j)} \right]^{\beta_j} \left[x_j + \exp\left(\frac{x_j}{4} \right) \right] - I(C_{\lim}) \right) \geq 0$$

$$\delta_{V_S}(1-\lambda) - \left(\sum_{j=1}^{N} v_j x_j^{\,2} - I(V_{\lim}) \right) \geq 0$$

$$\delta_{Q_S}(1-\lambda) - \left(\sum_{j=1}^{N} q_j x_j \exp\left(\frac{x_j}{4} \right) - I(Q_{\lim}) \right) \geq 0 \tag{8.20}$$

$$0.5 \leq R_{j,\min} \leq R_j \leq 1,\ 0 \leq R_s \leq 1, \text{ for } j = 1,2,\ldots,N.$$

$$\lambda \in [0, w],\ \sum_{p=1}^{2} W_p = 1.$$

where $I(C_{\lim})$, $I(V_{\lim})$, $I(Q_{\lim})$ denotes the integral value of the resources cost, volume, and weight, respectively.

8.6 NUMERICAL EXAMPLE

A two-stage reliability redundancy allocation problem with entropy as an additional objective function is considered for numerical exposure. The problem is as follows:

$$\text{Maximize } R_s(R) = \prod_{j=1}^{2} \left\{ 1 - (1 - R_j)^{x_j} \right\}$$

$$\text{Maximize } H_N(R) = -\sum_{j=1}^{2} \left(\frac{x_j}{\sum_{j=1}^{2} x_j} \right) \ln \left(\frac{x_j}{\sum_{j=1}^{2} x_j} \right)$$

$$\text{Subject to, } C_S(R) = \sum_{j=1}^{2} \alpha_j \left[\frac{-t}{(m+1)\ln(R_j)} \right]^{\beta_j} \left[x_j + \exp\left(\frac{x_j}{4} \right) \right] \le C_{\lim}$$

$$V_S(R) = \sum_{j=1}^{2} v_j x_j^2 \le V_{\lim}$$

$$Q_S(R) = \sum_{j=1}^{2} q_j x_j \exp\left(\frac{x_j}{4} \right) \le Q_{\lim}$$

$$0.5 \le R_{j,\min} \le R_j \le 1,\ 0 \le R_s \le 1,\ \text{for}\ j = 1,2.$$

$$x_j \ge 1,\ j = 1,2,\ldots, x_j \text{ being integer \& } m = 0. \tag{8.21}$$

Input data for the problem 8.21 are given in the next table.

Now according to step 2 the pay-off matrix is formulated as follows:

	$R_S(R)$	$H_N(R)$
R_1^+	0.9963024	0.6761873
R_2^+	0.8814947	0.6931472
R_1^-	0.9949039	0.6765330
R_2^-	0.8814947	0.6931472

Here $R_S^*(R_1^+) = 0.9963024$, $H_N^*(R_2^+) = 0.6931472$, and $R_S^{\min} = 0.8814947$, $H_N^{\min} =$ 0.6761873 are identified.

Considering the GTrFN for the resources $C_S(R)$, $V_S(R)$ and $Q_S(R)$ of the MOFROP and taking the following fuzzy input data instead of the crisp coefficient, the other data remain the same as given in Tables 8.2 and 8.3.

Comparative Analysis: Table 8.4

Discussion: It is observed that in all three approaches when the DM gives more preference to the entropy function than the reliability function, then $R_S^*(R)$ is maximum and $H_N^*(R)$ is minimum and when the decision maker supplies more preference

TABLE 8.2
Input data for model (6.1)

R_1	R_2	α_1	α_2	β_1	β_2	v_1	v_2	q_1	q_2	$C_{\lim}$	$V_{\lim}$	$Q_{\lim}$	t
0.85	0.95	0.1	0.1	0.15	0.15	3	4	7	7	100	45	200	1000

TABLE 8.3
Input Data for MOFROP

$C_{\lim}$	$V_{\lim}$	$Q_{\lim}$
(80,90,110,120; 0.7)	(35, 40, 50, 55; 0.8)	(175,200,210,215;0.8)

TABLE 8.4
Optimal Solutions for Different Weightages of System Reliability (W_1) and Entropy Functions (W_2) for MOFROP (for k = 0.5)

Method	Weight	x_1^*	x_2^*	R_S^*	H_N^*
Fuzzy multi-objective nonlinear programming (FMONLP) method	$W_1 = 0.8$ $W_2 = 0.2$	1	1	0.9099473	0.6931472
	$W_1 = 0.5$ $W_2 = 0.5$	2	2	0.9772956	0.6905676
	$W_1 = 0.2$ $W_2 = 0.8$	2	2	0.9788409	0.6816326
Fuzzy multi-objective goal programming (FMOGP) method	$W_1 = 0.8$ $W_2 = 0.2$	2	2	0.9579275	0.6931472
	$W_1 = 0.5$ $W_2 = 0.5$	3	2	0.9829743	0.6914693
	$W_1 = 0.2$ $W_2 = 0.8$	2	2	0.9864241	0.6801339
Interactive fuzzy multi-objective decision-making (IFMODM) method	$W_1 = 0.8$ $W_2 = 0.2$	2	2	0.9645298	0.6931472
	$W_1 = 0.5$ $W_2 = 0.5$	3	3	0.9934568	0.6927268
	$W_1 = 0.2$ $W_2 = 0.8$	3	2	0.9960115	0.6804166

to the reliability function than the entropy function, then $R_S^*(R)$ is minimum and $H_N^*(R)$ is maximum. It is also observed that, in each case, the obtained reliability of the system in the proposed IFMODM approach is much better than the result obtained in the FMONLP and FMOGP approaches.

The optimal solutions of MOFROP with equal weights by the IFMODM method for different values of k are presented in Table 8.5. It can be seen that as k decreases, the reliability R_S^* and the entropy H_N^* also decreases.

Discussion: From the above discussion it can be concluded that the reasons for considering entropy as an additional objective function and then solving the above-mentioned multi-objective optimization problem using an interactive method are as follows:

TABLE 8.5
Optimal Solutions of MOFROP with Equal Weights by IFMODM Method for Different Values of k

Test	x_1^*	x_2^*	R_S^*	H_N^*
Optimistic i.e. $k = 1$	3	3	0.9953258	0.6930030
About optimistic i.e. $k = 0.7$	3	3	0.9943126	0.6928532
Moderate i.e. $k = 0.5$	3	3	0.9934568	0.6927268
About pessimistic i.e. $k = 0.2$	3	2	0.9918161	0.6924845
Pessimistic i.e. $k = 0$	3	2	0.9904097	0.6922767

(1) For a redundancy allocation problem, entropy acts as a measure of dispersal of allocation between stages. So it will be more realistic and potentially useful if we have the maximum amount of entropy and maximum system reliability.
(2) The proposed method is an efficient and modified optimization technique and gives a highly reliable system than other existing methods. When solving a nonlinear programming problem, the interactive method considers a large variety of situations that the decision maker (DM) might encounter. In this method the DM may modify the original model continuously to obtain a satisfactory solution until the decision maker is satisfied with the obtained result at each stage.
(3) And also comparing the result obtained in the proposed interactive method with the result obtained in the FMONLP and FMOGP approaches, it is observed that the interactive method gives a better reliable system than the other methods. Thus we think that the proposed method is more efficient, effective, and a powerful tool to solve the above discussed multi-objective problem.

8.7 CONCLUSIONS

In this chapter, a reliability redundancy allocation problem of a series-parallel system with entropy as an additional objective function is considered. The problem is then solved by the IFMODM method. Here the available cost, weight, and volume of each component of the system are taken as generalized trapezoidal fuzzy numbers and the fuzzy number is ranked by the integral value method. Here the result obtained from the proposed interactive method is compared with the result obtained from FMONLP and FMOGP approaches. The optimal solutions are also presented here due to the different preferences on objective functions by the decision maker.

It is also discussed that when the number of redundant components increases, the system reliability increases and the entropy of the system decreases. The aim of this chapter is to find the optimum number of redundant components of the proposed entropy-based fuzzy reliability optimization problem and to maximize the entropy amount subject to the available resources.

FUTURE RESEARCH

From this chapter a lot of scope may arise for further research work.

- All the multi-objective models developed in this thesis contain only two objective functions, viz. system reliability and cost (or system reliability and entropy). In the future, one may extend the reliability models for higher objectives and then solve them by the proposed optimization technique.
- In this chapter, we have considered all the available resources like cost, weight, and volume of each component of the system as a trapezoidal fuzzy number (TrFN), but in the future, consideration of other types of fuzzy numbers like pentagonal fuzzy number, hexagonal fuzzy number, etc. and also intuitionistic and neutrosophic fuzzy number may be considered for evaluation of the reliability model.
- Also these problems can be solved by various search algorithms like genetic algorithm, evolutionary algorithms, particle swam optimization, etc.
- All the methods introduced here are quite general and can be applied to the multi-objective problem in other areas of engineering sciences and operation research, like, assignment problems, structural optimization, inventory problems, transportation problems, etc.

ACKNOWLEDGMENT

The authors are thankful to the University of Kalyani for providing financial assistance through DST-PURSE (Phase-II) Programme and UGC SAP (Phase-II).

REFERENCES

Allella, F., Chiodo, E. & Lauria, D. 2005. Optimal reliability allocation under uncertain conditions, with application to hybrid electric vehicle design. *International Journal of Quality and Reliability Management* 22(6): 626–641.

Bellman, R.E. & Zadeh, L.A. 1970. Decision-making in a fuzzy environment. *Management Science* 17(4): B141–B164.

Boland, P.J. & EL-Neweihi, E. 1995. Component redundancy vs. system redundancy in the hazard rate ordering. *IEEE Transactions on Reliability* 44(4): 614–619.

Bulfin, R.L. & Liu, C.Y. 1985. Optimal allocation of redundant components for a large systems. *IEEE Transactions on Reliability* 34(3): 241–247.

Caserta, M. & Nodar, M.C. 2009. A cross-entropy based algorithm for reliability problems. *Journal of Heuristics* 15(5): 479–501.

Chen, S.H. 1985. Operations on fuzzy members with function principal. *Tamkang Journal of Management Science* 6(1): 13–25.

Chen, S.H. 1985. Ranking fuzzy numbers with maximizing set and minimizing set. *Fuzzy Sets and Systems* 17(2): 113–129.

Chen, A., Yang, H., Lo, H.K. & Tang, W.H. 1999. A capacity related reliability for transportation networks. *Journal of Advanced Transportation* 33(2): 183–200.

Chern, M.S. & Jan, R.H. 1985. Parametric programming applied to reliability optimization problems. *IEEE Transactions on Reliability* 34(2): 165–170.

Chern, M.S. 1992. On the computational complexity of reliability redundancy allocation in a series system. *Operations Research Letters* 11(5): 309–315.

Coit, D.W. & Konak, A. 2006. Multiple weighted objectives heuristic for the redundancy allocation problem. *IEEE Transactions on Reliability* 55(3): 551–558.

Coit, D.W. & Smith, A.E. 1998. Redundancy allocation to maximize a lower percentile of the system time-to-failure distribution. *IEEE Transactions on Reliability* 47(1): 79–87.

Dancese, M., Abbas, F. & Ghamry, E. 2014. Reliability and cost analysis of a series system model using fuzzy parametric geometric programming. *International Journal of Innovative Science, Engineering and Technology* 1(8).

Garg, H. 2013. An approach for analyzing fuzzy system reliability using particle swarm optimization and intuitionistic fuzzy set theory. *Journal of Multiple-Valued Logic and Soft Computing* 21(3): 335–354.

Garg, H. 2013. Fuzzy multi-objective reliability optimization problem of industrial system using particle swarm optimization. *Journal of Industrial Mathematics* 2013: Article id: 872450.

Garg, H. 2015. Multi-objective optimization problem of system reliability under intuitionistic fuzzy set environment using cuckoo Search algorithm. *Journal of Intelligent and Fuzzy Systems* 29(4): 1653–1669.

Garg, H. 2016. A novel approach for analyzing the reliability of series-parallel system using credibility theory and different types of intuitionistic fuzzy numbers. *Journal of the Brazillian Society of Mechanical Science and Engineering* 38(3): 1021–1035.

Garg, H. 2017. Performance analysis of an industrial system using soft computing based hybridized technique. *Journal of the Brazillian Society of Mechanical Science and Engineering* 39(4): 1441–1451.

Garg, H. & Rani, M. 2013. An approach for reliability analysis of industrial systems using PSO and IFS technique. *ISA Transaction* (Elsevier) 52(6): 701–710.

Garg, H., Rani, M. & Sharma, S.P. 2013. Reliability analysis of the engineering system using intuitionistic fuzzy set theory. *International Journal of Quality and Reliability Engineering, Hindawi.* Article id: 943972. 10 pages.

Garg, H., Rani, M. & Sharma, S.P. 2014. An approach for analyzing the reliability of industrial systems using soft-computing based technique. *Experts Systems with Applications* 41(2): 489–501.

Garg, H., Rani, M., Sharma, S.P. & Viswakarma, Y. 2014. Bi-objective optimization of the reliability redundancy allocation problem for series-parallel system. *Journal of Manufacturing Systems* 33(3): 335–347.

Garg, H., Rani, M., Sharma, S.P. & Viswakarma, Y. 2014. Intuitionistic fuzzy optimization technique for solving multi-objective reliability optimization problem in interval environment. *Expert Systems with Applications* 41(7): 3157–3167.

Garg, H. & Sharma, S.P. 2013. Multi-objective reliability-redundancy allocation problem using particle swarm optimization. *Computers and Industrial Engineering* 64(1): 247–255.

Ghare, P.M. & Taylor, R.E. 1969. Optimal redundancy for reliability in series systems. *Operations Research* 17(5): 838–847.

Gong, B., Chen, X. & Hu, C. 2012. Fuzzy entropy clustering approach to evaluate the reliability of emergency logistics system. *Energy Procedia* 16: 278–283.

Islam, S. & Roy, T.K. 2006. A new fuzzy multi-objective programming: Entropy based geometric programming and its application of transportation problems. *European Journal of Operational Research* 173(2): 387–404.

Islam, S. & Roy, T.K. 2010. Multi-objective transportation problem with an additional entropy objective function in fuzzy environment. *Journal of Fuzzy Math* 18(2): 1–24.

Kang, H.Y. & Kwak, B.M. 2009. Application of maximum entropy principle for reliability-based design optimization. *Structural and Multidisciplinary Optimization* 38(4): 331–346.

Kapur, J.N. 1993. *Maximum-Entropy Models in Science and Engineering.* Wiley Eastern Limited, New Delhi.

Kim, H., Bae, C. & Park, D. 2006. Reliability-redundancy optimization using simulated annealing algorithms. *Journal of Quality in Maintenance Engineering* 12(4): 354–363.

Kim, J. & Yum, B. 1993. A heuristic method for solving redundancy optimization problems in complex systems. *IEEE Transactions on Reliability* 42(4): 572–578.

Kundu, T. & Islam, S. 2019. A new interactive approach to solve entropy based fuzzy reliability optimization model. *International Journal of Interactive Design and Manufacturing* 13(1): 137–146.

Kuo, W. & Prasad, V.R. 2000. An annotated overview of system-reliability optimization. *IEEE Transactions on Reliability* 49(2): 176–187.

Kuo, W., Prasad, V.R., Tillman, F.A. & Hwang, C. 2001. *Optimal Reliability Design – Fundamentals and Applications.* Cambridge University Press, Cambridge, United Kingdom.

Kuo, W. & Wan, R. 2007. Recent advances in optimal reliability allocation. *Computational Intelligence in Reliability Engineering* 39: 1–36.

Lai, Y.J. & Hwang, C.L. 1992. Interactive fuzzy linear programming. *Fuzzy Sets and Systems* 45(2): 169–183.

Lai, Y.J. & Hwang, C.L. 1994. *Fuzzy Multiple Objective Decision Making – Methods and Application.* Springer-Verlag, Berlin, Heidelberg.

Liang, Y. & Chen, Y. 2007. Redundancy allocation of series-parallel systems using a variable neighborhood search algorithm. *Reliability Engineering and System Safety* 92(3): 323–331.

Liang, Y. & Smith, A.E. 2004. An ant colony optimization algorithm for the redundancy allocation problem (RAP). *IEEE Transactions on Reliability* 53(3): 417–423.

Liou, T.S. & Wang, M.J.J. 1992. Ranking fuzzy numbers with integral value. *Fuzzy Sets and Systems* 50(3): 247–255.

Mahapatra, G.S. 2009. Reliability optimization of entropy based series-parallel system using global criterion method. *Intelligent Information Management* 1(3): 145–149.

Mahapatra, G.S., Mahapatra, B.S. & Roy, P.K. 2011. Fuzzy decision-making on reliability of series system: Fuzzy geometric programming approach. *Annals of Fuzzy Mathematics and Informatics* 1(1): 107–118.

Mettas, A. 2000. Reliability allocation and optimization for complex systems. In *Proceedings: Annual Reliability and Maintainability Symposium, Los Angeles, California, USA* 2000, January 24–27, IEEE.

Misra, K.B. 1975. Optimal reliability design of a system containing mixed redundancies. *IEEE Transactions on Power Apparatus Systems (PAS)* 94: 983–993.

Niwas, R. & Garg, H. 2018. An approach for analyzing the reliability and profit of an industrial system based on cost free warranty policy. *Journal of the Brazilian Society of Mechanical Science and Engineering (Springer)* 40: 1–9.

Onishi, J., Kimura, S., James, R.J.W. & Nakagawa, Y. 2007. Solving the redundancy allocation problem with a mix of components using the improved surrogate constraint method. *IEEE Transactions on Reliability* 56(1): 94–101.

Park, K.S. 1987. Fuzzy apportionment of system reliability. *IEEE Transaction on Reliability* 36(1): 129–132.

Ramirez-Marquez, J.E. & Coit, D.W. 2004. A heuristic for solving the redundancy allocation problem for multi-state series- parallel systems. *Reliability Engineering and System Safety* 83(3): 341–349.

Ridder, A.D. 2005. Importance sampling simulations of Markovian reliability systems using cross entropy. *Annals of Operation Research* 134(1): 119–136.

Sakawa, M. 1978. Multiobjective reliability and redundancy optimization of a series-parallel system by the surrogate worth trade off methods. *Microelectronics Reliability* 17(4): 465–467.

Sakawa, M. & Yano, H. 1987. An interactive satisficing method for multi-objective nonlinear problems with fuzzy parameters. *Fuzzy Sets and Systems* 30: 221–238.

Tillman, F.A., Hwang, C.L. & Kuo, W. 1977. Optimization technique for system reliability with redundancy: A review. *IEEE Transactions on Reliability* 26: 148–155.

Tillman, F.A., Hwang, C.L. & Kuo, W. 1977. Determining component reliability and redundancy for optimum system reliability. *IEEE Transactions on Reliability* 26(3): 162–165.

Wang, G.B., Huang, H.Z., Liu, Y., Zhang, X. & Wang, Z. 2009. Uncertainty estimation of reliability redundancy in complex system based on the cross-entropy method. *Journal of Mechanical Science and Technology* 23(10): 2612–2623.

Zadeh, L.A. 1965. Fuzzy sets. *Information and Control* 8(3): 338–353.

Zangiabadi, M. & Maleki, H.R. 2007. A method for solving linear programming problems with fuzzy parameters based on multiobjective linear programming technique. *Asia-Pacific Journal of Operational Research* 24(4): 557–573.

Zimmermann, H.J. 1996. *Fuzzy Set Theory – And Its Applications.* Allied Publishers Limited, New Delhi.

9 Multiple Constrained Reliability-Redundancy Optimization under Triangular Intuitionistic Fuzziness Using a Genetic Algorithm

R. Paramanik, S. K. Mahato, N. Bhattacharyee, P. Supakar, and B. Sarkar

CONTENTS

9.1 INTRODUCTION

The reliability theory was originally developed to meet the demands of modern technology, particularly due to the experiences with complex military systems in World War II. In military systems the problems that were encountered earlier were machine maintenance and system reliability. Reliability is included in almost every engineering system design,. Different reliability system designs were studied by Tillman et al. [1], Misra [2], Kuo et al.[3], and many other researchers. The aim was to achieve and establish the reliability of those systems or components as high as possible. Due to some technical errors, failures of the equipment or systems can lead to severe damages along with loss of property as well as life, and to overcome such situations it is desirable to ensure the proper functioning of the system up to the targeted duration. The aim, therefore, is to maximize the reliability of the system under consideration.

The reliability of a system can be maximized in many ways such as increasing the reliabilities of the components or adding some identical extra components to the system as well as allowing the maximum bounds of the constraints in terms of volume, cost, and weight. Redundancy allocation is very effective and also popular in the optimization of the reliability of a system. In redundancy allocation, some extra component(s) are installed in the system to maximize the system reliability by considering the constraints of the system.

Most of the researchers considered the design parameters/variables as having fixed values, indicating that the probability can be determined perfectly. The complete information/data of the system and the components in terms of their performance are generally available, although, in reality, in most of the cases, no sufficient statistical data are available, particularly for a new system/problem. For many problems, only some partial information is known because it is not possible to cover all dimensions of the performance of the system or components in terms of statistical data. So it is more justifiable to consider the reliability of a component of a system as an imprecise number. The impreciseness can be assumed in various ways such as interval number, fuzzy number, intuitionistic fuzzy number, stochastic approaches, and a combination of these.

The objective of this chapter is to study a reliability system in crisp and imprecise environments and to utilize genetic algorithms to solve it. The study also analyzes the use of several crispification methods for intuitionistic fuzzy numbers and the use of the penalty function method to reduce the constrained combinatorial optimization problem to an unconstrained one. Finally the optimal redundancy allocation is found so that the system reliability is maximum.

In this chapter, we discuss the optimization of system reliability for an n-stage series system with redundant units in parallel. The corresponding problem is formulated in crisp and intuitionistic fuzzy environments. In the intuitionistic fuzzy environment, we consider two situations. In the first case, only the parameters are taken as triangular intuitionistic fuzzy numbers while the reliability of each component of the system is considered as precise numbers. In the second case, the component reliabilities and all the parameters are taken as triangular intuitionistic fuzzy numbers. Thus, we obtain three types of models, viz., the crisp, the intuitionistic fuzzy (model-1), and the fully intuitionistic fuzzy (model-2). The intuitionistic fuzzy models 1 and 2 are more realistic in the sense of uncertainty of the real-life phenomena. We use the three different crispification methods, viz., the (α, β)-cut method, the Graded

Mean Integration Value (GMIV) method, and the ranking function method after extending them to intuitionistic fuzzy numbers. Then these problems are formulated as unconstrained integer programming problems with the help of the Big-M penalty technique. The transformed problems are solved by the real coded elitist genetic algorithm (RCEGA) for integer variables, tournament selection, intermediate crossover, and one-neighborhood mutation. Comparative studies of the results obtained are presented and also the sensitivity studies are carried out and presented in a tabular form as well as graphically.

9.2 BACKGROUND/LITERATURE REVIEW

Many researchers studying reliability engineering have tried to solve the reliability–redundancy allocation problems (RAP) which mainly occur as nonlinear integer or mixed integer programming problems [4–11].

The reliabilities at the component level and/or redundancy allocation can be considered as the decision variables of the problem. Several researchers have implemented different methods to solve several types of reliability optimization problems.

In the initial stage of development, various deterministic methods such as heuristic methods [12,13], reduced gradient method [14], linear programming approach [12], dynamic programming method [8], and branch and bound method [15] are used for solving this type of redundancy allocation problem. Later, evolutionary algorithms were being used to solve these problems. Evolutionary algorithms are more flexible in terms of assumptions on the objective function and on the constraints. Evolutionary algorithms are found to perform well both in the discrete and continuous search space.

Several works in reliability–redundancy optimization in a fuzzy environment have been encountered in the literature. Bourezg and Meglouli [16] have presented the reliability analysis of a power distribution system using a disjoint path-set algorithm. Huang [17] has presented the optimization of a series system in a fuzzy environment considering the concept of multi-objective decision making. Also, Chen [18] has reported on a fuzzy reliability model of the bridge system. Mahapatra and Roy [19] have also used the fuzzy multi-objective technique on reliability optimization. Mahapatra and Roy [20] have extended their work for reliability evaluation using intuitionistic fuzzy numbers. Mahapatra and Roy [21] have presented the optimal redundancy allocation using a generalized fuzzy number in a series-parallel system. Mahapatra and Roy [9] have also considered a bridge system with fuzzy reliability of the components using interval nonlinear programming. Mahato et al. [22] have discussed fuzzy reliability-redundancy optimization with the signed distance method for defuzzification.

Reliability optimization problems are reported in several ways, such as component reliability allocation, redundancy allocation, cost minimization, bi-objective modeling, and multi-objective modeling in several environments like fuzzy, interval, stochastic, intuitionistic fuzzy, etc. Garg et al. [23] formulated the bi-objective reliability–redundancy allocation problem. Garg and Sharma [24] formulated and solved the multi-objective reliability–redundancy allocation problem using particle swarm optimization.

Several methods and techniques are utilized to solve different reliability optimization problems. Soft computing techniques have worked successfully in reliability–redundancy allocation problems. Dolatshahi-Zand and Khalili-Damghani [25], Khalili-Damghani et al. [26], Garg and Rani [27], Garg and Sharma [24], Garg et al. [28] and Garg [29] have used the particle swarm optimization technique to solve reliability design problems. The genetic algorithm is used for reliability optimization by Gupta et al. [30], Mahato et al. [22,31], and Sahoo et al. [32,33]. Garg et al. [34] implemented the ABC algorithm for analyzing the reliability of the industrial system.

The reliability optimization models with intuitionistic fuzzy type uncertainty are yet to be studied to a great extent. However, some commendable works are reported in the literature [27,29,35–38]. Garg [35] presented the reliability analysis for different types of intuitionistic fuzzy numbers using the credibility theory. The PSO and IFS technique is presented for reliability analysis by Garg and Rani [27]. Again, Garg [29] used PSO and the intuitionistic fuzzy set theory for a fuzzy system. The intuitionistic fuzzy set theory is used in reliability analysis in the work of Garg et al. [36]. Later, Garg [37] implemented the Cuckoo search algorithm for multi-objective reliability optimization in an intuitionistic fuzzy environment. Prior to this work, Garg et al. [38] presented the intuitionistic fuzzy technique for multi-objective reliability optimization with interval type uncertainty.

Though several works are available in different environments like fuzzy, interval, and stochastic for reliability optimization, only few researchers have attempted

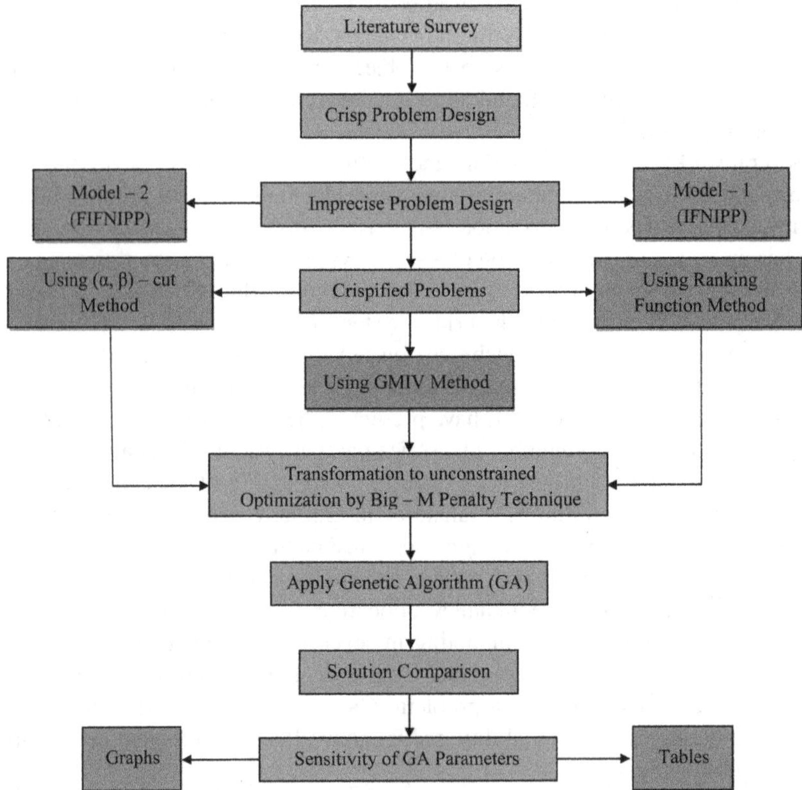

FIGURE 9.1 Flowchart of the contribution of the chapter.

the reliability–redundancy optimization in an intuitionistic fuzzy environment. Triangular intuitionistic fuzzy numbers are considered in reliability optimization through the arithmetic operations of the TIFN [20]. In this chapter we developed some new techniques of crispification of intuitionistic fuzzy numbers. We are the first to report these crispification techniques, and these methods have the advantage of not requiring the direct arithmetic operations of IFNs.

The rest of the chapter is described in the various sections whose interrelated connections are shown in the form of the flowchart given in Figure 9.1.

9.3 NOMENCLATURES AND ASSUMPTIONS

The nomenclatures and assumptions used in this chapter are given below.

9.3.1 Nomenclatures

n	number of subsystems
x_{ij}	number of redundant components of the design alternative j in stage i
R_{ij}	reliability of the design alternative j in stage i
$\tilde{R}_{ij}$	intuitionistic fuzzy reliability of the design alternative j in stage i
$R_S(x), \tilde{R}_S(x)$	crisp, intuitionistic fuzzy system reliability (objective function)
c_{ij}	cost of the design alternative j in stage i
$\tilde{c}_{ij}$	intuitionistic fuzzy cost of the design alternative j in stage i
w_{ij}	weight of the design alternative j in stage i
$\tilde{w}_{ij}$	intuitionistic fuzzy weight of the design alternative j in stage i
C_0	cost boundary
$\tilde{C}_0$	intuitionistic fuzzy cost boundary
W_0	weight boundary
$\tilde{W}_0$	intuitionistic fuzzy weight boundary
l_i	number of components in stage i
l_{ij}	lower bound of the redundant components of design alternative j in stage i
u_{ij}	upper bound of the redundant components of design alternative j in stage i
S	feasible region
$prob_{cross}$	probability of crossover
$prob_{mute}$	probability of mutation
max_{gen}	maximum number of generation
pop_{size}	population size

9.3.2 Assumptions

- The components' reliabilities are precise/intuitionistic fuzzy valued depending on the model.
- The failures of the components are independent of each other.
- All the redundancies are active and non-repairable.
- The cost coefficients are intuitionistic fuzzy valued.

9.4 MATHEMATICAL MODEL

9.4.1 The Crisp Model

An n-stage series system is considered in which the redundant units are connected in parallel as shown in Figure 9.2. Different types of components can be used as design alternatives in each stage. Then, a nonlinear integer programming problem is formulated as:

$$\text{Maximize } R_S = \prod_{i=1}^{n}\left[1-\prod_{j=1}^{l_i}\left(1-R_{ij}\right)^{x_{ij}}\right]$$

$$\text{Subject to } \sum_{i=1}^{n}\sum_{j=1}^{l_i} c_{ij}x_{ij} \le C_0 \qquad (9.1)$$

$$\sum_{i=1}^{n}\sum_{j=1}^{l_i} w_{ij}x_{ij} \le W_0$$

$x_{ij} \ge 0$ and are integers, $1 \le i \le n, 1 \le j \le l_i$.

9.4.2 Intuitionistic Fuzzy Model

Generally, it is assumed before solving a reliability optimization problem that the coefficients of components are precisely known in advance.

But in real-world situations, several diverse situations occur such as uncertain judgments, unpredictable conditions or human errors, incomplete knowledge and information, etc., due to which it is not possible to get relevant precise data for the reliability system. Such impreciseness of any data can be represented in different ways. Among these, one way is to represent the imprecise data by an intuitionistic fuzzy number. Therefore, problem (9.1) can be reformulated as an intuitionistic fuzzy nonlinear integer programming problem as follows.

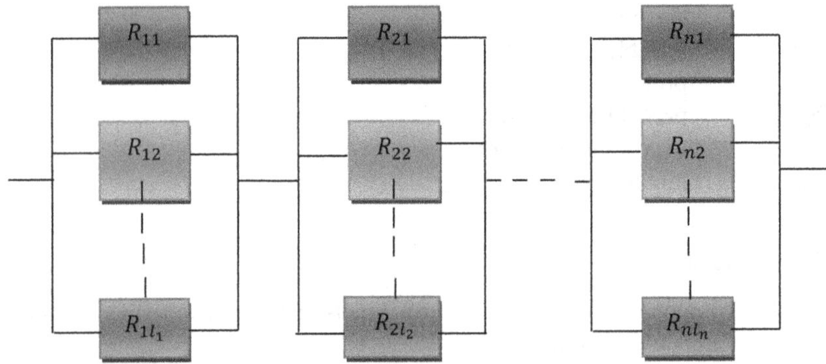

FIGURE 9.2 n-stage series system.

Model-1: Intuitionistic Fuzzy Nonlinear Integer Programming Problem (IFNIPP)

In this model, the cost and weight of the design alternative j in stage i, the system cost, and the system weight boundaries are considered as imprecise and assumed as TIFNs.

$$\text{Maximize } R_S = \prod_{i=1}^{n}\left[1-\prod_{j=1}^{l_i}\left(1-R_{ij}\right)^{x_{ij}}\right] \tag{9.2}$$

$$\text{subject to } \sum_{i=1}^{n}\sum_{j=1}^{l_i}\left(c_{ij}^{1},c_{ij}^{2},c_{ij}^{3};c_{ij}^{4},c_{ij}^{2},c_{ij}^{5}\right)x_{ij} \le \left(C_0^{1},C_0^{2},C_0^{3};C_0^{4},C_0^{2},C_0^{5}\right)$$

$$\sum_{i=1}^{n}\sum_{j=1}^{l_i}\left(w_{ij}^{1},w_{ij}^{2},w_{ij}^{3};w_{ij}^{4},w_{ij}^{2},w_{ij}^{5}\right)x_{ij} \le \left(W_0^{1},W_0^{2},W_0^{3};W_0^{4},W_0^{2},W_0^{5}\right)$$

$$x_{ij} \ge 0 \text{ and } x_{ij} \text{ are integers, } 1 \le i \le n, 1 \le j \le l_i.$$

Model-2: Fully Intuitionistic Fuzzy Nonlinear Integer Programming Problem (FIFNIPP)

In this model, the reliability components, the cost and weight of design alternative j in stage i, the system cost, and the system weight boundaries are considered as imprecise in terms of TIFNs.

$$\text{Maximize } \tilde{R}_S = \prod_{i=1}^{n}\left[1-\prod_{j=1}^{l_i}\left(1-\left(R_{ij}^{2},R_{ij}^{1},R_{ij}^{3};R_{ij}^{4},R_{ij}^{2},R_{ij}^{5}\right)^{x_{ij}}\right)\right]$$

$$\text{subject to } \sum_{i=1}^{n}\sum_{j=1}^{l_i}\left(c_{ij}^{1},c_{ij}^{2},c_{ij}^{3};c_{ij}^{4},c_{ij}^{2},c_{ij}^{5}\right)x_{ij} \le \left(C_0^{1},C_0^{2},C_0^{3};C_0^{4},C_0^{2},C_0^{5}\right)$$

$$\sum_{i=1}^{n}\sum_{j=1}^{l_i}\left(w_{ij}^{1},w_{ij}^{2},w_{ij}^{3};w_{ij}^{4},w_{ij}^{2},w_{ij}^{5}\right)x_{ij} \le \left(W_0^{1},W_0^{2},W_0^{3};W_0^{4},W_0^{2},W_0^{5}\right)$$

$$x_{ij} \ge 0 \text{ and } x_{ij} \text{ are integers, } 1 \le i \le n, 1 \le j \le l_i. \tag{9.3}$$

9.4.2.1 (α, β)-Cut Method

Using the method described in the Appendix, Section III, the problems (9.2) and (9.3) are reduced to the problems (9.4) and (9.5) which are nonlinear integer programming problems with interval valued constraints. Further, we see that the problem (9.5) also has the interval valued objective function. Hence, problem (9.5) is a fully interval optimization problem.

Model-1: Intuitionistic Fuzzy Nonlinear Integer Programming Problem (IFNIPP)

$$\text{Maximize}\, R_S = \prod_{i=1}^{n}\left[1-\prod_{j=1}^{l_i}\left(1-R_{ij}\right)^{x_{ij}}\right] \tag{9.4}$$

$$\text{subject to} \sum_{i=1}^{n}\sum_{j=1}^{l_i}\left[{}^{\alpha}\tilde{c}^{L}, {}^{\alpha}\tilde{c}^{U}\right]x_{ij} \le \left[{}^{\alpha}\tilde{C}_0{}^{L}, {}^{\alpha}\tilde{C}_0{}^{U}\right]$$

$$\sum_{i=1}^{n}\sum_{j=1}^{l_i}\left[{}^{\alpha}\tilde{w}^{L}, {}^{\alpha}\tilde{w}^{U}\right]x_{ij} \le \left[{}^{\alpha}\tilde{W}_0{}^{L}, {}^{\alpha}\tilde{W}_0{}^{U}\right]$$

$$x_{ij} \ge 0 \text{ and } x_{ij} \text{ are integers}, 1 \le i \le n, 1 \le j \le l_i.$$

Model-2: Fully Intuitionistic Fuzzy Nonlinear Integer Programming Problem (FIFNIPP)

$$\text{Maximize}\, \tilde{R}_S = \left[{}^{\alpha}\tilde{R}_S{}^{L}, {}^{\alpha}\tilde{R}_S{}^{U}\right] = \left[\prod_{i=1}^{n}[1-\prod_{j=1}^{l_i}(1-{}^{\alpha}\tilde{R}_{ij}^{L})^{x_{ij}}], \prod_{i=1}^{n}[1-\prod_{j=1}^{l_i}(1-{}^{\alpha}\tilde{R}_{ij}^{U})^{x_{ij}}]\right]$$

$$\sum_{i=1}^{n}\sum_{j=1}^{l_i}\left[{}^{\alpha}\tilde{c}^{L}, {}^{\alpha}\tilde{c}^{U}\right]x_{ij} \le \left[{}^{\alpha}\tilde{C}_0{}^{L}, {}^{\alpha}\tilde{C}_0{}^{U}\right]$$

$$\sum_{i=1}^{n}\sum_{j=1}^{l_i}\left[{}^{\alpha}\tilde{w}^{L}, {}^{\alpha}\tilde{w}^{U}\right]x_{ij} \le \left[{}^{\alpha}\tilde{W}_0{}^{L}, {}^{\alpha}\tilde{W}_0{}^{U}\right]$$

$$x_{ij} \ge 0 \text{ and } x_{ij} \text{ are integers}, 1 \le i \le n, 1 \le j \le l_i. \tag{9.5}$$

For solving problems (9.4) and (9.5), it is indeed necessary to use the definitions of interval order relations to compare the values of the objective function as well as satisfy the constraints with the help of the interval order relations described in the Appendix Section VI ; the best possible values of the objective function is selected after satisfying the constraints up to the desired degree of accuracy.

9.4.2.2 Graded Mean Integration Value (GMIV) Procedure

We also used the method described in the Appendix, Section IV, to reduce the problems given in (9.2) and (9.3) to the crispified forms given in (9.6) and (9.7), respectively. We may use the usual combinatorial techniques to solve problems (9.6) and (9.7).

Model-1: Intuitionistic Fuzzy Nonlinear Integer Programming Problem (IFNIPP)

$$\text{Maximize } R_S = \prod_{i=1}^{n}\left[1-\prod_{j=1}^{l_i}\left(1-R_{ij}\right)^{x_{ij}}\right] \tag{9.6}$$

subject to

$$\sum_{i=1}^{n}\sum_{j=1}^{l_i}\text{GMIV}\left(c_{ij}^{1},c_{ij}^{2},c_{ij}^{3};c_{ij}^{4},c_{ij}^{2},c_{ij}^{5}\right)x_{ij} \le \text{GMIV}\left(C_0^{1},C_0^{2},C_0^{3};C_0^{4},C_0^{2},C_0^{5}\right)$$

$$\sum_{i=1}^{n}\sum_{j=1}^{l_i}\text{GMIV}\left(w_{ij}^{1},w_{ij}^{2},w_{ij}^{3};w_{ij}^{4},w_{ij}^{2},w_{ij}^{5}\right)x_{ij} \le \text{GMIV}\left(W_0^{1},W_0^{2},W_0^{3};W_0^{4},W_0^{2},W_0^{5}\right)$$

$x_{ij} \ge 0$ and x_{ij} are integers, $1 \le i \le n, 1 \le j \le l_i$.

Model-2: Fully Intuitionistic Fuzzy Nonlinear Integer Programming Problem (FIFNIPP)

$$\text{Maximize } \tilde{R}_S = \prod_{i=1}^{n}\left[1-\prod_{j=1}^{l_i}\left(1-\left(\text{GMIV}(\tilde{R}_{ij})\right)\right)^{x_{ij}}\right]$$

subject to

$$\sum_{i=1}^{n}\sum_{j=1}^{l_i}\text{GMIV}\left(c_{ij}^{1},c_{ij}^{2},c_{ij}^{3};c_{ij}^{4},c_{ij}^{2},c_{ij}^{5}\right)x_{ij} \le \text{GMIV}\left(C_0^{1},C_0^{2},C_0^{3};C_0^{4},C_0^{2},C_0^{5}\right)$$

$$\sum_{i=1}^{n}\sum_{j=1}^{l_i}\text{GMIV}\left(w_{ij}^{1},w_{ij}^{2},w_{ij}^{3};w_{ij}^{4},w_{ij}^{2},w_{ij}^{5}\right)x_{ij} \le \text{GMIV}\left(W_0^{1},W_0^{2},W_0^{3};W_0^{4},W_0^{2},W_0^{5}\right)$$

$x_{ij} \ge 0$ and x_{ij} are integers, $1 \le i \le n, 1 \le j \le l_i$.

(9.7)

9.4.3 Ranking Function Method

Using the method described in the Appendix, Section V, we reduced the problems given in (9.2) and (9.3) to the crispified forms given in (9.8) and (9.9) respectively. Here, we may also use the usual combinatorial techniques to solve problems (9.8) and (9.9).

Model-1: Intuitionistic Fuzzy Nonlinear Integer Programming Problem (IFNIPP)

$$\text{Maximize } R_S = \prod_{i=1}^{n}\left[1-\prod_{j=1}^{l_i}\left(1-R_{ij}\right)^{x_{ij}}\right] \tag{9.8}$$

subject to

$$\sum_{i=1}^{n}\sum_{j=1}^{l_i}\text{Rank}\left(c_{ij}^{1},c_{ij}^{2},c_{ij}^{3};c_{ij}^{4},c_{ij}^{2},c_{ij}^{5}\right)x_{ij} \le \text{Rank}\left(C_0^{1},C_0^{2},C_0^{3};C_0^{4},C_0^{2},C_0^{5}\right)$$

$$\sum_{i=1}^{n}\sum_{j=1}^{l_i}\text{Rank}\left(w_{ij}^{1},w_{ij}^{2},w_{ij}^{3};w_{ij}^{4},w_{ij}^{2},w_{ij}^{5}\right)x_{ij} \le \text{Rank}\left(W_0^{1},W_0^{2},W_0^{3};W_0^{4},W_0^{2},W_0^{5}\right)$$

$x_{ij} \ge 0$ and x_{ij} are integers, $1 \le i \le n, 1 \le j \le l_i$.

Model-2: Fully Intuitionistic Fuzzy Nonlinear Integer Programming Problem (FIFNIPP)

$$\text{Maximize}\,\tilde{R}_S = \prod_{i=1}^{n}\left[1-\prod_{j=1}^{l_i}\left(1-\left(\text{Rank}\left(\tilde{R}_{ij}\right)^{x_{ij}}\right)\right)\right]$$

subject to

$$\sum_{i=1}^{n}\sum_{j=1}^{l_i}\text{Rank}\left(c_{ij}^{1},c_{ij}^{2},c_{ij}^{3};c_{ij}^{4},c_{ij}^{2},c_{ij}^{5}\right)x_{ij} \le \text{Rank}\left(C_0^{1},C_0^{2},C_0^{3};C_0^{4},C_0^{2},C_0^{5}\right)$$

$$\sum_{i=1}^{n}\sum_{j=1}^{l_i}\text{Rank}\left(w_{ij}^{1},w_{ij}^{2},w_{ij}^{3};w_{ij}^{4},w_{ij}^{2},w_{ij}^{5}\right)x_{ij} \le \text{Rank}\left(W_0^{1},W_0^{2},W_0^{3};W_0^{4},W_0^{2},W_0^{5}\right)$$

$x_{ij} \ge 0$ and x_{ij} are integers, $1 \le i \le n, 1 \le j \le l_i$.

(9.9)

9.5 SOLUTION PROCEDURE

This is to be noted that problems (9.1)–(9.3) are constrained optimization problems and so the existing techniques may be used to solve them. In this work, we used the penalty function technique in which the constrained optimization problem is converted into an unconstrained optimization problem. The easiest and the most effective Big-M penalty technique [30,39,32] is used in this work. Hence, the unconstrained optimization problems corresponding to problems (9.1)–(9.3) are as follows:

$$\text{Maximize } R_s(x) = \begin{cases} R_s(x) & \text{when } x \in S_1 \\ -M & \text{when } x \notin S_1 \end{cases} \tag{9.10}$$

where,

$$S_1 = \left\{x : \sum_{i=1}^{n}\sum_{j=1}^{l_i} c_{ij}x_{ij} \le C_0,\ \sum_{i=1}^{n}\sum_{j=1}^{l_i} w_{ij}x_{ij} \le W_0,\ 1 \le i \le n, 1 \le j \le l_i \text{ and } 1 \le l_{ij} \le x_{ij} \le u_{ij},\ x_{ij} \in Z^{+}\right\}$$

$$\text{Maximize } R_s(x) = \begin{cases} R_s(x) & \text{when } x \in S_2 \\ -M & \text{when } x \notin S_2 \end{cases} \tag{9.11}$$

where,

$$S_2 = \left\{ x : \sum_{i=1}^{n} \sum_{j=1}^{l_i} \tilde{c}_{ij} x_{ij} \le \tilde{C}_0, \sum_{i=1}^{n} \sum_{j=1}^{l_i} \tilde{w}_{ij} x_{ij} \le \tilde{W}_0, 1 \le i \le n, 1 \le j \le l_i \text{ and } 1 \le l_{ij} \le x_{ij} \le u_{ij}, x_{ij} \in Z^+ \right\}$$

$$\text{Maximize } \tilde{R}_s(x) = \begin{cases} \tilde{R}_s(x) & \text{when } x \in S_3 \\ -M & \text{when } x \notin S_3 \end{cases} \tag{9.12}$$

where,

$$S_3 = \left\{ x : \sum_{i=1}^{n} \sum_{j=1}^{l_i} \tilde{c}_{ij} x_{ij} \le \tilde{C}_0, \sum_{i=1}^{n} \sum_{j=1}^{l_i} \tilde{w}_{ij} x_{ij} \le \tilde{W}_0, \ 1 \le i \le n, 1 \le j \le l_i \text{ and} 1 \le l_{ij} \le x_{ij} \le u_{ij}, \ x_{ij} \in Z^+ \right\}$$

Evolutionary algorithms may be used to solve this problem. In this work, we developed the real coded elitist genetic algorithm for solving the abovementioned problem.

The crispified problems given in Equations 9.4–9.9 are then transformed to unconstrained problems by using the Big-M penalty method described in the Appendix.

9.6 GENETIC ALGORITHM

We used the Genetic Algorithm (GA) to solve the reliability optimization problems described in this chapter. GA is a stochastic search and optimization technique based on the evolutionary principle "survival of the fittest" and natural genetics [12,22,39,40]. Gen and Cheng [41] described the applications of GA to combinatorial problems including reliability optimization problems. GA has the following basic features:

(i) it works with a coding of solution set, not the solution itself;
(ii) it searches over a population of solutions, not a single solution;
(iii) it uses payoff information, not derivatives or other auxiliary knowledge;
(iv) it applies stochastic transformation rules, not deterministic.

The main components considered for implementing the genetic algorithm are the GA parameters (population size, maximum number of generations, crossover, and mutation), Chromosome representation, initialization of population, evaluation of the fitness function, selection process, genetic operators (crossover, mutation, and elitism), and termination criteria.

In our work, the value of the objective function of the transformed unconstrained optimization problems corresponding to the chromosome is considered as the fitness value of that chromosome. The tournament selection of size two, intermediate crossover for integer variables, one-neighborhood mutation for integer variables, and the termination condition as the maximum number of generations are used in this work.

The GA Parameters

The GA parameters which are used to implement the GA code are presented below.

***Population Size* ($\mathbf{pop_{size}}$)** It determines the amount of information stored by the GA. Generally, it is dependent on the dimension of the problem.

***Maximum Number of Generations* ($\mathbf{max_{gen}}$)** It is also dependent upon the number of genes (variables) of a chromosome of the problem and it controls the termination criterion for the convergence of the solution.

***Probability of Crossover* ($\mathbf{prob_{cross}}$)** It is the main search operator in GA. It is used to thoroughly investigate the search process. By this operator, the genetic information among the individuals is mixed to create new individuals. Its normal range is taken as [0.60, 0.95].

***Probability of Mutation* ($\mathbf{prob_{mute}}$)**This operator plays a crucial role in the genetic algorithm as it prevents the solution from being trapped in local optimal. Mutation is done after the crossover. This operator randomly changes the offspring that results after the crossover. The mutation rate generally lies in [0.05, 0.20].

Chromosome Representation

The real coded representation for the chromosomes is very popular. In real coded representation, a chromosome is coded in the form of a vector/matrix of integer/floating point or a combination of both the numbers. Every component of that chromosome represents a decision variable of the problem, and each chromosome is encoded as a vector of integer numbers, with the same component as the vector of decision variables of the problem. This representation is accurate and more efficient. A chromosome denoted as $c_l\left(l = 1,2,\ldots,pop_{size}\right)$ is the ordered list of n genes, $c_l = \left\{c_{l1}, c_{l1}, c_{lpop_{size}}\right\}$.

Initialization of the Population

It is necessary to initialize the participating chromosomes in artificial genetics after chromosome representation. The independent variables and their ranges are identified. Then the initialization process produces pop_{size} number of chromosomes, in which every component for each chromosome is randomly generated within the ranges of the corresponding decision variable. In this work, the following algorithm is applied to select an integer random number for this purpose.

An integer random number between a and b can be generated as either $I = I_1 + I_r$ or $I = I_1 - I_r$, where I_r is a random integer between 1 and $|I_1 - I_2|$.

Evaluation of Fitness Function

The evaluation of fitness function is the same for the natural evolution process in the biological and physical environments. After initialization of the chromosomes of a potential solution, we need to see how relatively good they are. Therefore, the fitness value for each chromosome has to be calculated. In our work, the value of the objective function of the reduced unconstrained optimization problems corresponding to the chromosome is considered as the fitness value of that chromosome.

Selection of Fitness Function

The selection operator, which is the first operator in artificial genetics, has an effective role in GA. This selection process is based on Darwin's principle on natural evolution “survival of the fittest”. The primary objective of this process is to select the above average individuals/chromosomes from the population according to the fitness value of each chromosome and eliminate the rest of the individuals/chromosomes. There are several methods to implement the selection process.

The tournament selection of size two is taken in this work.

Genetic Operators

After the selection process, other genetic operators like crossover and mutation are applied to the surviving chromosomes. Crossover is an operator that creates new individuals/chromosomes (offspring) by combining the features of both parent solutions. It operates on two or more parent solutions at a time and produces the offspring for the next generation. The intermediate crossover for integer variables is used.

The aim of the mutation operator is to introduce the random variations into the population used to prevent the search process from converging to the local optima. This operator helps to regain the information lost in earlier generations and is responsible for the fine tuning capabilities of the system and is applied to a single individual only. Usually, its rate is very low, because, otherwise, it would defeat the order building generated through the selection and crossover operations. Here, one-neighborhood mutation is used.

9.7 NUMERICAL EXAMPLES

To illustrate the proposed solution methodology, we consider a numerical example from Chern and Jan [4]. In this problem, we take a three-stage series system with redundant units in parallel (1-out-of-3:G stage) and assume that in each stage, different types of components can be used as design alternatives. The system budget and system weight for the crisp model are $C_0 = 30$ units and $W_0 = 17$ units and for the intuitionistic fuzzy model are $\widetilde{C_0} = (29.15, 30, 31.25; 28.99, 30, 32.19)$ and $\widetilde{W_0} = (16, 17, 18; 15.8, 17, 18.5)$. The complete data required to formulate the problem are given in Tables 9.1 through 9.3 (Figures 9.4 through 9.7).

TABLE 9.1
Input Data for Crisp Model

		i		
j	**Parameters**	**1**	**2**	**3**
1	C	4	8	11
	W	2	3	4
	R	0.99	0.98	0.98
2	C	13	3	5
	W	3	3	6
	R	0.95	0.8	0.92
3	C	7	3	
	W	5	9	
	R	0.92	0.90	
l_i		3	3	2

$C_0 = 30$, $W_0 = 17$

TABLE 9.2
Input Data for IFNIPP with TIFN Parameters (Type-1)

		i		
j	**Parameters**	**1**	**2**	**3**
1	$\tilde{C}$	(2,4,5;1,4,6)	(6,8,9;5,8,10)	(9,11,12;8,11,13)
	$\tilde{W}$	(1,2,4;0.95,2,4.5)	(2,3,5;1,3,6)	(3,4,6;2,4,7)
	R	0.99	0.98	0.98
2	$\tilde{C}$	(11,13,14;10,13,15)	(1,2,4;0.5,2,5)	(3,5,6;2,5,7)
	$\tilde{W}$	(2,3,5;1,3,6)	(2,3,6;1,3,7)	(5,6,8;4,6,9)
	R	0.95	0.8	0.92
3	$\tilde{C}$	(5,7,8;4,7,9)	(1,3,4;0.5,3,4.5)	–
	$\tilde{W}$	(4,5,7;3,5,8)	(8,9,11;7,9,12)	
	R	0.92	0.90	
		3	3	2

$\widetilde{C_0} = (29.15, 30, 31.25; 28.99, 30, 32.19), \widetilde{W_0} = (16, 17, 18; 15.8, 17, 18.5).$

9.8 DISCUSSION

The results for both the models are presented in Table 9.4. It is seen that our methodology gives slightly better results compared to Chern and Jan [42] for the crisp model with some different redundancy allocations. Also, in this table all the outcomes for different methods of crispifications for both the models are presented. It is to be noted that the (α, β)-cut method of crispification produces the best allocation of redundancy with maximum system reliability for both the models. The ranking function method gives a better result for model-1. It is observed that the GMIV method produces the same result as that of the ranking function method with the same redundancy allocations. However, for model-2, the ranking function method gives a better result than the GMIV method with the same redundancy vector. Therefore, it may be concluded that the (α, β)-cut method is the best method of crispification with regard to the problem under consideration.

The sensitivities of the GA parameters for the problem using the (α, β)-cut method are presented graphically (Figures 9.2 through 9.5) for model-1 and in tabular form for model-2 (Tables 9.5 through 9.8). From Figure 9.2 it is clear that the optimal allocation is reached at a population size of 50 and found to be stable for higher values also. It is observed from Figure 9.3 that optimum system reliability is achieved around the value of 50 for maximum number of generations and it is stable. Figure 9.4 depicts the stability of system reliability with respect to the probability

TABLE 9.3
Input Data for FIFNIPP with TIFN Parameters (Type-2)

		i		
j	**Parameters**	**1**	**2**	**3**
1	$\tilde{C}$	(2,4,5;1,4,6)	(6,8,9;5,8,10)	(9,11,12;8,11,13)
	$\tilde{W}$	(1,2,4;0.95,2,4.5)	(2,3,5;1,3,6)	(3,4,6;2,4,7)
	$\tilde{R}$	(0.95,097,0.98; 0.90,0.97,0.99)	(0.90,0.96,0.98; 0.85,0.96,0.99)	(0.88,0.95,0.98; 0.83,0.95,1.00)
2	$\tilde{C}$	(11,13,14;10,13,15)	(1,2,4;0.5,2,5)	(3,5,6;2,5,7)
	$\tilde{W}$	(2,3,5;1,3,6)	(2,3,6;1,3,7)	(5,6,8;4,6,9)
	$\tilde{R}$	(0.85,0.95,0.99; 0.80,0.95,1.00)	(0.77,0.80,0.85; 0.70,0.80,0.95)	(0.85,0.92,0.97; 0.80,0.92,1.00)
3	$\tilde{C}$	(5,7,8;4,7,9)	(1,3,4;0.5,3,4.5)	---------
	$\tilde{W}$	(4,5,7;3,5,8)	(8,9,11;7,9,12)	
		(0.88,0.92,0.95; 0.80,0.92,1.00)	(0.85,0.90,0.97; 0.80,0.90,1.00)	
l_i		3	3	2

$\widetilde{C_0} = (29.15, 30, 31.25; 28.99, 30, 32.19), \widetilde{W_0} = (16, 17, 18; 15.8, 17, 18.5).$

TABLE 9.4
Optimal Solutions

Model/method	**Type**	$x = (x_{11}, x_{12}, x_{13}, x_{21}, x_{22}, x_{23}, x_{31}, x_{32})$	R_s
Crisp	---	(2,0,0,1,1,0,1,1)	0.98799087
	In [4]	(2,0,0,0,3,0,1,0)	0.9779422
(α, β)-cut method	1	(2,0,0,1,1,0,1,1)	0.99430696
	2	(2,0,0,1,1,0,1,1)	[0.97534941, 0.99390139] Mid-value = 0.98462540
Ranking function method	1	(2,0,0,1,1,0,1,0)	0.97598239
	2	(2,0,0,1,1,0,1,0)	0.96897374
GMIV method	1	(2,0,0,1,1,0,1,0)	0.97598239
	2	(2,0,0,1,1,0,1,0)	0.93055733

of the crossover. In Figure 9.5 it can be seen that the maximum system reliability is achieved with suitable redundancy allocation with stability with respect to the probability of mutation. Further, it is quite clear from Tables 9.5 through 9.8 that for model-2 using the (α, β)-cut method, the optimal redundancy is reached with stable system reliability with respect to all GA parameters.

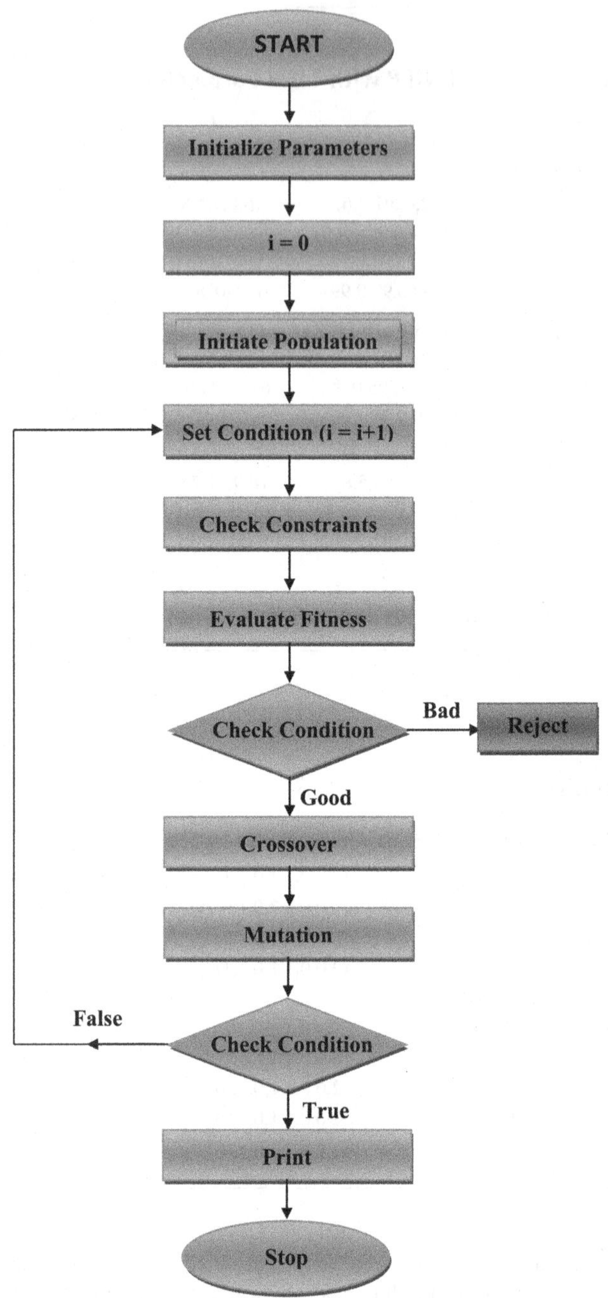

FIGURE 9.3 Flowchart of GA.

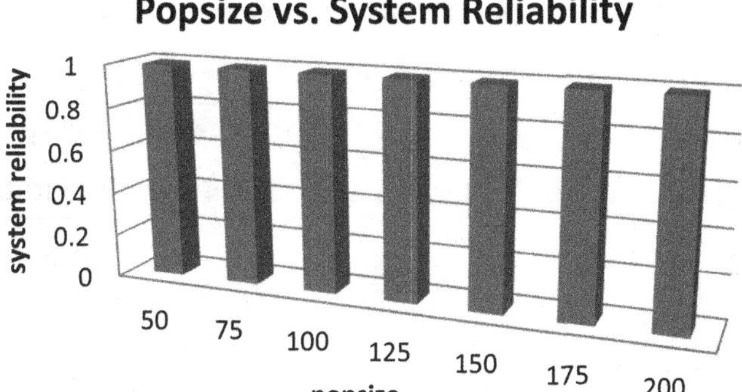

FIGURE 9.4 Sensitivity w. r. t. population size for model-1 problem using (α, β)-cut method.

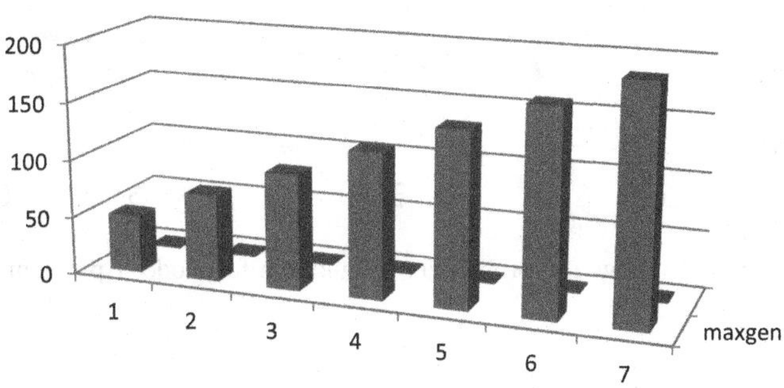

FIGURE 9.5 Sensitivity w. r. t. maximum no. of generations for model-1 problem using the (α, β)-cut method.

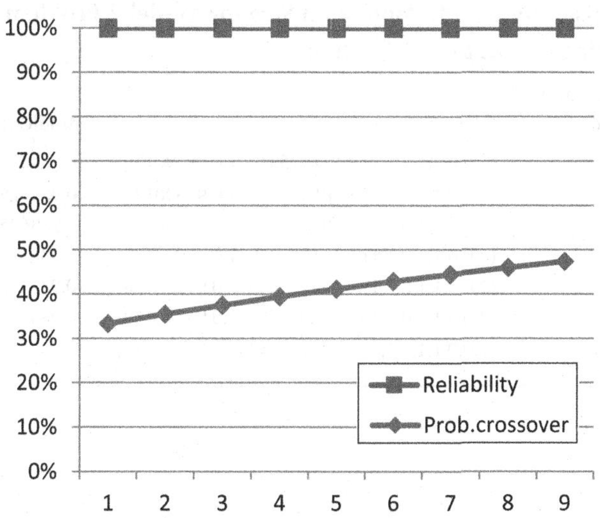

FIGURE 9.6 Sensitivity w. r. t. probability of crossover for model-1 problem using the (α, β)-cut method.

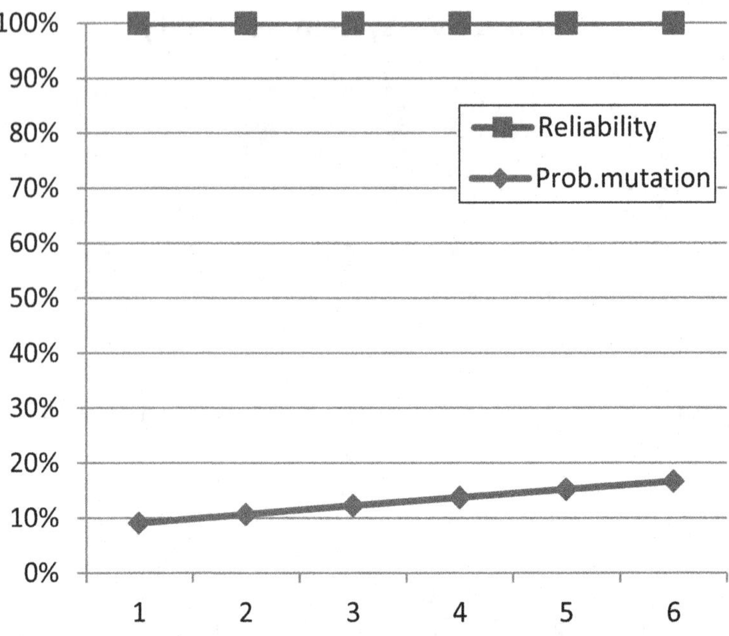

FIGURE 9.7 Sensitivity w. r. t. probability of mutation for model-1 problem using the (α, β)-cut method.

TABLE 9.5
Sensitivity w. r. t. Population Size for Model-2 Problem Using the (α, β)-Cut Method

Population size	x	R_s	Middle value
50	(2,0,0,1,1,0,1,1)	[0.97534941,0.99390139]	0.98462540
75	(2,0,0,1,1,0,1,1)	[0.97534941,0.99390139]	0.98462540
100	(2,0,0,1,1,0,1,1)	[0.97534941,0.99390139]	0.98462540
125	(2,0,0,1,1,0,1,1)	[0.97534941,0.99390139]	0.98462540
150	(2,0,0,1,1,0,1,1)	[0.97534941,0.99390139]	0.98462540
175	(2,0,0,1,1,0,1,1)	[0.97534941,0.99390139]	0.98462540
200	(2,0,0,1,1,0,1,1)	[0.97534941,0.99390139]	0.98462540

TABLE 9.6
Sensitivity w. r. t. Maximum No. of Generations for Model-2 Problem Using (α, β)-Cut Method

Maximum generation	x	R_s	Middle value
50	(2,0,0,1,1,0,1,1)	[0.97534941,0.99390139]	0.98462540
75	(2,0,0,1,1,0,1,1)	[0.97534941,0.99390139]	0.98462540
100	(2,0,0,1,1,0,1,1)	[0.97534941,0.99390139]	0.98462540
125	(2,0,0,1,1,0,1,1)	[0.97534941,0.99390139]	0.98462540
150	(2,0,0,1,1,0,1,1)	[0.97534941,0.99390139]	0.98462540
175	(2,0,0,1,1,0,1,1)	[0.97534941,0.99390139]	0.98462540
200	(2,0,0,1,1,0,1,1)	[0.97534941,0.99390139]	0.98462540

TABLE 9.7
Sensitivity w. r. t. Probability of Crossover for Model-2 Problem Using (α, β)-Cut Method

Probability of crossover	x	R_s	Middle value
0.50	(2,0,0,1,1,0,1,1)	[0.97534941,0.99390139]	0.98462540
0.55	(2,0,0,1,1,0,1,1)	[0.97534941,0.99390139]	0.98462540
0.60	(2,0,0,1,1,0,1,1)	[0.97534941,0.99390139]	0.98462540
0.65	(2,0,0,1,1,0,1,1)	[0.97534941,0.99390139]	0.98462540
0.70	(2,0,0,1,1,0,1,1)	[0.97534941,0.99390139]	0.98462540
0.75	(2,0,0,1,1,0,1,1)	[0.97534941,0.99390139]	0.98462540
0.80	(2,0,0,1,1,0,1,1)	[0.97534941,0.99390139]	0.98462540

TABLE 9.8
Sensitivity w. r. t. Probability of Mutation for Model-2 Problem Using the (α, β)-Cut Method

Probability of mutation	x	R_s	Middle value
0.10	(2,0,0,1,1,0,1,1)	[0.97534941,0.99390139]	0.98462540
0.12	(2,0,0,1,1,0,1,1)	[0.97534941,0.99390139]	0.98462540
0.14	(2,0,0,1,1,0,1,1)	[0.97534941,0.99390139]	0.98462540
0.16	(2,0,0,1,1,0,1,1)	[0.97534941,0.99390139]	0.98462540
0.18	(2,0,0,1,1,0,1,1)	[0.97534941,0.99390139]	0.98462540
0.20	(2,0,0,1,1,0,1,1)	[0.97534941,0.99390139]	0.98462540
0.22	(2,0,0,1,1,0,1,1)	[0.97534941,0.99390139]	0.98462540

9.9 CONCLUSION

An n-stage series system with redundancy in parallel is formulated as a nonlinear integer programming problem (NIPP) in crisp and intuitionistic fuzzy environments. Then the intuitionistic fuzzy models are transformed after crispifying the TIFN parametric values by using the crispification methods, viz., the (α, β)-cut method, the GMIV method, and the ranking function method. The Big-M technique is used in each case to convert the constrained optimization problem into an unconstrained one. Interval order relations are used in the case of solving the problems in the (α, β)-cut method. Then the problems are solved by using a real coded genetic algorithm with tournament selection, one-neighborhood mutation for integer variables, and intermediate crossover. The results are presented for the numerical examples, and the sensitivity analyses are shown in tabular form and graphically for the (α, β)-cut method.

9.10 FUTURE RESEARCH DIRECTIONS

This chapter provides new scope for using crispification methods for the intuitionistic fuzzy numbers developed in this chapter. Several decision-making problems in operations research and in optimization can be considered in an intuitionistic fuzzy environment to make them more realistic, and such problems can be solved using these crispification methods along with genetic algorithms and interval order relations. Several other soft computing techniques can be utilized to solve similar problems and a comparative study can also be carried out. Other types of impreciseness such as neutrosophic fuzzy numbers, type-2 fuzzy number, etc., for better approximation of the reality can be taken into account.

ACKNOWLEDGMENTS

The authors are grateful to the anonymous referees for their constructive comments toward the development of the work. The second and fourth authors are thankful to the Department of Science & Technology and Biotechnology, West Bengal, for the financial support through the research project [Memo No. 30(Sanc)/ST/P/S&T/16G-43/2017 Dated 12/06/2018].

APPENDIX

I Intuitionistic Fuzzy Number [43,29,36–38]

An intuitionistic fuzzy number $\tilde{A}^i$ is an extension of fuzzy number [44] and can be described as

a) An intuitionistic fuzzy subset of the real line.
b) Normal i.e., there is any $x_0 \in R$ such that $\mu_{\tilde{A}^i}(X_0) = 1$ and $v_{\tilde{A}^i}(X_0) = 0$.
c) Convex for the membership function $\mu_{\tilde{A}^i}(X)$ i.e., $\mu_{\tilde{A}^i}(\lambda x_1 + (1-\lambda)x_2) \geq \min(\mu_{\tilde{A}^i}(x_1), \mu_{\tilde{A}^i}(x_2))$ $\forall x_1, x_2 \in R$, $\lambda \in [0,1]$.
d) Concave for non-membership function $v_{\tilde{A}^i}(x)$ i.e., $v_{\tilde{A}^i}(\lambda x_1 + (1-\lambda)x_2) \leq \max(v_{\tilde{A}^i}(x_1), v_{\tilde{A}^i}(x_2))$ $\forall x_1, x_2 \in R$, $\lambda \in [0,1]$ Figure 9A.1.

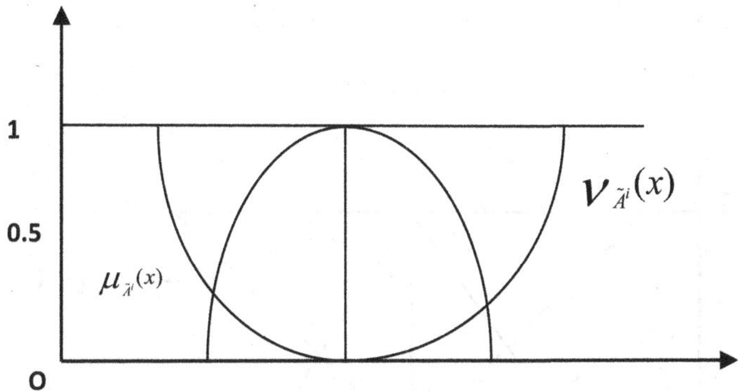

FIGURE 9A.1 Membership and non-membership functions of $\tilde{A}^i$.

II Triangular Intuitionistic Fuzzy Number (TIFN) [20]

A triangular intuitionistic fuzzy number $\tilde{A}^i$ is an intuitionistic fuzzy set in R with the following membership function ($\mu_{\tilde{A}^i}(x)$) and non-membership function ($\nu_{\tilde{A}^i}(x)$)

$$\text{and } \nu_{\tilde{A}^i} = \begin{cases} \dfrac{a_2 - x}{a_2 - a_1'}, & \text{for } a_1' \le x \le a_2 \\ \dfrac{x - a_2}{a_3' - a_2}, & \text{for } a_2 \le x \le a_3' \\ 1, & \text{otherwise} \end{cases}$$

where, $a_1' \le a_1 \le a_2 \le a_3 \le a_3'$ and $\mu_{\tilde{A}^i}(x), \nu_{\tilde{A}^i}(x) \le 0.5$ for $0 \le \mu_{\tilde{A}^i}(x) + \nu_{\tilde{A}^i}(x) \le 1, \forall x \in R$

This TIFN is denoted by $\tilde{A}^i = (a_1, a_2, a_3;\, a_1', a_2, a_3')$.

Property: Transformation rules [20] for the TIFN, $\tilde{A}^i = (a_1, a_2, a_3;\, a_1', a_2, a_3')$ to

(i) triangular fuzzy number TFN, $\tilde{A} = (a_1, a_2, a_3)$ is that $a_1 = a_1'$, $a_3 = a_3'$ and $\nu_{\tilde{A}^i}(x) = 1 - \mu_{\tilde{A}^i}(x)$,
(ii) crisp interval $[a_1, a_3]$ is $a_1' = a_1$ and $a_3 = a_3'$,
(iii) a real number "a" is $a_1' = a_1 = a_2 = a_3 = a_3' = a$ Figure 9A.2.

III (α, β)-Cuts of Triangular Intuitionistic Fuzzy Number [20]

If $\tilde{A}^i$ is a TIFN, (α, β)-level intervals or (α, β)-cut [12] is given by

$$\tilde{A}^i_{\alpha,\beta} = \left\{\left(x, \mu_{\tilde{A}^i}(x), \nu_{\tilde{A}^i}(x)\right) : x \in X, \mu_{\tilde{A}^i}(x) \ge \alpha, \nu_{\tilde{A}^i}(x) \le \beta, \alpha, \beta \in [0,1]\right\}$$

$$= \left\{\left[A_1(\alpha), A_2(\alpha)\right], \left[A_1'(\beta), A_2'(\beta)\right], \alpha + \beta \le 1, \alpha, \beta \in [0,1]\right\}$$

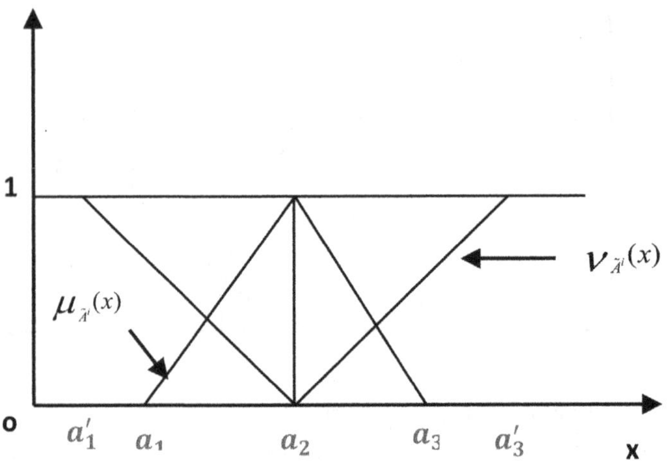

FIGURE 9A.2 Membership and non-membership functions of TIFN.

Where, $A_1(\alpha)$, $A_2'(\beta)$ and $A_2(\alpha), A_1'(\beta)$ will be the increasing and decreasing function of α, β, respectively, with $A_1(1) = A_2(1)$ and $A_1'(0) = A_2'(0)$.

Then, for the TIFN, $\tilde{A}^i = (a_1, a_2, a_3; a_1', a_2, a_3')$, we have

$\tilde{A}^i_{\alpha,\beta} = \{[a_1 + \alpha(a_2 - a_1), a_3 - \alpha(a_3 - a_2)]; [a_2 - \beta(a_2 - a_1'), a_2 + \beta(a_3' - a_2)]\}$, α, $\beta \in [0,1]$, where $\alpha + \beta \leq 1$.

Now, taking $\alpha = \beta$ with $\alpha \leq 0.5$, we have

$$\tilde{A}^t_{\alpha,\alpha} = \{[a_1 + \alpha(a_2 - a_1), a_3 - \alpha(a_3 - a_2)]; [a_2 - \alpha(a_2 - a_1'), a_2 + \alpha(a_3' - a_2)]\}.$$

Now, taking the mean of the α-level intervals for the membership and non-membership functions of the TIFN $\tilde{A}^i$ we have,

$$\alpha(\tilde{A}^i) = \left[\left(\frac{a_1 + a_2}{2}\right) - \alpha\left(\frac{a_1 - a_1'}{2}\right), \left(\frac{a_2 + a_3}{2}\right) + \alpha\left(\frac{a_3' - a_3}{2}\right)\right] = \left[{}^{\alpha}\tilde{A}^L, {}^{\alpha}\tilde{A}^U\right]$$

IV Graded Mean Integration Value (GMIV) of a Triangular Intuitionistic Fuzzy Number [45,46]

For the TIFN $\widetilde{A}^i$ with membership function $\mu_{\tilde{A}^i}(x)$, the graded mean integral value is $\text{GMIV}_\mu(\tilde{A}^i) = \dfrac{\int_0^1 x\{(1-w)L^{-1}(x) + wR^{-1}(x)\}dx}{\int_0^1 xdx}$, where $w \in [0,1]$ is the pre-assigned parameter which refers to the degree of optimism [31,45,46].

Similarly, for the non-membership function $\nu_{\tilde{A}^i}(x)$, the graded mean integral value $\text{GMIV}_\nu\left(\tilde{A}^i\right) = \frac{\int_0^1 x\left\{(1-w)L_1^{-1}(x) + wR_1^{-1}(x)\right\}dx}{\int_0^1 xdx}$.

Then, for the TIFN, $\tilde{A}^i = \left(a_1, a_2, a_3;\ a_1', a_2, a_3'\right)$, we have

$$\text{GMIV}_\mu\left(\tilde{A}^i\right) = \frac{a_1 + 4a_2 + a_3}{6}$$

and

$$\text{GMIV}_\nu\left(\tilde{A}^i\right) = \frac{a_1' + 4a_2 + a_3'}{6}$$

Taking the average, we have the GMIV of the TIFN as

$$\text{GMIV}\left(\tilde{A}^i\right) = \frac{a_1 + 8a_2 + a_3 + a_1' + a_3'}{12}.$$

V Ranking Function of a Triangular Intuitionistic Fuzzy Number [47]

Let $\tilde{A}^i = \left(a_1, a_2, a_3;\ a_1', a_2, a_3'\right)$ be a TIFN, then the ranking functions [47,48] for the membership and the non-membership functions are given by

$$R_\mu\left(\tilde{A}^i\right) = \frac{a_1 + 2a_2 + a_3}{4}$$

and

$$R_\nu\left(\tilde{A}^i\right) = \frac{a_1' + 2a_2 + a_3'}{4}$$

Taking the average of these, we have the ranking function of the TIFN as

$$\text{Rank}\left(\tilde{A}^i\right) = \frac{a_1 + 4a_2 + a_3 + a_1' + a_3'}{8}$$

VI Order Relations of Interval Numbers [33,49,50]

In this chapter, we solved reliability optimization problems with interval valued objectives and constraints with the help of a genetic algorithm. In this algorithm, interval order relations are used to compare two interval numbers.

Any two distinct interval numbers $A = [a_L, a_R]$ and $B = [b_L, b_R]$ can be categorized as below:

Category 1: These two intervals are disjoint.
Category 2: These two intervals are partially overlapping.
Category 3: One of the intervals includes the other one.

Optimistic decisions are taken in some cases of decision making for intervals of Category 3 [49] and irrespective of optimistic and pessimistic [50].

Definition 1: Let $A = [a_L, a_R] = \langle a_c, a_w \rangle$ and $B = [b_L, b_R] = \langle b_c, b_w \rangle$ be two interval numbers. Then for maximization problems, the order relation $>_{\max}$ (reflexive and transitive but not symmetric) in which interval A is accepted is

$$A >_{\max} B \Leftrightarrow \begin{cases} a_c > b_c \text{ for Category 1 and Category 2 intervals and} \\ \text{either } a_c \geq b_c \wedge a_w < b_w \text{ or } a_c \geq b_c \wedge a_R > b_R \text{ for Category 3 intervals.} \end{cases}$$

Definition 2: Let $A = [a_L, a_R] = \langle a_c, a_w \rangle$ and $B = [b_L, b_R] = \langle b_c, b_w \rangle$ be two interval numbers. Then for minimization problems, the order relation $>_{\max}$ (reflexive and transitive but not symmetric) in which interval A is accepted is

$$A <_{\min} B \Leftrightarrow \begin{cases} a_c < b_c \text{ for Category 1 and Category 2 intervals and} \\ \text{either } a_c \leq b_c \wedge a_w < b_w \text{ or } a_c \leq b_c \wedge a_L < b_L \text{ for Category 3 intervals.} \end{cases}$$

VII The Big-M Penalty Method

The Big-M penalty method can be used to handle constraints of the problem under consideration. In this method, the fitness function is penalized by a large positive/negative number depending on the type of the problem (minimization/maximization). In this work, the value of M is taken as 999999, while utilizing the Big-M penalty method. The method is described below.

Let us consider a constrained optimization problem

$$\text{Maximize } f(x) \tag{9A.1}$$

Subject to the constraints

$$g_j(x) \leq b_j, j = 1, 2, \ldots, m.$$

Thereby, using the Big-M method, the problem of (A1) is converted into:

$$\text{Maximize } \hat{f}(x) = \begin{cases} f(x) & \text{if } x \in S \\ M & \text{if } x \notin S \end{cases} \tag{9A.2}$$

and $S = \{x : g_j(x) \leq b_j, j = 1, 2, \ldots, m \text{ and } l \leq x \leq u\}$.

The problem (Equation 9A.2) is a nonlinear unconstrained optimization problem with an imprecise objective value.

After using the Big-M penalty method, the problem (4)–(9) is reduced to the unconstrained form and we use GA to solve these problems.

REFERENCES

1. I.Tillman, C.Hwang and W.Kuo, *Optimization of System Reliability*, Marcel Dekker, New York.
2. K.Misra, On optimal reliability design: A review, *Systems Science*, 12(4), 5–30, 1986, K.Park, Fuzzy apportionment of system reliability, *IEEE Transactions on Reliability R*, 36(1), 129–132, 1987.
3. W.Kuo, V.Prasad, F.Tillman and C.Hwang, *Optimal Reliability Design Fundamentals and Applications*. Cambridge University Press, Cambridge LINGO User Guide (2013) Lindo Systems Inc., Chicago, IL, 2001.
4. M. S.Rao and V.Naikan, Reliability analysis of repairable systems using system dynamics modeling and simulation, *Journal of Industrial Engineering International*, 10(3), 1–10, 2014.
5. R. O. T.Sasaki and S.Shingai, A new technique to optimize system reliability, *IEEE Transactions on Reliability*, 32, 175–182, 1983.
6. C. S.Sung and Y. K.Cho, Reliability optimization of a series system with multiple-choice and budget constraints, *European Journal of Operational Research*, 127(1), 159–171, 2000.
7. W.Kuo and V. R.Prasad, An annotated overview of system-reliability optimization, *IEEE Transactions on Reliability*, 49, 176–187, 2011.
8. W.Kuo, V. R.Prasad, F. A.Tillman and C. L.Hwang, *Optimal Reliability Design: Fundamentals and Application*. Cambridge University Press, Cambridge, 2001.
9. G. S.Mahapatra and T. K.Roy, Reliability evaluation of bridge system with fuzzy reliability of components using interval nonlinear programming, *Electronic Journal of Applied Statistical Analysis*, 5(2), 151–163, 2012.
10. F. A.Tillman, C. L.Hwang and W.Kuo, Optimization technique for system reliability with redundancy: A review, *IEEE Transactions on Reliability*, 26, 148–155, 1977a.
11. I.Yusuf, Comparative analysis of profit between three dissimilar repairable redundant systems using supporting external device for operation, *Journal of Industrial Engineering International*, 10(4), 199–207, 2014.
12. J. H.Kim and B. J.Yum, A heuristic method for solving redundancy optimization problems in complex systems, *IEEE Transactions on Reliability*, 42(4), 572–578, 1993.
13. Y.Nakagawa and K.Nakashima, A heuristic method for determining optimal reliability allocation, *IEEE Transactions on Reliability*, 26(3), 156–161, 1977.
14. C. L.Hwang, F. A.Tillman and W.Kuo, Reliability optimization by generalized Lagrangian-function based and reduced-gradient methods, *IEEE Transactions on Reliability*, 28, 316–319, 1979.
15. X. L.Sun and D.Li, Optimization condition and branch and bound algorithm for constrained redundancy optimization in series system, *Optimization and Engineering*, 3(1), 53–65, 2002.
16. A. Bourezg and H. Meglouli, Reliability assessment of power distribution systems using disjoint path-set algorithm, *Journal of Industrial Engineering International*, 11(1), 45–57, 2015.
17. H.Huang, Fuzzy multi-objective optimization decision-making of reliability of series system, *Microelectronics Reliability*, 37(3), 447–449, 1996.
18. D. X.Chen, Fuzzy reliability of the bridge circuit system, *Systems Engineering-Theory and Practice*, 11, 109–112, 1977.
19. G. S.Mahapatra and T. K.Roy, Fuzzy multi-objective mathematical programming on reliability optimization model, *Applied Mathematics and Computation*, 174(1), 643–659, 2006.

20. G. S.Mahapatra and T. K.Roy, Reliability evaluation using triangular intuitionistic fuzzy numbers arithmetic operations, *World Academy of Science, Engineering and Technology*, 50, 574–581, 2009.
21. G. S.Mahapatra and T. K.Roy, Optimal redundancy allocation in series-parallel system using generalized fuzzy number, *Tamsui Oxford Journal of Information and Mathematical Sciences*, 27(1), 1–20, 2011.
22. S. K.Mahato, N.Bhattacharyee and R. Paramanik, Fuzzy reliability redundancy optimization with signed distance method for defuzzification using genetic algorithm, *International Journal of Operation Research (IJOR)*, 37(3), 307–323, 2020.
23. H.Garg, M.Rani, and Y.Vishwakarma, Bi-objective optimization of the reliability-redundancy allocation problem for series-parallel system, *Journal of Manufacturing Systems*, 33(3), 335–347, 2014.
24. H.Garg and S. P.Sharma, Multi-objective reliability-redundancy allocation problem using particle swarm optimization, *Computers and Industrial Engineering*, 64(1), 247–255, 2013.
25. A.Dolatshahi-Zand and K.Khalili-Damghani, Design of scada water resource management control center by a bi-objective redundancy allocation problem and particle swarm optimization, *Reliability Engineering and System Safety*, 133, 11–21, 2015.
26. K.Khalili-Damghani, A. R.Abtahi and M.Tavana, A new multi objective particle swarm optimization method for solving reliability redundancy allocation problems, *Reliability Engineering and System Safety*, 111, 58–75, 2013.
27. H.Garg and M.Rani, An approach for reliability analysis of industrial systems using PSO and IFS technique, *ISA Transactions*, Elsevier, 52(6), 701–710, 2013.
28. H.Garg, Performance analysis of an industrial system using soft computing based hybridized technique, *Journal of the Brazilian Society of Mechanical Sciences and Engineering*, 39(4), 1441–1451, April 2017.
29. H.Garg, An approach for analyzing fuzzy system reliability using particle swarm optimization and intuitionistic fuzzy set theory, *Journal of Multiple-Valued Logic and Soft Computing*, 21(3–4), 335–354, 2013.
30. R.Gupta, A. K.Bhunia and D.Roy, A GA based penalty function technique for solving constrained redundancy allocation problem of series system with interval valued reliabilities of components, *Journal of Computational and Applied Mathematics*, 232(2), 275–284, 2009.
31. S. K.Mahato, L.Sahoo and A. K.Bhunia, Effects of defuzzification methods in redundancy allocation problem with fuzzy valued reliabilities *via* genetic algorithm, *International Journal of Information and Computer Science*, 2(6), 106–115, 2013.
32. L.Sahoo, A. K.Bhunia and S. K.Mahato, Optimization of system reliability for series system with fuzzy component reliabilities by genetic algorithm, *Journal of Uncertain Systems*, 8(2), 136–148, 2014.
33. L.Sahoo, A. K.Bhunia and D.Roy, Reliability optimization with high and low level redundancies in interval environment *via* genetic algorithm, *International Journal of Systems Assurance Engineering and Management*, 2013. doi: 10.1007/s13198-013-0199-9.
34. H.Garg, M.Rani and S. P.Sharma, An approach for analyzing the reliability of industrial systems using soft-computing based technique, *Expert Systems with Applications*, 41(2), 489–501, 1 February 2014.
35. H.Garg, A novel approach for analyzing the reliability of series-parallel system using credibility theory and different types of intuitionistic fuzzy numbers, *Journal of Brazilian Society of Mechanical Sciences and Engineering*, 38(3), 1021–1035, 2016.
36. H.Garg, M. Rani and S. P. Sharma, Reliability analysis of the engineering systems using intuitionistic fuzzy set theory, *International Journal of Quality and Reliability Engineering, Hindawi*, 2013, 10 pages, Article ID 943972, 2013.

37. H.Garg, Multi-objective optimization problem of system reliability under intuitionistic fuzzy set environment using cuckoo Search algorithm, *Journal of Intelligent and Fuzzy Systems*, 29(4), 1653–1669, 2015.
38. H.Garg, M.Rani, S. P.Sharma and Y.Vishwakarma, Intuitionistic fuzzy optimization technique for solving multi-objective reliability optimization problems in interval environment, *Expert Systems with Applications*, 41(7), 3157–3167, 2014.
39. Z.Michalewicz, *Genetic Algorithms + Data Structure = Evolution Programs*. Springer-Verlag, Berlin, 1996.
40. D. E.Goldberg, *Genetic Algorithms in Search, Optimization, and Machine Learning*. Addison Wesley, Reading, MA, 1989.
41. M.Gen and R.Cheng, *Genetic Algorithm and Engineering Design(1st ed.)*. John Wiley & Sons, New York, NY, USA, 1997.
42. M. S.Chern and R. H.Jan, Reliability optimization problems with multiple constraints, *IEEE Transactions on Reliability*, 35(4), 431–436, 1986.
43. E.Baloui Jamkhaneh, System reliability using generalized intuitionistic fuzzy exponential lifetime distribution, *International Journal of Soft Computing and Engineering (IJSCE)*, 7, 2231–2307, 2017.
44. L. A.Zadeh, Fuzzy sets, *Information and Control*, 8(3), 338–353, 1965.
45. S. Chen and C. Hsieh, Graded mean integration representation of generalized fuzzy numbers, *Journal of the Chinese Fuzzy Systems Association*, 5(2), 1–7, 1999.
46. C.Hsieh and S.Chen, Similarity of generalized fuzzy numbers with graded mean integration representation. *Proceedings of 8th International Fuzzy System Association World Congress*, 2, 551–555, 1999.
47. T. S.Liou and M. J.Wang, Ranking fuzzy numbers with integral value, *Fuzzy Sets and Systems*, 50(3), 247–255, 1992.
48. G. Bortolana and R. Degani, A review of some methods for ranking fuzzy numbers, *Fuzzy Sets and Systems*, 15(1), 1–19, 1985.
49. S. K.Mahato and A. K.Bhunia, Interval-arithmetic-oriented interval computing technique for global optimization, *Applied Mathematics Research Express*, 2006, 1–19, Article ID 69642, 2006.
50. L.Sahoo, A. K. Bhunia and P. K. Kapur, Genetic algorithm based multi-objective reliability optimization in interval environment, *Computer and Industrial Engineering*, 62(1), 152–160, 2012.

10 A Framework for Assessing the Stability of Groundwater Quality Using Reliability and Resilience

Deepesh Machiwal, Adlul Islam, and Priyanka Sharma

CONTENTS

10.1 INTRODUCTION

Groundwater, an important component of the hydrologic cycle, plays a vital role in chemical cycles (solute transport) and biochemical cycles (biosphere) and is influenced by changes in the carbon cycle (climate change) (Van der Gun, 2012). This largest source of freshwater currently contributes to about one-third of global freshwater consumption, satisfying the needs of two billion people (Kundzewicz & Döll, 2009; Famiglietti, 2014; Gorelick & Zheng, 2015; Richey et al., 2015). Owing to its relatively stable yield, groundwater has emerged as an extremely important water resource for meeting all kinds of water needs in various sectors such as domestic, agricultural, industrial, and environmental (Howard, 2015). Groundwater fulfills the domestic needs of about half of the world's population and sustains global food production by providing 38% of the required water for irrigation (Zektser & Everett, 2004; Siebert et al., 2010; Rodell et al., 2018). As a result, this vast and hidden water resource is being depleted at a high rate of 800 km^3/year all over the world by supplying for agriculture (67%), domestic consumption (22%), and industrial needs (11%) (Van der Gun, 2012; Burek et al., 2016).

It is expected that in the 2050s the depletion rate may reach as high as 1100 km^3/year, and about 67% of global groundwater abstraction is carried out in India, the United States, China, Iran, and Pakistan (UNESCO, 2018). The net groundwater depletion rate has doubled since 1990 (Konikow, 2011), leaving one-third of the world's largest groundwater systems in distress (Richey et al., 2015). Depletion of groundwater reservoirs leads to several risks and problems, e.g. increased pumping cost, reduced crop yields, degraded water quality, seawater intrusion, land subsidence, wetland and ecosystem degradation, among others (Van der Gun, 2012). Groundwater depletion in the long run may ultimately result in the exhaustion of the aquifer.

Groundwater quality and quantity are both equally important because the degradation of groundwater quality leads to the reduction in its usable quantity. Two major sources affecting groundwater quality are (i) geogenic (natural) processes and (ii) anthropogenic (human) activities (Machiwal et al., 2018a). Geogenic processes mostly occur beneath the soil surface under the vadose (unsaturated) zone and the saturated zone, and depend on the geological, hydrochemical, and hydrogeologic conditions of the aquifer. On the other hand, anthropogenic activities result in pollution of shallow and young groundwater. In contrast, deeper and old groundwater is more vulnerable to geogenic contamination due to their longer contact time with geological formation (Machiwalet al., 2018b). In India, national groundwater quality data published by the Central Ground Water Board (CGWB) have revealed the concentrations of fluoride, arsenic, nitrate, iron, and heavy metals in excess of the Bureau of Indian Standards (BIS) permissible limits in isolated pockets in various parts of the country. Beyond the permissible limits the following elements are found in several different parts of the country: fluoride in 335 districts of 20 states, nitrate in 386 districts of 21 states, arsenic in 153 districts of 21 states, iron in 301 districts of 25 states, lead in 93 districts of 14 states, cadmium in 24 districts of 9 states, and chromium in 30 districts of 10 states. The salinity (electrical conductivity>3000 micro-mhoscm^{-1}) beyond the permissible limit is documented in 212 districts of 15 states (CGWB, 2018; GoI, 2018). Hence, management of groundwater quality is of a major concern in the regions experiencing heavy groundwater depletion, especially in the arid and semi-arid areas lacking perennial water bodies. Globally, it is imperative to address the key issues of depleting groundwater reservoirs and degrading groundwater quality to ensure the sustainability of groundwater resources (Van der Gun, 2012). Groundwater can be sustainably managed, and its degradation can be kept in check by comprehensive evaluation and characterization of groundwater quality (Machiwal et al., 2011).

Groundwater quality can be evaluated by using a wide variety of available tools and techniques, including graphical, statistical, artificial intelligence (e.g. fuzzy logic, and artificial neural network), and geochemical modeling methods (Machiwal et al., 2018a). Popular graphical methods include Radial/Vector diagram, Stiff diagram, Pie chart, Piper diagram, Schoeller diagram, United States Salinity Laboratory (USSL) diagram, Wilcox diagram, Gibbs diagram, among others. Likewise, widely used statistical methods comprise histogram/frequency plot, box and whisker plot, correlation matrix, Student's t-test, time series modeling, geostatistical modeling, and multivariate analysis techniques such as principal component analysis, cluster analysis, and discriminant analysis. These methods are helpful in better understanding the role of geochemical and hydrogeochemical processes causing spatial and temporal variations in groundwater quality of the aquifer system. Hence, several researchers have employed these methods to characterize groundwater quality around the world – for example, Wunderlin et al. (2001), Kumar et al. (2007), Cloutier et al. (2008), Güler et al. (2012), Machiwal and Jha (2015), and Davies and Crosbie (2018). However, it is challenging to provide adequate scientific and technical knowledge of the geochemical processes to groundwater managers in simple terms by integrating groundwater-quality issues within the groundwater sustainability action framework. To address this challenge, researchers felt the need for developing indices to assess groundwater quality.

A water quality index is a simplified approach to aggregate the scientific geochemical knowledge and communicate this knowledge to the novice end users in plain and simple language (Machiwal et al., 2018a). Different agencies and researchers have proposed several types of water quality indices across the literature over the past 55 years. These indices were initially developed solely for assessing surface water quality, and later on, few of them were extended to assess groundwater quality. In his pioneering work, Horton (1965) developed an index by weighing eight water quality parameters according to their relative importance. This index was further improved by the National Sanitation Foundation (NSF) of the United States (Brown et al., 1970; Deininger & Maciunas, 1971). Lumb et al. (2011) presented a review of different water quality indices developed and used in different parts of the world. In 1998, several attempts were made to develop for the first time a water quality index to specifically characterize groundwater quality (Backman et al., 1998; Melloul & Collin, 1998). Afterward, the pace of applying the indices for groundwater-quality evaluation increased, and as a result, many different indices have evolved to evaluate groundwater quality; the salient indices are enlisted in Machiwal et al. (2018a, 2019). Expressing water quality through a single index was not possible due to the complexities of deriving an index suitable for the intended water use (e.g. drinking, agricultural, irrigation, livestock, and industrial), the spatial and temporal variability in water quality parameters, and the choice of water quality variables (e.g. chemical, physical, microbiological, and radiological) for deriving the index (Abbasi & Abbasi, 2012). These and other complexities led to the development of several such indices. For example, Chanapathi and Thatikonda (2019) developed a Mamdani-type fuzzy-based regional water quality index (FRWQI) consisting of ten water quality parameters such as dissolved oxygen (DO), fecal coliforms (FC), biological oxygen demand (BOD), pH, nitrogen, suspended solids (SS), alkalinity, turbidity, chemical oxygen demand (COD), and electrical conductivity (EC) for evaluating the water quality indices of various river basins across the world. Chanapathi et al. (2019) further developed the Mamdani-type fuzzy-based groundwater sustainable index (FGSI) with a centroid defuzzification framework containing five components, covering five dimensions of sustainability (i.e. environmental, social, economic, mutual trust, and institutional) and 24 indicators. The FGSI model was evaluated for selected Asian cities, and it was observed that overall groundwater sustainability index in the city of Hyderabad, India, was high, and that in Lahore (Pakistan), Bangkok (Thailand), Ho Chi Minh (Vietnam), and Yangon (Myanmar), it was moderate. They also reported that the fuzzy inference system provided more accurate information than conventional methods that used the continuous fuzzy surface.

Computation of a water quality index involves four basic steps: (i) selection of water-quality dataset, (ii) transformation of water-quality parameters to define them on a common scale or obtaining sub-index values, (iii) assignment of suitable weights to parameters, and (iv) aggregation of sub-indices (Machiwal et al., 2018a). None of the developed indices account for the temporal variability of groundwater quality as computation of indices requires time-specific water quality dataset values recorded within a monitoring network. Thus, the indices are not capable of handling the multi-year or multi-season dataset to depict the temporal dynamics of groundwater quality of an aquifer system. Also, these indices do not deal with the probabilistic assessment of groundwater quality. Probabilistic indicators assess three important properties of groundwater quality, i.e. reliability, resilience, and vulnerability, in addition to the aggregated index. Groundwater-quality datasets monitored at multiple time points over a network of sampling sites may be used to explore the dynamics of the geochemical processes responsible for sustainability of groundwater quality (Babiker et al., 2007).

This chapter proposes a framework for assessing a novel stability index of groundwater quality using the reliability–resilience–vulnerability (RRV) concept. It first provides an overview of the RRV concept and presents a review on the application of the RRV concept in hydrology and water resources engineering. It then introduces the reader to a framework for computing the groundwater-quality stability using the RRV concept. Furthermore, it demonstrates the application of the developed framework through a case study undertaken at Jaipur district, Rajasthan, India.

10.2 OVERVIEW OF RRV CONCEPT

The proposed novel probability-based groundwater stability index considers temporal variability of groundwater quality, which involves the criteria of "system-robustness" and "system sustainability" (Hashimoto et al., 1982; Loucks, 1997).

The developed index integrates two probabilistic indicators, i.e. reliability and resilience. "System robustness", in economic terms of water resources systems and as defined by Hashimoto et al. (1982), is the possible deviation between the actual costs of a proposed project and of the project. However, "system sustainability" of the water resource systems, as explained by Loucks (1997), is examined by their planning and management that contribute to the current and future objectives of society, while maintaining their ecological, environmental, and hydrological integrity.

The World Conservation Strategy introduced the concept of sustainable development about 40 years ago (IUCN, 1980). Initially, a sustainability index was proposed to assess the performance of alternative policies from the perspectives of both water users and environment. The sustainability index was defined as a measure of a system's adaptive capacity to reduce its vulnerability. Hence, when implementation of a policy makes a system more sustainable, its sustainability index indicates that the system has a large adaptive capacity. Sustainability indices are generally computed by integrating non-weighted or weighted combinations of reliability, resilience, and vulnerability indicators of various economic, environmental, ecological, and social criteria (Loucks, 1997). The concept of sustainability index was, for the first time, defined using reliability (R), resilience (R), and vulnerability (V) as the performance criteria to evaluate and compare water management policies (Loucks, 1997). Thereafter, many researchers used the RRV-based index for various research purposes (e.g. McMahon et al., 2006; Ray et al., 2010; Sandoval-Solis et al., 2011; Machiwal et al., 2016). In general, the RRV-based sustainability concept is used to evaluate the performance of water resource systems (Loucks, 1997; Kay, 2000; Ajami et al., 2008; Sandoval-Solis et al., 2011).

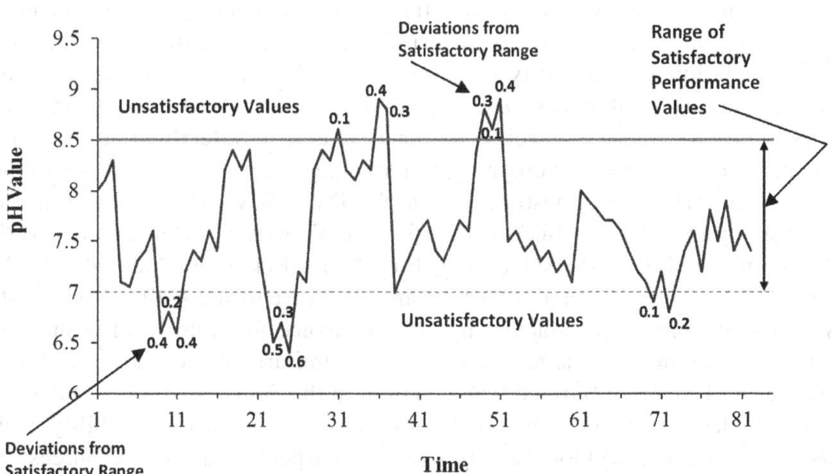

FIGURE 10.1 Time series of pH value of the groundwater illustrating the concepts of satisfactory and unsatisfactory values needed in estimating reliability, resilience, and vulnerability indicators.

The RRV concept is illustrated by considering a time series plot for 82 arbitrary pH values of groundwater as shown in Figure 10.1. Here, the range of pH value from 7 to 8.5 on the y-axis represents the area of satisfactory values which meet the criterion of safe drinking water prescribed by the World Health Organization (WHO). pH values above and below this range are considered unsatisfactory values, indicating that the water is nonpotable. Reliability is indicated by the fraction of the total 82 values that remain within the range of satisfactory values during the entire time span. Resilience is derived by the number of times the pH of the groundwater changes from unsatisfactory to satisfactory values divided by the total number of unsatisfactory pH values. Vulnerability is measured by the average deviations of pH values from the satisfactory range.

10.3 APPLICATION OF RRV IN WATER RESOURCES PLANNING AND MANAGEMENT

The concept of RRV framework was introduced by Hashimoto et al. (1982), who were among the first to propose the use of the terms reliability, resilience, and vulnerability in the area of water resources management to describe the performance of a multipurpose reservoir system. These risk-based performance indicators of reliability, resilience, and vulnerability have been used either individually or in combination as a comprehensive index for quantifying risk and improving decision making for the management of eco-hydrological systems. These performance indicators have a wide range of application in the selection of system capacities, configurations, operational policies, and targets. Cai et al. (2002) developed a long-term modeling framework incorporating quantified sustainability criteria in a long-term optimization model of a river basin, ensuring risk minimization in water supply, environmental conservation, equity in water allocation, and economic efficiency in water infrastructure development. The model included quantitative measures of risk characteristics, namely, reliability (the frequency of system failure), reversibility (the time required for a system to return from failure), and vulnerability (the severity of system failure). Model outputs included proposals for long-term reservoir operations, water supply, facility improvements, irrigation development, and cropping pattern changes. Applicability of this modeling framework was demonstrated in the Syr Darya River Basin of central Asia and was found to be an effective tool for policy analysis in the context of the river basin. Almost at the same time, First-Order Reliability Method (FORM) was developed by Maier et al. (2001) as an alternative to simulating probabilistic estimates of reliability, vulnerability, and resilience for water quality management problems in Willamette River, Oregon. Similarly, performance of Tampa Bay Water's Enhanced Surface Water System in the Southeast United States was studied under varying climatic conditions using three metrics (reliability, resilience, and vulnerability) for evaluating different aspects of a water resources system (Asefa et al., 2014). They demonstrated the implementation of a seemingly intractable ensemble size (1000 likely future realizations) and length of simulation period (300 years) in a distributed (a cluster of 52 distributed computers) computing environment and highlighted the benefits of comprehensive system performance metrics that are easy to understand by decision makers and stake holders. In their study, Asefa et al. (2014) did not aggregate three metrics to define the sustainability index as it requires additional subjective weights across metrics.

Loucks (1997) expressed sustainability as the product of RRV, and this sustainability criterion has been applied to water resource systems by several researchers (e.g. Jain, 2010; Sandoval-Solis et al., 2011; Vieira & Sandoval-Solis, 2018). Kjeldsen and Rosbjerg (2004) investigated the most appropriate combination of RRV for use in connection with a multi-objective risk assessment of a water resource system. They observed that sample estimates of probabilistic indicators based on average values of failure duration and magnitude for entire length of time series are not appropriate as the sample estimates should be non-monotonic. They also reported that combinations of probabilistic indicators based on the average statistics of the entire time series should not be used to avoid any overlapping in time series values. It might be beneficial to abandon either the resilience or the vulnerability index from a multicriteria analysis. Alternatively, both could be used while abandoning the reliability index altogether. Jain (2010) employed reliability, resilience, and vulnerability indices to assess the performance of the Dharoi reservoir located in a semiarid region of Gujarat, India. They reported that time and volume reliability and maximum volume vulnerability are suitable indicators for an assessment of reservoir performance. They also measured the sustainability index based on time reliability, vulnerability, and storage to mean annual runoff ratio as a composite indicator of reservoir performance assessment. Sandoval-Solis et al. (2011) and Vieira and Sandoval-Solis (2018) used the sustainability index (SI) with measures of reliability, resilience, vulnerability and maximum deficit to evaluate and compare alternative plans for future water availability and water supply in the Rio Verde Grande Basin (RVGB), Brazil. SI is defined as a geometric mean of the performance criteria.

In recent years, there is growing interest in the use of RRV-based performance indices to evaluate droughts (Hazbavi et al., 2018; Yilmaz, 2018), to assess watershed health (Hoque et al., 2014a; Hazbavi & Sadeghi, 2017; Alilou et al., 2019), and to provide safe (reliable) water management in a resilient and sustainable manner (Butler et al., 2017). Maity et al. (2013) observed that assessing drought using a univariate measure (severity or reliability) is inadequate and proposed joint probability distribution of reliability, resilience, and vulnerability (RRV) to characterize droughts. They used soil moisture data to characterize the drought. As reliability and resilience exhibit a well-defined monotonic nonlinear relationship, the joint probability distribution of only resilience and vulnerability was considered for developing the drought management index (DMI) for long-term drought characterization of the Malaprabha River basin of Karnataka, India. They also suggested the use of a five-year period satisfactory for assessment of DMI variability. Similarly, Chanda et al. (2014) developed the drought management index (DMI) using the reliability–resilience–vulnerability (RRV) rationale with the assumption that depletion of soil moisture across a vertical soil column is equivalent to the decline of water surface in a water supply reservoir. The Permanent Wilting Point (PWP) was used as the threshold value, and joint cumulative distribution functions (CDF) of resilience and vulnerability were used for estimating the DMI. Reconnaissance Drought Index (RDI) is one of the widely used drought indices all over the world as it takes into account both the cumulative precipitation and the potential evapotranspiration. Yilmaz (2018) proposed a modified version of the Reconnaissance Drought Index (mRDI) using the reliability–resilience–vulnerability (RRV) concept. The RRV approach was used with the RDI time series by considering the relative frequency of the drought event (reliability), the drought recovery period (resilience), and the probability of drought severity during drought occurrences (vulnerability). An additional performance indicator was used in the mRDI formula to evaluate the drought extremity (maximum extent). A modified version of the RDI (mRDI), computed as a geometric average of performance indicators, has been successfully used in four different provinces of Turkey with different precipitation and temperature parameters.

Application of a conceptual framework of the RRV has also been successfully demonstrated for a comprehensive assessment of watershed health through different approaches and by utilizing different datasets such as water quality and precipitation. Hoque et al. (2012) developed a method for assessing watershed health by employing measures of reliability, resilience, and vulnerability (RRV) using the water quality data of stream water with different monitoring points within the Cedar Creek watershed in north-east Indiana (USA). They reported that RRV indicators are useful for assessing the health of a watershed, and highlighted the importance of accounting data uncertainty when conducting RRV analysis using the reconstructed time series, particularly in the case of sediment analysis. Hoque et al. (2014a) suggested risk-based analyses coupled with deterministic methods for a more comprehensive assessment of the health of the Wildcat Creek watershed in Indiana, USA, under projected scenarios. However, the methodology is sensitive to water quality constituent threshold values. While studying the effect of a spatial scale on risk-based watershed health assessment using RRV indices for five agricultural watersheds in the mid-western United States, it was observed that RRV indicators do change with the spatial scale. However, to achieve stable values, these indicators require a representative threshold value, which is the ratio of the contributing upland area to the area required for channel initiation. Further, scaling with the Strahler stream order is feasible if the watershed possesses a tree-like stream network. They also studied the role of BMPs (best management practices) within an agricultural watershed, and observed that implementation of BMPs based on optimizing a cost–benefit ratio may not significantly change watershed risk measures (reliability and resilience) but are likely to cause a significant reduction in vulnerability. Further, if primarily upland BMPs are implemented in a diffuse manner throughout the watershed, there might not be a significant change in the scaling behavior of RRV values (Hoque et al., 2014b). Similarly, the aggregate time series data of water quality at multiple locations were found to be useful for providing a composite picture of watershed health (Hoque et al., 2016) and also enabled comparison between locations with different types of water quality data.

For watershed health assessment in the Shazand Watershed of Markazi Province, Iran, Hazbavi and Sadeghi (2017) conceptualized and customized the RRV framework by using criteria of the standardized precipitation index (SPI), low and high flow discharges (LFD and HFD), and suspended sediment concentration (SSC). The threshold values of 0.1, 0.16 $m^3 s^{-1}$, 12.63 $m^3 s^{-1}$, and 25 mg l^{-1} day^{-1} were selected for SPI, LFD, HFD, and SSC, respectively. The susceptibility of watershed health was estimated by combining all four factors (SPI, HFD, LFD, and SSC) using the geometric mean as a sustainability index. The hydrological watershed health index (HWHI) was also calculated by standardized values of SPI, LFD, HFD, and SSC. The integrated HWHI of 0.16 ± 0.11 obtained from the geometric mean of the RRV indices showed a decreasing trend for the Shazand Watershed health during the study period. Sadeghi and Hazbavi (2017) further analyzed the temporal and spatial variability of reliability, resilience, and vulnerability in connection with the standardized precipitation index (SPI) for 24 sub-watersheds in the Shazand Watershed of Markazi Province, Iran. The RRV framework, expressed as the geometric mean of RRV, indicated the need for assessing the sensitivity of the RRV framework to different SPI threshold values. Hazbavi et al. (2018) conducted a study examining the sensitivity of the RRV framework, using SPI variability as an indicator of drought in three watersheds with varying climatic conditions in Europe and Asia. They observed that the climatic gradient significantly influenced the watershed vulnerability and drought-based RRV index, but not the SPI and reliability and resilience indicators. Also, they suggested the use of multiple statistical means (at least two) by combining the RRV indicators and compared the results, while investigating the responses of the RRV indicators to different SPI threshold values Hazbavi et al., 2018). Almost a similar research was conducted by Veettil et al. (2018) to characterize droughts in watersheds located in the contiguous United States (CONUS) using the Standardized Precipitation Evapotranspiration Index (with an accumulation period of 3 months) (SPEI03) as a drought index.

An extensive literature search revealed that very few studies were reported on the application of RRV for groundwater management. The very first study, reported by Rodak et al. (2014), proposed RRV indices as viable tools for the assessment of time-dependent health risks due to contaminated groundwater. Studies suggest that the time-dependent health risk method, with the aid of the RRV criteria, provides insight into the complex temporal relationship between health risk, population variability, regulatory risk threshold, and time-dependent concentration of the contaminants in the well. They also noted that the concept of resilience may be of limited utility for chronic health risk due to groundwater contamination events requiring transport from a source to the well. Recently, Machiwal et al. (2019) developed and applied a groundwater-quality stability index (GQSI), which considers a probabilistic estimate of reliability and resilience through a case study conducted in quaternary alluvial and quartzite aquifer systems of Jaipur district, Rajasthan, India. This study utilized 14 groundwater-quality parameters, monitored over a network of 250 sites for a period of 12 years (2001–2012). The results indicate that GQSI is a robust tool to assess steadiness of the groundwater-quality parameters over time and to identify factors responsible for changes in groundwater quality.

A key advantage of the RRV framework approach lies in its ability to combine and quantify risk indicators, viz., reliability, resilience, and vulnerability. Development of composite frameworks is gaining more emphasis nowadays to quantify the dimensions of different ecosystem using different drivers. The most critical factor of the RRV framework is choosing an adequate threshold value for different hydro-climatic regions (Hazbavi et al., 2018).

10.4 FRAMEWORK FOR ASSESSMENT OF GROUNDWATER QUALITY AND STABILITY

Prior to explaining the framework for the assessment of groundwater quality and stability, it is imperative to understand two terms, i.e., the satisfactory (success) state and the unsatisfactory (failure) state of a groundwater-quality time series at a particular time, depending on the condition that the values of the groundwater-quality parameter at certain times are above or below a threshold level. The threshold level may be chosen as the desirable and permissible limits of a given parameter prescribed for safe drinking water by the World Health Organization (WHO) (WHO, 2017). Suppose the time series of a water quality parameter is denoted by $X_t = \{x_1, x_2, x_3, \ldots, x_t, \ldots, x_n\}$., the value of that parameter at a certain time "t" is considered to be in a satisfactory state (success domain, S) if its value x_t at a particular time "t" is found to be less than or equal to the threshold value, i.e. the WHO limit (Hashimoto et al., 1982). On the other hand, if the value of that parameter x_t at a particular time "t" is more than the threshold value of the parameter, the value at that time is said to be in an unsatisfactory state (failure domain, F). The terms "success domain'" and "failure domain" are used in defining three probabilistic indicators, i.e. reliability (R_y), resilience (R_e), and vulnerability (V_y), as described below (Hashimoto et al., 1982).

10.4.1 Reliability

Reliability (R_y) is defined as the probability of finding a value (concentration) of a groundwater-quality parameter less than or equal to the desirable/permissible limit prescribed for drinking purpose by WHO. The R_y value for a groundwater-quality parameter is computed by dividing the total number of values in a satisfactory state (f_{SE}) by the total number of values (n) in the time series of that particular groundwater-quality parameter, as expressed below (Hashimoto et al., 1982):

$$R_y = \Pr\left(X_t \in S\right) = f_{SE} / n \tag{10.1}$$

The R_y value varies from 0 to 1, where no reliability (R_y=0) occurs when the value of a groundwater-quality parameter exceeds the threshold value and the highest reliability (R_y= 1) occurs when the value of groundwater-quality parameter is within the threshold value. Sometimes, reliability is expressed as the opposite of risk (probability) of failure and is defined as $1 - R_y$.

10.4.2 Resilience

Resilience (R_e) of a groundwater-quality parameter is defined by the probability that its value being in an unsatisfactory state (F) at time "t" occurs in a satisfactory state (S) at time "$t+1$". In other words, if the concentration of a parameter exceeds the threshold value at any time "t" then the concentration of that parameter should become less than the threshold value during the next time "$t+1$". In other words, an R_e value is computed as the ratio of number of times satisfactory state (S) values in a time series of a groundwater-quality parameter follows unsatisfactory state (F) values ($f_{\text{FE-SE}}$) to the total number of unsatisfactory state values (f_{FE}). Basically, the R_e value of a groundwater-quality parameter is measured by its capacity to adapt to changing conditions such as climate variability and climate change over time (WHO, 2009). R_e is expressed by the following equation (Hashimoto et al., 1982):

$$R_e = \frac{\Pr\{X_t \in F \text{ and } X_{t+1} \in S\}}{\Pr\{X_t \in F\}} = \Pr\{X_{t+1} \in S \mid X_t \in F\} = f_{\text{FE-SE}} \Big/ f_{\text{FE}} \quad (10.2)$$

The value of R_e ranges between 0 and 1, where no resilience ($R_e = 0$) is observed when the value of groundwater-quality parameter always exceeds the threshold value, and the highest resilience ($R_e = 1$) occurs when the value of the groundwater-quality parameter never exceeds the threshold value.

10.4.3 Vulnerability

The system vulnerability (V_L) to a failure domain or an unsatisfactory state (F), when the system remains under an unsatisfactory state, is quantified by assigning a numerical indicator of the severity, s_j, to each discrete failure state $X_j \epsilon F$. Also, consider e_j to be the probability that X_j corresponding to s_j is the most unsatisfactory and severe outcome that occurs into the set of unsatisfactory states F. The overall system vulnerability is the expected maximum severity of failure states into the set of unsatisfactory value (Hashimoto et al., 1982):

$$V_L = \sum_{j \epsilon F} s_j e_j \quad (10.3)$$

Vulnerability (V_y) of a groundwater-quality parameter is a probabilistic measure that indicates the severity of the deviations of the concentration values of the parameter under an unsatisfactory state (S) from the threshold values based on the desirable/permissible limits set by WHO. The probability of exceedance to vulnerability (V_y) is computed by calculating the difference between the parameter concentration (x_t) at time "t" and threshold values (x^T), and then dividing the difference by unsatisfactory state values (f_{FE}) as shown below (Hashimoto et al., 1982):

$$V_y = \sum_{i=1}^{n} \text{difference}\left(x^T - x_t\right) \Big/ f_{\text{FE}} \sqrt{b^2 - 4ac};\ t = 1, 2, \ldots n \quad (10.4)$$

where $\sum_{i=1}^{n} \text{difference}\left(x^T - x_t\right) =$ sum of absolute values of ($x^T - x_t$) for unsatisfactory state values.

The vulnerability of a groundwater-quality parameter may be treated as the magnitude of instability of groundwater quality with respect to that particular parameter. Vulnerability measured in groundwater stability studies is as important as the magnitude of a trend in trend-assessment studies.

The framework developed in this study calculates vulnerability of the groundwater-quality parameters by obtaining the differences between the parameters' observed values and their desirable or permissible limits for safe drinking water (threshold values). Obviously, the unit of the concerned parameter is the unit for the V_y values, for example, mg l^{-1} for the major cations and anions.

10.4.4 Stability Index of Groundwater Quality

The framework developed in this study for evaluating the stability of groundwater quality requires estimates of three probabilistic indicators, i.e. R_y, R_e, and V_y. The developed stability index explores the dynamics of the groundwater-quality parameters with respect to their threshold values based on the desirable or permissible limits prescribed for safe drinking water by WHO. It is worth mentioning that the stability index is similar to the sustainability index that assesses the performance of the water resource systems (Loucks, 1997). The sustainability index is defined through a multiplicative form of R_y, R_e, and V_y indicators. In the literature, a few alternative forms of these three probabilistic indicators have also been used to define sustainability. However, the developed stability index (SI_{GWQ}) for evaluating groundwater quality considers a multiplicative form of only R_y and R_e indicators and is given by the following expression:

$$SI_{GWQ} = R_y \times R_e \tag{10.5}$$

The value of SI_{GWQ} varies from 0 (no stability) to 1 (highest stability). The multiplicative form of R_y and R_e indicators is supported by the fact that a positive correlation between R_y and R_e values is generally found significant. Thus, the small value of one indicator associates with the small value of another indicator, and vice-versa. Also, the approach of SI_{GWQ} involves an implicit weighting as it adds the least weight to the probabilistic indicator having the least impact on stability of groundwater quality. It is worth noting that the exclusion of the V_y indicator from the definition of SI_{GWQ} is justifiable as V_y involves a unit of measurement, whereas R_y and R_e do not have any unit. A flowchart depicting a step-by-step procedure to compute the stability index of groundwater quality is shown in Figure 10.2.

Values of SI_{GWQ} for individual groundwater-quality parameters can be aggregated by averaging, and a composite stability index (CSI_{GWQ}) is computed using the following expression:

$$CSI_{GWQ} = \frac{\sum_{i=1}^{n} SI_{GWQi}}{n} \tag{10.6}$$

where, SI_{GWQi} represents the stability index for *i*th groundwater-quality parameter and *n* indicates the total number of groundwater-quality parameters.

10.5 A CASE STUDY

In this section, the developed framework for evaluating the stability of groundwater quality is applied through a case study of quaternary alluvial and quartzite aquifer systems of Jaipur district, Rajasthan, India.

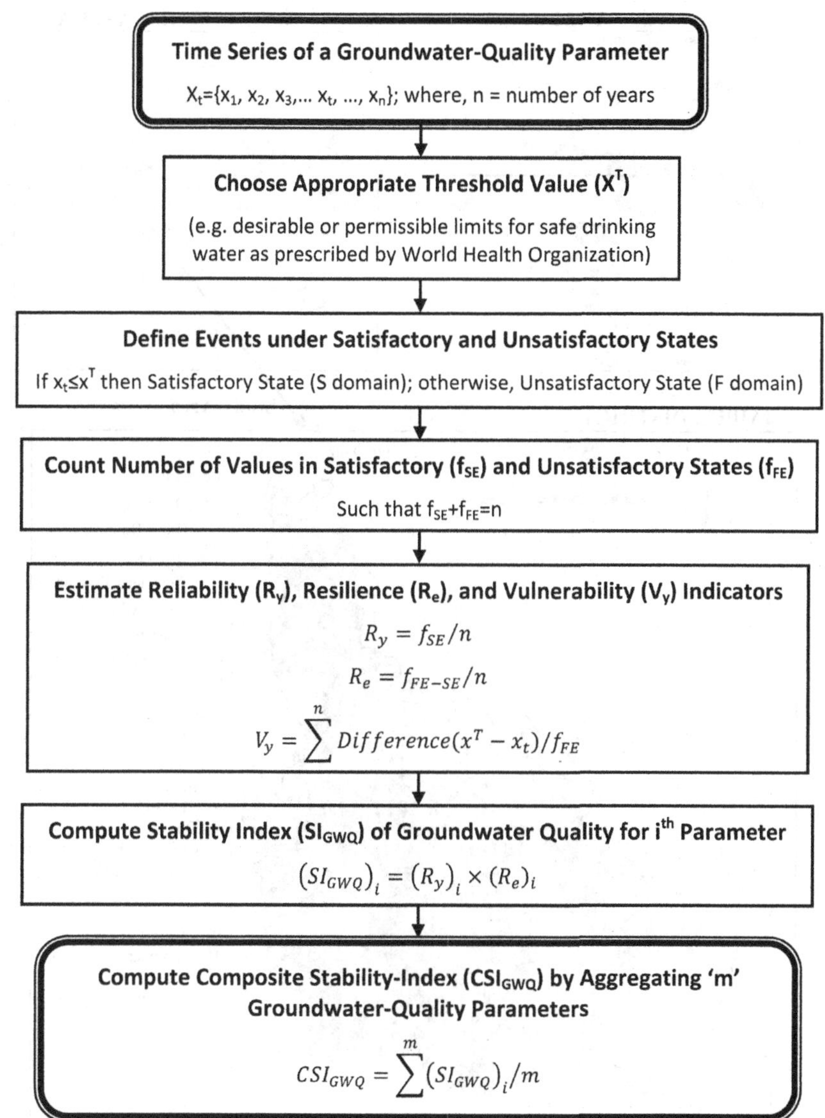

FIGURE 10.2 Flowchart depicting a step-by-step procedure for computing stability index of groundwater quality using time series of groundwater-quality parameters.

10.5.1 Study Area Description

Jaipur district, the capital of the largest and driest state (Rajasthan) of India, is situated between latitudes 26°25′ and 27°51′ N and longitudes 74°55′ and 76°10′ E (Figure 10.3). The extent of the study area is about 10,878 km^2, which is 3.23% of the total area of Rajasthan. The average annual rainfall of the study area is 548 mm, and annual potential evapotranspiration is 1745 mm (CGWB, 2007). The study area experiences

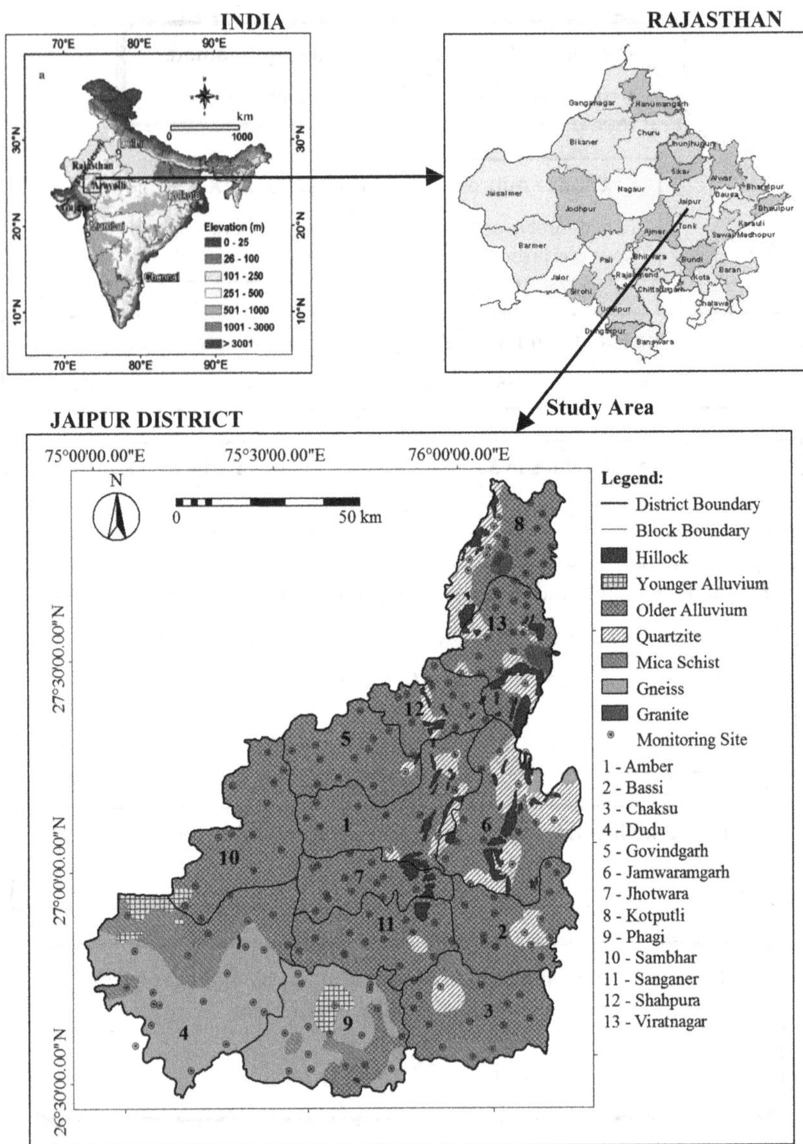

FIGURE 10.3 Location map of study area showing blocks of Jaipur district and sites of 196 monitoring wells.

a dry and semi-arid climate with extreme cold and heat at distinct places. The minimum and maximum temperatures are 3°C and 45°C, respectively, with the mean temperature of 24°C. The entire area is divided into 13 blocks or sub-divisions, mainly for administrative purposes (Figure 10.3). Major landscapes of the study area are hillocks, pediments, undulating fluvial plains, aeolian dune fields, ravines, and paleochannels. Structural hills, composed of Delhi quartzite and trending NNE-SSW, are present in the north and northeast portions of the study area. Pediments with thin to thick soil cover are mainly found around Dudu, Phagi, and Chaksu blocks that form gneissic outcrops. Aeolian sand-dunes covering Sambhar, Jobner, and Renwal are present in the western region. The major soils of the area are loamy sand, sandy loam, sandy clay loam, sandy clay, wind-blown sand, and river sand.

The major source of surface water resources in the area is reservoirs, e.g. Ramgarh, Champarwara, Kalakh, Hingonia, Buchara, and Mansagar. The ephemeral rivers of Banganga, Bandi, Dhund, Mendha, Mashi, Sota, and Sabi drain the area. Groundwater occurs both in unconsolidated and in consolidated formations. Well yield of the underlying aquifer varies from 100 m^3d^{-1} to 500 m^3d^{-1}. Wells are shallow in the alluvial aquifer, with depths ranging from 50 m to 100 m, and they are deep in consolidated formations, with depths varying from 50 m to 200 m. The specific capacity of wells ranges between 58 lpm m^{-1} and 500 lpm m^{-1}. Values of the storage coefficient and transmissivity vary from 4.70×10^{-5} to 1.05×10^{-3} and from 10 m^2 d^{-1} to 850 m^2 d^{-1}, respectively (CGWB, 2017). A total of 120,471 dug wells and dug-cum-tube wells are operational for irrigation purposes in the area. On the other hand, 27,378 hand pumps and dug-cum-tube wells are in use for domestic and industrial purposes (CGWB, 2013). A recent estimate indicates that groundwater is the major source of water, supplying 1178.92×10^6 m^3 for irrigation and 315.96×10^6 m^3 for drinking and industrial purposes in the area (CGWB, 2017).

10.5.2 Data Collection

A total of nine groundwater-quality parameters namely pH, total dissolved solids (TDS), calcium (Ca), magnesium (Mg), sodium (Na), chloride (Cl), sulfate (SO_4), nitrate (NO_3), and total hardness (TH) were collected for the 250 sites in the study area. A twelve-year (2001–2012) groundwater-quality dataset was obtained from the Ground Water Department, Jaipur, Rajasthan, India. The data were collected during the pre-monsoon season (April–May) and were checked for absence of anomalies and presence of regularity. After screening out the anomalies, the dataset of 196 sites was used for the analysis.

10.5.3 Methodology

The developed framework for computing the stability of groundwater quality was applied, and satisfactory and unsatisfactory states for 12 ordered values (2001–2012) of 9 groundwater-quality time series were computed using MS Excel. Thereafter, values of R_y, R_e, and V_y were calculated using Equations 10.1–10.4, and subsequently, values of SI_{GWQ} were computed using Equation 10.5 for all the individual nine parameters. Finally, the value of CSI_{GWQ} was computed using Equation 10.6.

10.5.4 Results and Discussion

10.5.4.1 R_y, R_e, and V_y Values for Groundwater-Quality Parameters

Values of reliability (R_y) for the 196 sites were categorized into three equal-interval groups: (i) "high" ($1.0 > R_y \geq 0.66$), (ii) "moderate" ($0.66 > R_y \geq 0.33$), and (iii) "low" ($0.33 > R_y \geq 0$). The number and the percentage of sites classified into low, moderate, and high values of reliability for the nine groundwater-quality parameters are given in Table 10.1. It is observed that reliability of calcium, sulfate, and nitrate is "high" at 196 (100%), 190 (97%), and 137 (70%) sites, which are relatively more than the sites under "high" reliability of other parameters. In contrast, reliability of total hardness and total dissolved solids is "low" at 186 (95%) and 93 (47%) sites, respectively, which are relatively larger than the sites under the category of "low" reliability of other groundwater-quality parameters. It is also observed that the sites having "low" reliability of total dissolved solids, sodium, and chloride are mainly present in the southern, southwest, and southeast parts where gneiss and mica-schist types of rocks exist. Hence, the "low" reliability is attributed to natural hydrogeologic and geochemical processes occurring in the hard-rock aquifer system that caused geogenic contamination of the groundwater. The reliability of pH and magnesium is "moderate" at 130 (66%) and 124 (63%) sites, respectively, and these two parameters have moderate reliability at the largest number of sites.

Similar to the classification scheme adopted for R_y, the R_e values of groundwater-quality parameters were categorized into three equal-interval probability groups: (i) "high" ($1.0 > R_e \geq 0.66$), (ii) "moderate" ($0.66 > R_e \geq 0.33$), and (iii) "low" ($0.33 > R_e \geq 0$). The number of sites and their percentages classified into three groups of "low", "moderate", and "high" resilience for nine parameters are presented in Table 10.1. It is observed that calcium, sulfate, and nitrate depict "high" resilience at 196 (100%), 186 (95%), and 125 (64%) sites, respectively, which are comparatively larger than the sites under the "high" resilience category of other parameters. It is further observed that sodium and chloride have a relatively large number of total sites under "high" to "moderate" resilience, i.e. 145 (74%) and 151 (77%) sites, respectively. On the other hand, pH, total dissolved solids, and magnesium showed "moderate" to "low" resilience at 148 (76%), 142 (73%), and 171 (87%) sites, respectively. It is found that the resilience of sodium and chloride is "low" in gneiss and mica-schist types of rocks that exist in the southern, southeast, and southwest parts of the study area.

The value of vulnerability (V_y) accompanies a unit of measurement of the parameter, i.e. concentration or magnitude, and hence, the range of V_y values is relatively small for pH and large for total dissolved solids in comparison with other parameters. Threshold values for nine groundwater-quality parameters were chosen depending upon their desirable or permissible limits as prescribed by WHO, and the same are presented in Table 10.2. Values of V_y for the nine groundwater-quality parameters were grouped into three categories of "low", "moderate" and "high" vulnerability, but the range of the three categories was not considered the same for all the nine parameters due to the association of the different units of measurement with their V_y values. The range considered for this case study for categorizing V_y values into three groups is given in Table 10.3. The number and the percentage of sites categorized into three groups of "low", "moderate" and "high" vulnerability for the nine parameters are provided in Table 10.1. It is observed that sulfate, calcium, and magnesium have "low" vulnerability at 178 (91%), 187 (95%), and 121 (62%) sites, respectively, which are relatively larger than the sites having "low" vulnerability for other parameters. On the contrary, the parameter of pH has the largest number of sites (106 sites, 54%) under the "high" vulnerability category. Also, total dissolved

TABLE 10.1
Number and Percentage of Sites Classified into Low, Moderate, and High Values of Reliability, Resilience and Vulnerability for Nine Groundwater-Quality Parameters in Study Area

Class	pH	TDS	Calcium	Magnesium	Sodium	Chloride	Sulfate	Nitrate	Total hardness
(a) Reliability									
Low	25 (13)	93 (47)	0 (0)	44 (22)	43 (22)	30 (15)	0 (0)	3 (2)	186 (95)
Moderate	130 (66)	64 (33)	0 (0)	124 (63)	63 (32)	69 (35)	6 (3)	56 (29)	9 (5)
High	41 (21)	39 (20)	196 (100)	28 (14)	90 (46)	97 (49)	190 (97)	137 (70)	1 (1)
(b) Resilience									
Low	54 (28)	78 (40)	0 (0)	86 (44)	51 (26)	45 (23)	0 (0)	14 (7)	91 (46)
Moderate	94 (48)	64 (33)	0 (0)	85 (43)	71 (36)	67 (34)	10 (5)	57 (29)	13 (7)
High	48 (24)	54 (28)	196 (100)	25 (13)	74 (38)	84 (43)	186 (95)	125 (64)	92 (47)
(c) Vulnerability									
Low	0 (0)	85 (43)	187 (95)	121 (62)	97 (49)	72 (37)	178 (91)	95 (48)	55 (28)
Moderate	90 (46)	63 (32)	9 (5)	58 (30)	70 (36)	53 (27)	17 (9)	55 (28)	102 (52)
High	106 (54)	48 (24)	0 (0)	17 (9)	29 (15)	71 (36)	1 (1)	46 (23)	39 (20)

Note: Values outside and inside parentheses are respectively, number and percentage of sites classified into low, moderate, and high indicator values.

TABLE 10.2
Desirable and Permissible Limits of Nine Groundwater-Quality Parameters Prescribed by the World Health Organization (WHO) for Safe Drinking Water

S. no.	Water quality parameter	Desirable limit	Permissible limit
1	pH	7–8.5	n.a.
2	Total dissolved solids	500 mg l^{-1}	1500 mg l^{-1}
3	Calcium	75 mg l^{-1}	200 mg l^{-1}
4	Magnesium	30 mg l^{-1}	150 mg l^{-1}
5	Sodium	200 mg l^{-1}	n.a.
6	Chloride	200 mg l^{-1}	600 mg l^{-1}
7	Sulfate	200 mg l^{-1}	400 mg l^{-1}
8	Nitrate	45 mg l^{-1}	n.a.
9	Total hardness	100 mg l^{-1}	500 mg l^{-1}

Source: WHO (2017); n.a.: not available.

TABLE 10.3
Range of "Low", "Moderate", and "High" Vulnerability Groups for Nine Groundwater-Quality Parameters

		Vulnerability class		
S. no.	Water quality parameter	Low	Moderate	High
1	pH	0–1.5	1.5–3.0	>3.0
2	Total dissolved solids (mg l^{-1})	0–500	500–1000	>1000
3	Calcium (mg l^{-1})	0–75	75–150	>150
4	Magnesium (mg l^{-1})	0–30	30–60	>60
5	Sodium (mg l^{-1})	0–200	200–400	>400
6	Chloride (mg l^{-1})	0–200	200–400	>400
7	Sulfate (mg l^{-1})	0–200	200–400	>400
8	Nitrate (mg l^{-1})	0–45	45–90	>90
9	Total hardness (mg l^{-1})	0–100	100–200	>200

solids, sodium, nitrate, and total hardness depicted "low" to "moderate" vulnerability at 148 (75%), 167 (85%), 150 (76%), and 157 (80%) sites, respectively. Sites under "high" vulnerability to total dissolved solids, sodium, chloride, nitrate, and total hardness are mostly located in the southern, southeast, and southwest parts that are covered by gneiss and mica-schist types of rocks. These sites revealed a relatively high concentration of groundwater-quality parameters, and hence, it is suggested that the vulnerability indicator helps in understanding the temporal dynamics of the geochemical processes occurring in the aquifer system.

10.5.4.2 Stability of Groundwater Quality

Values of a composite probability-based stability index (CSI_{GWQ}) were computed for groundwater quality at 196 sites in the study area using R_y and R_e values. The CSI_{GWQ} values of the 196 sites were categorized into three groups of "low stability", "moderate stability" and "high stability", and the spatial distribution of the categorized sites in the study area is shown in Figure 10.4.

The stability of groundwater quality is found to be "low", "moderate", and "high" at 92 (47%), 80 (41%), and 24 (12%) sites, respectively. It is observed that the sites depicting "low" stability of groundwater quality are mainly present in southern, southwest, and southeast parts, where gneiss and mica-schist types of rocks exist. The presence of hard-rock terrain and "low" groundwater-quality stability indicate the presence of natural hydrogeochemical processes occurring in the aquifer system, which caused the geogenic contamination of groundwater (Machiwal & Jha, 2015). On the other hand, the "moderate" stability of groundwater quality is apparent in the central, eastern, and northeast parts of the area. In addition, stability of groundwater quality is found to be "high", mainly in the central and north-central portions where older alluvial rocks exist.

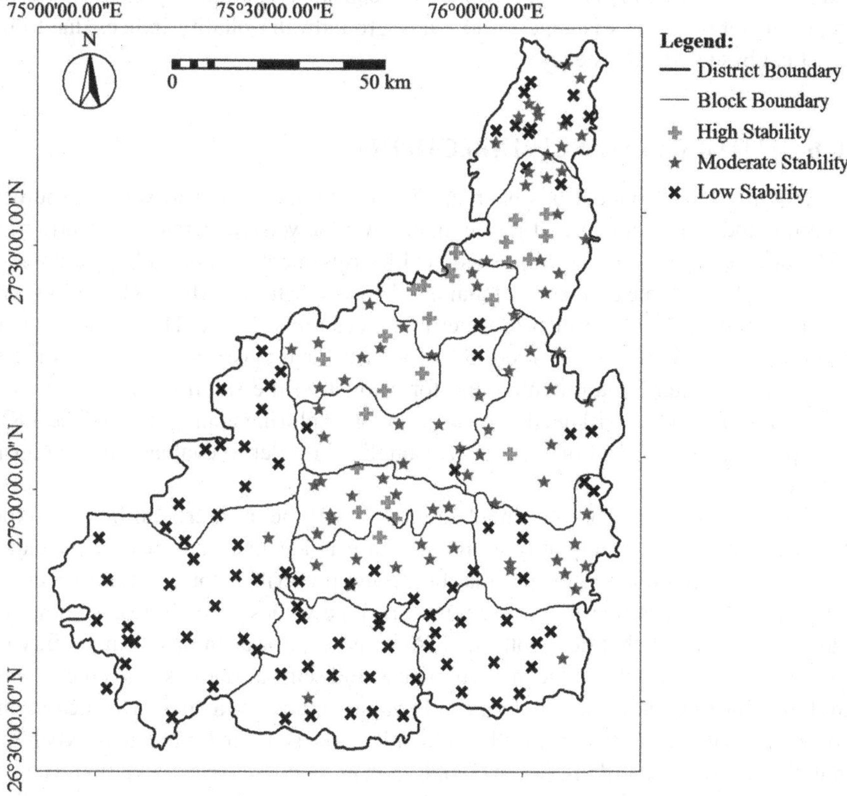

FIGURE 10.4 Spatially distributed sites categorized into "low", "moderate", and "high" stability of groundwater quality in the study area.

10.5.5 Conclusions of the Case Study

This study demonstrated the application of a novel framework for assessing the temporal stability of groundwater quality using two probabilistic indicators, i.e. reliability and resilience. The stability of groundwater quality is evaluated in quaternary alluvial and quartzite aquifer systems of Jaipur district, Rajasthan, India, using nine groundwater-quality parameters. In addition, vulnerability of the nine groundwater-quality parameters is computed, which suggests deviations of parameters' values from the threshold values based on their desirable or permissible limits prescribed for safe drinking water. The results revealed that the groundwater-quality stability is "low" at relatively more sites (47% sites) in comparison with sites categorized under "moderate" (41% sites) and "high" (24% sites) stability groups. The "low" stability of groundwater quality is mostly found under the gneiss and mica-schist types of hard-rock geologic terrain that exists in the southern, southeast, and southwest parts of the study area. It is most likely that the natural hydrogeochemical processes occurring in the aquifer systems are responsible for geogenic contamination of the groundwater in the area. In contrast, groundwater quality is categorized as having "high" stability under the older alluvium type of geologic terrain. The stability index developed and applied in this study is comparable with the groundwater quality indices, but the proposed stability index is superior to many groundwater-quality indices due to its statistical base and consistency.

10.6 FUTURE RESEARCH DIRECTIONS

Recently the RRV concept has been applied in studies related to water resources planning and management and researchers have showed great interest in utilizing RRV for drought evaluation, watershed health assessment, and providing safe water supply in a sustainable and resilient manner. However, studies where RRV is applied for the management of subsurface water resources are still rare. Hardly one or two studies exist in the literature where RRV is used for the assessment of health risk due to groundwater contamination and for evaluating the stability of groundwater quality. Future studies will need to focus more on exploring applications of the RRV concept in the assessment of groundwater quality in order to evaluate its long-term stability.

Choosing an appropriate threshold value is one of the most critical factors in the RRV framework. In water quality studies, the threshold values may be easily decided based on the desirable and/or permissible limits prescribed for drinking water by WHO. However, selection of the optimum threshold values is not quite easy in water management studies that deal with the RRV concept. For example, vulnerability of drought estimates would be meaningful in arid and semi-arid areas when the annual rainfall is lower than the selected threshold value, indicating a "negative" deviation. Conversely, vulnerability of rainfall would only make sense in humid areas when the annual rainfall is more than the threshold, indicating a "positive" deviation. Hence, in future research applications of RRV, more emphasis will be needed on deciding appropriate threshold values, depending upon the climate of the areas as well as the type of parameters under consideration.

This study considered the stability index in a multiplicative form of only two probabilistic indicators, i.e. reliability and resilience. However, it should be noted that a few studies on RRV-based drought assessment included vulnerability as the third parameter in the expression of the RRV index. Likewise, few recent studies excluded the reliability indicator due to the similarity between resilience and reliability. Thus, the behaviors of all three probability indicators and the linear and nonlinear relationships among them should be investigated in future studies for different types of variables and hydro-climatic regions.

A total of nine groundwater quality parameters are considered in the case study. Selection of the parameters is dependent upon the availability of their desirable or permissible limits for the drinking water, and consequently, certain groundwater-quality parameters having the maximum impact on groundwater-quality stability are not chosen for the computation of the stability index. This limitation may be overcome in the future by developing adequate criteria for choosing reasonable values of the thresholds for the water quality parameters whose desirable or permissible limits for the drinking water are not available. Moreover, the geographic information system (GIS) and the geostatistical modeling are two powerful techniques that may be easily integrated with the point data of three probabilistic indicators of reliability, resilience, and vulnerability. Hence, it is strongly suggested that GIS be used in conjunction with geostatistical modeling to compute the stability index over the space.

10.7 CONCLUDING REMARKS

Groundwater protection from pollution involves several actions, and groundwater-quality evaluation is one of the basic and important steps that help in adequate planning and management of the groundwater resources. Of the several methods available for evaluating groundwater quality, water quality indices help to easily communicate scientific and technical knowledge to end users involved for taking appropriate actions to ascertain sustainable groundwater protection. Formulation of a water quality index follows four steps of selecting parameters, their transformations, assigning weights, and aggregation. However, it is found that the existing indices generally use the groundwater-quality dataset recorded through a monitoring network at a chosen point of time, and hence, none of the indices account for the temporal variability of groundwater quality.

The proposed framework uses RRV statistical indicators of reliability (R_y), resilience (R_e), and vulnerability (V_y) and aggregates the two indicators, i.e. reliability and resilience, to develop a novel index of groundwater-quality stability. The RRV criterion has been earlier utilized in the performance assessment of water resources systems, drought management, and assessment of watershed health. However, unlike the earlier studies, the multiplicative form of only R_y and R_e indicators is found efficient in evaluating groundwater-quality stability. The RRV-based stability index is superior to other indices because the former amalgamates the temporal variability with deviations of groundwater quality from the standards prescribed by the World Health Organization (WHO). Use of the RRV stability index avoids use of other measures for capturing the temporal variability or the dynamics of groundwater quality such as coefficient of variation. The stability index can be satisfactorily applied to a multi-year or a multi-season database of groundwater quality. It is worth mentioning that computation of the stability index does not require assigning arbitrary weights to different quality parameters, and hence, the index is free from subjectivity or bias. The stability index is a useful, preferable, excellent indicator of the groundwater-quality stability and has the potential to prove its applicability in different aquifer systems of the world.

The stability index may be integrated with the modern tools and techniques such as the geographic information system and the geostatistical modeling to depict spatial distribution of the groundwater-quality stability over an area. Usability of the index may be proved under different hydrogeologic settings in future studies for better planning and management of water resources so that problems arising due to polluted aquifer systems and degraded groundwater quality could be avoided.

REFERENCES

Abbasi, T. and Abbasi, S.A. (2012). *Water Quality Indices.* Elsevier, 384 pp.

Ajami, N.K., Hornberger, G.M. and Sunding, D.L. (2008). Sustainable water resource management under hydrological uncertainty. *Water Resources Research*, 44(11): 1–10.

Alilou, H., Rahmati, O., Singh, V.P., Choubin, B., Pradhan, B., Keesstra, S., Ghiasi, S.S. and Sadeghi, S.H. (2019). Evaluation of watershed health using Fuzzy-ANP approach considering geo-environmental and topo-hydrological criteria. *Journal of Environmental Management*, 232: 22–36.

Asefa, T., Clayton, J., Adams, A. and Anderson, D. (2014). Performance evaluation of a water resources system under varying climatic conditions: Reliability, resilience, vulnerability and beyond. *Journal of Hydrology*, 508: 53–65.

Babiker, I.S., Mohamed, M.A.A. and Hiyama, T. (2007). Assessing groundwater quality using GIS. *Water Resources Management*, 21(4): 699–715.

Backman, B., Bodiš, D., Lahermo, P., Rapant, S. and Tarvainen, T. (1998). Application of a groundwater contamination index in Finland and Slovakia. *Environmental Geology*, 36(1–2): 55–64.

Brown, R.M., McClelland, N.I., Deininger, R.A. and Tozer, R.G. (1970). A water quality index - Do we dare? *Water and Sewage Works*, 117(10): 339–343.

Burek, P., Satoh, Y., Fischer, G., Kahil, M.T., Scherzer, A., Tramberend, S., Nava, L.F., Wada, Y., Eisner, S., Flörke, M., Hanasaki, N., Magnuszewski, P., Cosgrove, B. and Wiberg, D. (2016). *Water Futures and Solution: Fast Track Initiative (Final Report).* IIASA Working Paper, International Institute for Applied Systems Analysis (IIASA), Laxenburg, Austria.

Butler, D., Ward, S., Sweetapple, C., Astaraie-Imani, M., Diao, K., Farmani, R. and Fu, G. (2017). Reliable, resilient and sustainable water management: The Safe &SuRe approach. *Global Challenges*, 1(1): 63–77.

Cai, X., McKinney, D.C. and Lasdon, L.S. (2002). A framework for sustainability analysis in water resources management and application to the Syr Darya basin. *Water Resources Research*, 38(6): 21-1–21-14. doi: 10.1029/2001WR000214.

CGWB (2007). *Groundwater Scenario: Jaipur District, Rajasthan.* District Groundwater Brochure, Central Ground Water Board (CGWB), Western Region, Jaipur, Ministry of Water Resources, Government of India, 25 pp.

CGWB (2013). *Ground Water Information, Jaipur District, Rajasthan.* Central Ground Water Board (CGWB), Western Region, Ministry of Water Resources, River Development and Ganga Rejuvenation, Government of India, 15 pp.

CGWB (2017). *Dynamic Groundwater Resources of India (as on 31st March 2013).* Central Ground Water Board (CGWB), Ministry of Water Resources, River Development and Ganga Rejuvenation, Bhujal Bhawan, Faridabad, 280 pp.

CGWB (2018). *Groundwater Quality in Shallow Aquifers of India (Selected Parameters).* Central Ground Water Board (CGWB), Ministry of Water Resources, River Development and Ganga Rejuvenation, Bhujal Bhawan, Faridabad, pp. 190.

Chanapathi, T. and Thatikonda, S. (2019). Fuzzy-based regional water quality index for surface water quality assessment. *Journal of Hazardous, Toxic, and Radioactive Waste*, 23(4): 04019010.

Chanapathi, T., Thatikonda, S., Pandey, V.P. and Shrestha, S. (2019). Fuzzy-based approach for evaluating groundwater sustainability of Asian cities. *Sustainable Cities and Society*, 44: 321–331.

Chanda, K., Maity, R., Sharma, A. and Mehrotra, R. (2014). Spatiotemporal variation of long-term drought propensity through reliability-resilience-vulnerability based drought management index. *Water Resources Research*, 50(10): 7662–7676.

Cloutier, V., Lefebvre, R., Therrien, R. and Savard, M.M. (2008). Multivariate statistical analysis of geochemical data as indicative of the hydrogeochemical evolution of groundwater in a sedimentary rock aquifer system. *Journal of Hydrology*, 353(3–4): 294–313.

Davies, P.J. and Crosbie, R.S. (2018). Mapping the spatial distribution of chloride deposition across Australia. *Journal of Hydrology*, 561: 76–88.

Deininger, R.A. and Maciunas, J.M. (1971). *Water Quality Index for Public Water Supplies.* Department of Environment and Industrial Health, University of Michigan, Ann Arbor.

Famiglietti, J.S. (2014). The global groundwater crisis. *Nature Climate Change*, 4(11): 945–948.

GoI (2018). *Study on Groundwater Contamination.* Press Information Bureau, Government of India Ministry of Water Resources, New Delhi. http://pib.nic.in/newsite/PrintRelease.aspx?relid=181183 (accessed on May 21, 2019).

Gorelick, S.M. and Zheng, C. (2015). Global change and the groundwater management challenge. *Water Resources Research*, 51(5): 3031–3051.

Güler, C., Kurt, M.A., Alpaslan, M. and Akbulut, C. (2012). Assessment of the impact of anthropogenic activities on the groundwater hydrology and chemistry in Tarsus coastal plain (Mersin, SE Turkey) using fuzzy clustering, multivariate statistics and GIS techniques. *Journal of Hydrology*, 414–415: 435–451.

Hashimoto, T., Stedinger, J.R. and Loucks, D.P. (1982). Reliability, resiliency and vulnerability criteria for water resource system performance evaluation. *Water Resources Research*, 18(1): 14–20.

Hazbavi, Z., Baartman, J.E., Nunes, J.P., Keesstra, S.D. and Sadeghi, S.H. (2018). Changeability of reliability, resilience and vulnerability indicators with respect to drought patterns. *Ecological Indicators*, 87: 196–208.

Hazbavi, Z. and Sadeghi, S.H.R. (2017). Watershed health characterization using reliability-resilience-vulnerability conceptual framework based on hydrological responses. *Land Degradation and Development*, 28(5): 1528–1537.

Hoque, Y.M., Raj, C., Hantush, M.M., Chaubey, I. and Govindaraju, R.S. (2014a). How do land-use and climate change affect watershed health? A scenario-based analysis. *Water Quality, Exposure and Health*, 6(1–2): 19–33.

Hoque, Y.M., Hantush, M.M. and Govindaraju, R.S. (2014b). On the scaling behavior of reliability-resilience-vulnerability indices in agricultural watersheds. *Ecological Indicators*, 40: 136–146.

Hoque, Y.M., Tripathi, S., Hantush, M.M. and Govindaraju, R.S. (2012). Watershed reliability, resilience and vulnerability analysis under uncertainty using water quality data. *Journal of Environmental Management*, 109: 101–112.

Hoque, Y.M., Tripathi, S., Hantush, M.M. and Govindaraju, R.S. (2016). Aggregate measures of watershed health from reconstructed water quality data with uncertainty. *Journal of Environmental Quality*, 45(2): 709–719.

Horton, R.K. (1965). An index number system for rating water quality. *Journal of Water Pollution Control Federation*, 37(3): 300–306.

Howard, K.W.F. (2015). Sustainable cities and the groundwater governance challenge. *Environmental Earth Sciences*, 73(6): 2543–2554.

IUCN (1980). *The World Conservation Strategy: Living Resource Conservation for Sustainable Development*. International Union for Conservation of Nature (IUCN), United Nations Environment Program, and World Wildlife Fund, Gland, Switzerland.

Jain, S.K. (2010). Investigating the behavior of statistical indices for performance assessment of a reservoir. *Journal of Hydrology*, 391(1–2): 90–96.

Kay, P.A. (2000). Measuring sustainability in Israel's water system. *Water International*, 25(4): 617–623.

Kjeldsen, T.R. and Rosbjerg, D. (2004). Choice of reliability, resilience and vulnerability estimators for risk assessments of water resources systems/Choixd'estimateurs de fiabilité, de résilienceet de vulnérabilité pour les analyses de risque de systèmes de ressources en eau. *Hydrological Sciences Journal*, 49(5): 755–767.

Konikow, L. (2011). Contribution of global groundwater depletion since 1900 to sea-level rise. *Geophysical Research Letters*, 38(17): 1–5.

Kumar, M., Kumari, K., Ramanathan, A.L. and Saxena, R. (2007). A comparative evaluation of groundwater suitability for irrigation and drinking purposes in two intensively cultivated districts of Punjab, India. *Environmental Geology*, 53(3): 553–574.

Kundzewicz, Z.W. and Döll, P. (2009). Will groundwater ease freshwater stress under climate change? *Hydrological Sciences Journal*, 54(4): 665–675.

Loucks, D.P. (1997). Quantifying trends in system sustainability. *Hydrological Sciences Journal*, 42(4): 513–530.

Lumb, A., Sharma, T.C. and Bibeault, J.-F. (2011). A review of genesis and evolution of water quality index (WQI) and some future directions. *Water Quality, Exposure and Health*, 3: 1–14.

Machiwal, D., Cloutier, V., Güler, C. and Kazakis, N. (2018a). A review of GIS-integrated statistical techniques for groundwater quality evaluation and protection. *Environmental Earth Sciences*, 77(19): 681. doi:10.1007/s12665-018-7872-x.

Machiwal, D., Jha, M.K., Singh, V.P. and Mohan, C. (2018b). Assessment and mapping of groundwater vulnerability to pollution: Current status and challenges. *Earth-Science Reviews*, 185: 901–927.

Machiwal, D., Islam, A. and Kamble, T. (2019). Trends and probabilistic stability index for evaluating groundwater quality: The case of quaternary alluvial and quartzite aquifer system of India. *Journal of Environmental Management*, 237: 457–475.

Machiwal, D. and Jha, M.K. (2015). Identifying sources of groundwater contamination in a hard-rock aquifer system using multivariate statistical analyses and GIS-based geostatistical modeling techniques. *Journal of Hydrology: Regional Studies*, 4(A): 80–110.

Machiwal, D., Jha, M.K. and Mal, B.C. (2011). GIS-based assessment and characterization of groundwater quality in a hard-rock hilly terrain of Western India. *Environmental Monitoring and Assessment*, 174(1–4): 645–663.

Machiwal, D., Kumar, S. and Dayal, D. (2016). Characterizing rainfall of hot arid region by using time-series modeling and sustainability approaches: A case study from Gujarat, India. *Theoretical and Applied Climatology*, 124(3–4): 593–607.

Maier, H.R., Lence, B.J., Tolson, B.A. and Foschi, R.O. (2001). First-order reliability method for estimating reliability, vulnerability, and resilience. *Water Resources Research*, 37(3): 779–790.

Maity, R., Sharma, A., Kumar, D.N. and Chanda, K. (2013). Characterizing drought using the reliability-resilience-vulnerability concept. *Journal of Hydrologic Engineering, ASCE*, 18(7): 859–869.

McMahon, T.A., Adeloye, A.J. and Sen-Lin, Z. (2006). Understanding performance measures of reservoirs. *Journal of Hydrology*, 324(1–4): 359–382.

Melloul, A.J. and Collin, M. (1998). A proposed index for aquifer water quality assessment: The case of Israel's Sharon region. *Journal of Environmental Management*, 54(2): 131–142.

Ray, P.A., Vogel, R.M. and Watkins, D.W. (2010). Robust optimization using a variety of performance indices. *Proceedings of the World Environmental and Water Resources Congress*, ASCE, Reston, VA.

Richey, A.S., Thomas, B.F., Lo, M.H., Reager, J.T., Famiglietti, J.S., Voss, K., Swenson, S. and Rodell, M. (2015). Quantifying renewable groundwater stress with GRACE. *Water Resources Research*, 51(7): 5217–5238.

Rodak, C., Silliman, S.E. and Bolster, D. (2014). Time-dependent health risk from contaminated groundwater including use of reliability, resilience, and vulnerability as measures. *Journal of the American Water Resources Association*, 50(1): 14–28.

Rodell, M., Famiglietti, J.S., Wiese, D.N., Reager, J.T., Beaudoing, H.K., Landerer, F.W. and Lo, M.-H. (2018). Emerging trends in global freshwater availability. *Nature*, 557(7707): 651–659.

Sadeghi, S.H. and Hazbavi, Z. (2017). Spatiotemporal variation of watershed health propensity through reliability-resilience-vulnerability based drought index (case study: Shazand Watershed in Iran). *Science of the Total Environment*, 587: 168–176.

Sandoval-Solis, S., McKinney, D. and Loucks, D. (2011). Sustainability index for water resources planning and management. *Journal of Water Resources Planning and Management, ASCE*, 137(5): 381–390.

Siebert, S., Burke, J., Faures, J.M., Frenken, K., Hoogeveen, J., Döll, P. and Portmann, F.T. (2010). Groundwater use for irrigation - A global inventory. *Hydrology and Earth System Sciences*, 14(10): 1863–1880.

UNESCO (2018). *Nature-Based Solutions for Water.* The United Nations World Water Development Report (WWDR), World Water Assessment Program, United Nations Educational, Scientific and Cultural Organization (UNESCO), Paris, France.

Van der Gun, J. (2012). *Groundwater and Global Change: Trends, Opportunities and Challenges.* SIDE Publication Series: 01,United Nations World Water Assessment Programme, United Nations Educational, Scientific and Cultural Organization (UNESCO), France, 38 pp.

Veettil, A.V., Konapala, G., Mishra, A.K. and Li, H.Y. (2018). Sensitivity of drought resilience-vulnerability-exposure to hydrologic ratios in contiguous United States. *Journal of Hydrology*, 564: 294–306.

Vieira, E.D.O. and Sandoval-Solis, S. (2018). Water resources sustainability index for a water-stressed basin in Brazil. *Journal of Hydrology: Regional Studies*, 19: 97–109.

WHO (2009). *Summary and Policy Implications Vision 2030: The Resilience of Water Supply and Sanitation in the Face of Climate Change.* World Health Organization (WHO), Geneva.

WHO (2017). *Guidelines for Drinking-Water Quality.* Fourth edition, Incorporating the First Addendum, World Health Organization (WHO), Geneva, 631 pp.

Wunderlin, D.A., delPilar, D.M., Valeria, A.M., Fabiana, P.S., Cecilia, H.A. and de los Angeles, B.M. (2001). Pattern recognition techniques for the evaluation of spatial and temporal variations in water quality. A case study: Suquía River Basin (Córdoba-Argentina). *Water Research*, 35(12): 2881–2894.

Yilmaz, B. (2018). Proposed index for drought assessment: Modified reconnaissance drought index (mRDI). *Journal of Environmental Protection and Ecology*, 19(4): 1796–1804.

Zektser, I.S. and Everett, L.G. (editors) (2004). *Groundwater Resources of the World and Their Use.* IHP-VI, Series on Groundwater No. 6, United Nations Educational, Scientific and Cultural Organization (UNESCO), Paris, France.

11 Big Data Analytics to Enhance the Reliability of Smart Cities

Kanika Jindal and Rajni Aron

CONTENTS

11.1 BIG DATA

This chapter discusses the fundamental concepts of big data (BD) and the importance of big data analytics in today's life.

11.1.1 Definition

Big data came into existence when traditional data mining and handling approaches proved incapable of uncovering the potential insights of the primary data. Initially, relational database systems were used for this purpose, but in this information age, data can be in an unstructured format or time-bound or large in size. Therefore, big data is an approach to analyze these types of data because it is capable of handling vast amount of data processing on readily available hardware. It has revolutionized modern living (Madden, 2012). The main principle behind this approach is to extract useful information from a huge amount of unstructured data for making informed decisions and gaining better results (O'Leary, 2016). Data processing is the building block for an organization to function efficiently and attain its goals (Elgendy & Elraga, 2014).

According to Gartner IT Glossary (2013),

"Big data is high-volume, high-velocity and high-variety information assets that demand cost-effective, innovative forms of information processing for enhanced insight and decision making".

11.1.2 Characteristics of Big Data

Initially, the attributes of big data were Volume, Velocity, and Variety (Laney, 2001), denoted by three Vs. But IBM recommended a fourth V of big data: Veracity. Together, these four Vs form the pillars of big data for defining its basic concept, and are depicted in Figure 11.1.

1) *Volume*: Volume refers to the amount of available data, which is growing day by day at very high speed. Initially, storing big data was challenging because of the high storage costs. However, with decreasing storage costs, this problem has been resolved.
2) *Velocity*: Velocity is the rate of growth of data. It is described as the speed with which data can be gathered for analysis. It is also defined as the rate at which data are being generated every day from different sources.
3) *Variety*: Variety provides information about different types of data, i.e. structured, unstructured, and semi-structured. Structured data are highly organized. But most data available these days are unstructured. Semi-structured data contains meta-data for data analysis and storage but lacks a static scheme like XML files, emails etc. Data are available in hundreds of formats, from documents to databases to excel tables to pictures, videos, and audios.
4) *Veracity*: Veracity refers to the trustworthiness of data. It deals with quality and accuracy of data. Data sources should be trustworthy, and data should be accurate for an automated decision-making process.

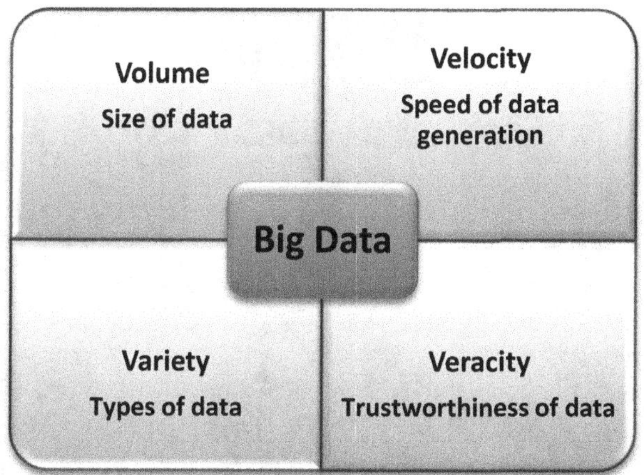

FIGURE 11.1 Big data four Vs.

11.1.3 Big Data Analytics (BDA)

The recent technology that has the potential to change the ways of our everyday functioning and livelihood is big data analytics. It is one step ahead of business intelligence and has the capability to manage these four Vs of big data (Oussous et al., 2018). BDA tools can analyze all the data along with historical information to predict the probability of possible events in the future. BDA employs various methods such as data mining tools, statistical algorithms, text analytics, social network analysis, and machine learning techniques (Chen & Zhang, 2014), which are helpful for data analysts, predictive modelers, statisticians, and data scientists. Therefore, big data analytics has prepared the society to enter into a new era of advanced technologies where people can handle the boundless challenges associated with an ever-growing big data ecosystem.

11.1.4 Importance of Big Data Analytics

Big data analytics is playing a prominent role in businesses and organizations (Mukherjee & Shaw, 2016). To increase their efficiency and predict outcomes, various industries like healthcare, entertainment, transportation, banking, and education are using analytics extensively (Sharma & Mangat, 2015). It can help organizations and businesses to collect information on customer preference and to forecast the demand for specific products. Big data analytics is providing various benefits, which are represented in Figure 11.2. Some benefits are:

1) *Cost Reduction*: BDA technologies help to reduce the cost by providing efficient means of business dealings (Mukherjee & Shaw, 2016). It is helpful in data storage and analysis at low costs, thereby saving capital for future prospects.
2) *Better and Timely Decision Making*: They enable businesses to analyze pertinent information gathered from heterogeneous sources and to make

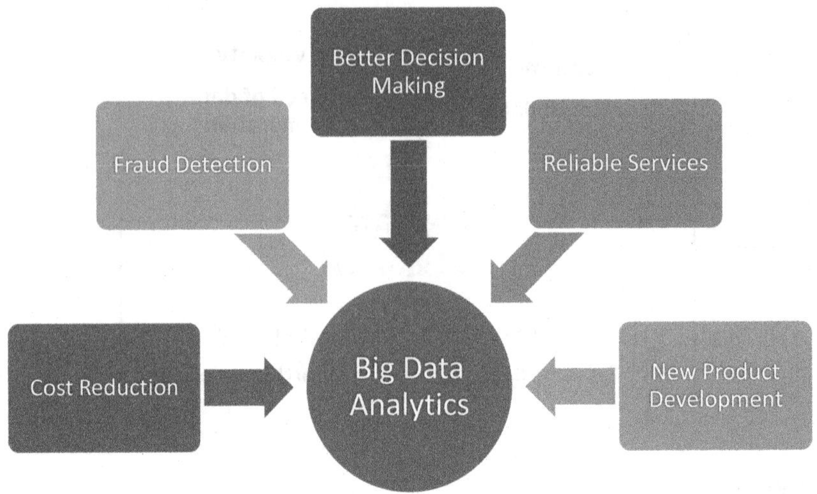

FIGURE 11.2 Benefits of big data analytics.

instant decisions (Tiwarkhede & Kakde, 2015). Thus, BDA helps authorities to take complex, effective, and timely decisions and to mitigate risks.

3) *Fraud Detection*: Data analytics techniques are playing a crucial role in fraud detection and monitoring (Elgendy & Elraga, 2014). Organizations are employing these methods along with human intuition to detect frauds at early stages for protection and security.
4) *Providing Reliable Services*: Nowadays, companies and businesses are interacting with customers in both online and offline modes. They are using data analytics to provide reliable services by customizing the product according to the customer's need (Oussous et al., 2018).
5) *New Product Development*: Customer requirements and satisfaction levels can be understood by using data acquired from social media platforms. Analytical tools can be applied to learn customer preferences, and based on the collected data, new products and services can be developed(Sharma & Mangat, 2015).

11.2 SMART CITY

In this section, the basic concepts related to Smart City (SC) are described. Further, the importance of BDA for smart cities and the need for improving the reliability of smart cities are also discussed.

11.2.1 Definition

A smart city is a city that has the basic infrastructure available to its citizens and that can provide them with a quality lifestyle. It combines information technology with basic functional infrastructure for effective resource management and better decision making (Harrison and Donnelly, 2011). A smart city can be referred to as

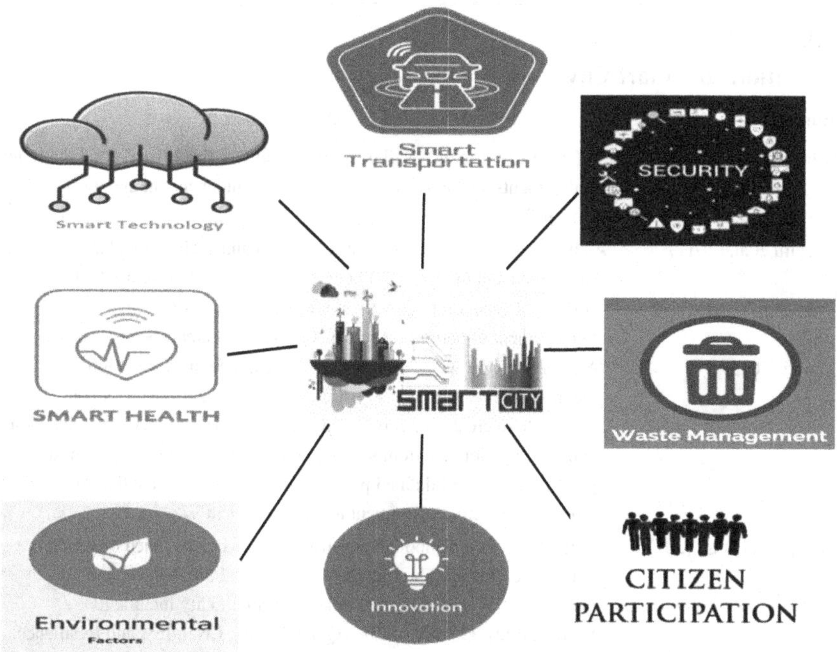

FIGURE 11.3 Smart city concepts.

an amalgamation of Artificial Intelligence (AI), Internet of Things (IoT), Machine Learning (ML), and big data (BD) technologies that help make human life easier and smarter than before. As shown in Figure 11.3, a smart city uses Information and Communication Technology (ICT) to serve various areas of society. Some working definitions of a smart city are discussed in Table 11.1.

11.2.2 Dimensions of a Smart City

Smart cities can be recognized by six main dimensions, i.e. Smart Economy, Smart Mobility, Smart Environment, Smart People, Smart Living, and Smart Governance (Giffinger et al., 2007), as illustrated in Figure 11.4.

1) *Smart Economy*: The fundamental features of a smart economy include: innovative and competitive spirit, increased productivity, cost reduction, digital knowledge, flexible labor market, and socially responsible citizens. All these factors will produce a sustainable and reliable urban ecosystem for citizens, business communities, and authorities.
2) *Smart Mobility*: Today's metropolitan cities are the result of modern urbanization and smart mobility practices. Smart mobility includes innovative, safe, and reliable transportation systems, effortless local or international transport accessibility, real-time traffic management, and car parking services. Therefore, all these aspects are necessary to facilitate smart mobility in cities and provide a secure and friendly environment to dwellers.

TABLE 11.1
Definitions of Smart City

Sources	Definition
Giffinger et al. (2007)	"A Smart City is a city well performing built on the 'smart' combination of endowments and activities of self-decisive, independent and aware citizens".
Caragliu et al. (2011)	A city is smart when investments in human and social capital and traditional (transport) and modern (ICT) communication infrastructure fuel sustainable economic growth and high quality of life, with a wise management of natural resources, through participatory governance.
Dameri (2013)	"A smart city is a well-defined geographical area, in which high technologies such as ICT, logistics, energy production, and so on, cooperate to create benefits for citizens in terms of well being, inclusion and participation, environmental quality, intelligent development; it is governed by a well-defined pool of subjects, able to state the rules and policy for the city government and development".
Ramaprasad et al. (2017)	A Smart City is a compound construct with two parts, each of which is a complex construct. The City is defined by its Stakeholders and the Outcomes. The desirable outcomes of a Smart City include its Sustainability, Quality of Life (QoL), Equity, Livability, and Resilience. The Stakeholders in a city include its Citizens, Professionals, Communities, Institutions, Businesses, and Governments. Thus, the effects on "citizens' QoL", "communities' equity", "businesses' resilience", and 27 ($6 \times 5 - 3$) other possible combinations of Stakeholder and Outcome, defines the smartness of a city.

3) *Smart Environment*: While developing a reliable smart city, governments and authorities should consider a few key parameters. Among them, sustainable management of resources and environment protection should be top priority. Smart and innovative solutions are essential for managing various environmental issues like pollution, water, and waste management and natural resource management.
4) *Smart People*: Key to ensuring a reliable and sustainable smart city are citizens' awareness of and acceptance towards growing technologies, revised government policies, and improved services. Only smart citizens can make a city smart by their creativity, strength of mind, and social contribution, and involvement in activities around them.
5) *Smart Living*: The prominent qualities of smart living in urban systems are providing residents with cultural facilities, better health conditions, smart home systems, and improved public safety. Educated and smart citizens will have social cohesion among them, which, in turn, will be the foundation of a reliable smart city.
6) *Smart Governance*: The main objective of smart governance is to provide public and social services and high-quality infrastructure as well as to take

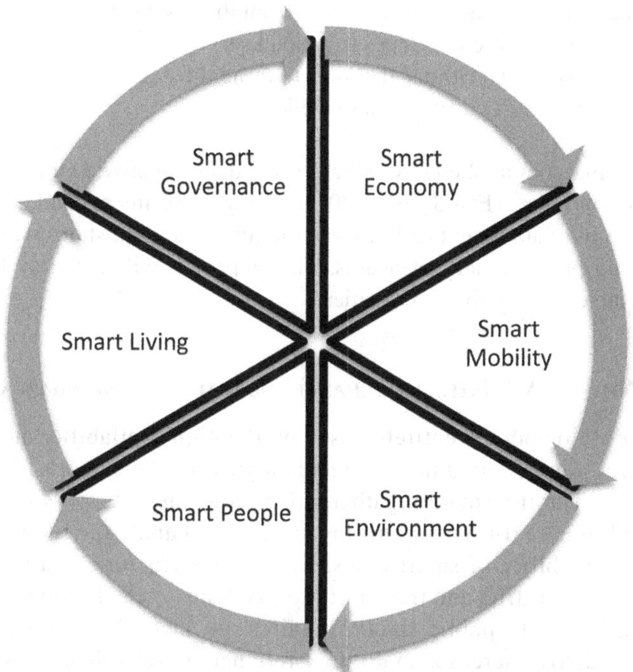

FIGURE 11.4 Smart city dimensions.

efficient and effective decisions for citizens' improved quality of life. Social media analytical techniques are facilitating citizens with a transparent governance system, which is making the cities more reliable and democratic.

11.2.3 Significance of Big Data Analytics for Smart Cities

Big data analytics plays a vital role in the sustainability and reliability of smart cities. Due to the worldwide growth of urbanization, smart devices are generating a huge amount of heterogeneous data every day. These data are stored, processed, and analyzed effectively and efficiently by using Information and Communications Technology (ICT) to generate useful information (Arroub et al., 2016). It can enhance the reliability of smart cities and can help decision makers to develop further smart city services and resources (Kumar & Prakash, 2014). Big data analytics has started to serve numerous sectors of smart cities to enhance their fundamental attributes. These characteristics include sustainability, infrastructure development, transit improvements, efficient public services, intelligent management of resources, and safety and protection of citizens. The major requirements for the smart city (Al Nuaimi et al., 2015) applications are:

1. Trustworthy data sources
2. Sensors to monitor the utilities

3. Interfaces to aggregate real-time data from these sensors
4. ICT to connect these interfaces to a cloud server
5. Wide variety of algorithms and tools to mine the data
6. Big data analytics: putting data to work

For example, analytics applied to weather related data can provide forecasts on perilous weather conditions (Fan & Bifet, 2013). Many governments and authorities are implementing big data analytics to assess the diverse circumstances of their cities and are taking precautionary measures. This approach will certainly lead to economically viable and reliable smart cities.

11.2.4 Why Do We Need to Enhance the Reliability of Smart Cities?

Smart city development is entirely based on the high availability and reliability of data sources and communication technologies for the smooth functioning of its operations. The information gathered from the smart sensors and devices is incorporated into real-time monitoring systems to handle the inefficiencies and improve the reliability of smart cities. Every infrastructure of a reliable smart city, ranging from traffic and transportation systems to electric grids, water, and waste management to public safety, requires advanced ICT. But as the city's population and size increase, so do their reliance on reliable data, sensors, and algorithms. According to Osman et al. (2017), all these factors are adding to other serious concerns of smart city development, such as lack of security, privacy, economic development, and reliability. Therefore, innovative ideas and efficient monitoring systems are required to provide improved quality of life to smart city inhabitants.

Corici et al. (2014) proposed an ETSI M2M communication framework for a reliable smart city platform by investigating various cases and possible solutions. Although much research has been made on the development of smart city applications by using ICT, their reliability has not yet been considered important until now (Tragos et al., 2014). Tragos et al. (2014) have discussed open reliability issues and have developed a framework named RERUM to address these issues. Uribe-Perez and Pous (2017) proposed a flexible communication architecture that can satisfy the requirements of a smart city. The proposed model has been stimulated in the human nervous system and developed using Smart Gateways. This work discusses ways of resolving reliability issues among smart cities. Nathali Silva et al. (2017) proposed a framework of the smart city based on big data analytics which consists of three levels. Even though the architecture presented in the study examines a variety of datasets, it is mainly focused on community development services and the report concludes that the reliability issue has not been fully resolved yet.

11.3 BACKGROUND

This section discusses the various smart city services and the available BDA platforms as well as the related work done in this direction.

11.3.1 Smart City Services

Big data analytics has the potential to provide many smart city services. It needs to be supported with an appropriate Information and Communication Technology (ICT) infrastructure (Al Nuaimi et al., 2015). The amalgamation of these two will always lead to a reliable and sustainable city for its citizens. It also helps to provide better customer experiences and business prospects. Some common applications of BDA in smart cities are discussed below and are presented in Figure 11.5 and their benefits are summarized in Table 11.2.

1) ***Smart Transportation and Traffic Management***
Due to the growth of urban population, traffic congestion in cities has become a major problem nowadays. It can be resolved by using intelligent transportation systems according to the needs of the physical infrastructure of smart cities (Debnath et al., 2014). For this purpose, appropriate data analysis should be made for improving the reliability of transport facilities in smart cities. Big data for the transportation sector can be collected through different sources like sensors, radars, global positioning systems (GPS), and navigation systems. Big data analytics has been widely used to resolve traffic congestion problems in London. It is used to plan the complete journey of passengers by generating accurate travel patterns. Traffic congestion is also avoided by providing alternative routes to customers through emails (Sager Weinstein, 2015).

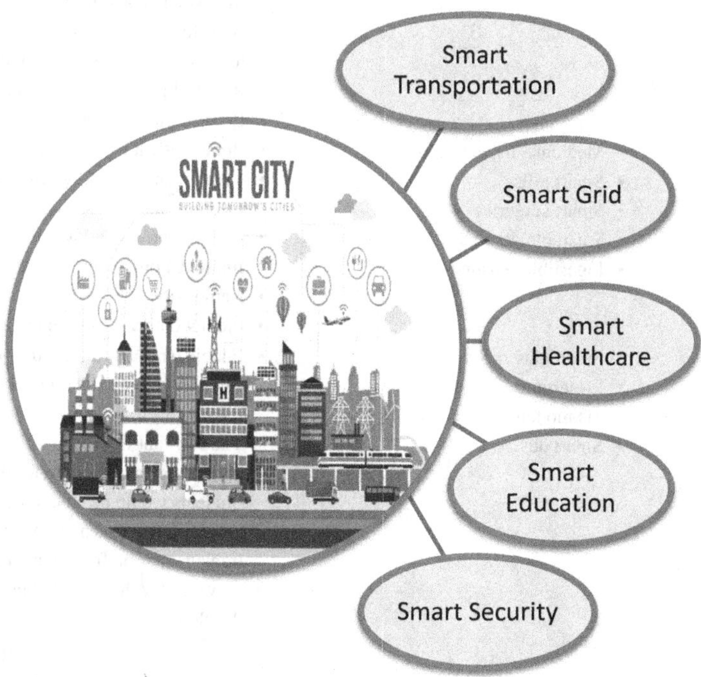

FIGURE 11.5 Key application areas for smart cities.

TABLE 11.2
Applications of Smart City

Applications	Technologies used	Benefits
Smart transportation	• Traffic signal system • Pedestrian signal • Intelligent road studs • Speed cameras • Red light cameras • Incident detection and management system • Traffic news broadcasting • Public transport information sharing system • Taxi booking system • Parking charge payment system • Electronic toll collection system	• Road safety improvement by adopting preventative measures • Real-time tracking • Automatic traffic management • Efficient route management • Less traffic congestion • Parking management
Smart grid	• Wireless smart meters • Phasor Measurement Units (PMU) • Sensors	• Efficient electricity transmission • Improved incorporation of large-scale renewable energy resources • Faster electricity restoration after power turbulence • Enhanced integration of customer and provider power generation systems • Reduced management and operational costs for services which eventually lead to lower electricity costs for consumers • Reduced peak demand of consumers
Smart healthcare	• Sensors • Medicine dispensation • Smart pills • Smart surgeries • Wearable devices • Ingestible sensors	• Lower costs of treatment • Better patient experience • Better management of drugs and medicine adherence • Reduced errors and waste • Improved outcomes of treatment • Remote medical assistance • Predict outbreaks of epidemics
Smart education	• Smart classrooms • E-learning • 3D models • Smart devices	• Enable teachers to meet the needs of students with specific modes of learning. • The constant flow of real-time knowledge and learning for the world. • Students would be able to perform analysis, experimentation, and implementation from what they study. • Advanced technology to help students develop.

(*Continued*)

TABLE 11.2 (CONTINUED)
Applications of Smart City

Applications	Technologies used	Benefits
		• Helping each student and giving them an equal opportunity to perform. • Inspiring students to become responsible citizens who can contribute towards the development of their society.
Smart security	• Sensors • Mobile apps • Intelligent Electronic Devices (IEDs) • Closed Circuit Television Cameras (CCTV)	• Remote accessibility • Provides night vision recording system • Doorbell cameras help to respond from anywhere

2) ***Smart Grid***

A smart grid refers to a two-way digital communication technology application used to supply electricity to its customers. When a system of computers, controls, automation tools, and technologies works together on an electric grid to manage the electricity supply chain, then it is termed as a smart grid. Smart cities are using smart meters to improve the reliability and efficiency of production and allocation of power supply (Munshi & Yasser, 2017). Many governments have started to incorporate smart grids to tackle the problems of emergency resilience and global warming. The analytics applied to smart grid data guides the users in the following ways:

I. To minimize energy consumption in case of heavy loads
II. To use low priority electric devices when electricity rates are higher
III. To decentralize power generation
IV. To predict the future needs of power supply

3) ***Smart Healthcare***

Big data analytics has played an important role in turning traditional healthcare systems into smart healthcare. Smart cities are using Information and Communication Technology (ICT) along with healthcare analytics to enhance the quality of life (Wang et al., 2018). This technology also helps in making better and timely diagnosis of diseases as well as providing improved treatment for patients. Smart healthcare includes the services of electronic health (eHealth) and mobile health (mHealth) in smart cities. These are used for information sharing between patients and health professionals, managing electronic health records, and providing portable monitoring devices for patients. Big data analytics in healthcare have the potential to predict the outbreak of serious diseases, minimize the costs of treatments, and improve the reliability and sustainability of smart citizens' lives (Belle et al., 2015).

4) ***Smart Education***

The traditional education system comprises face-to-face communication between teachers and students. It includes group discussions and conducting tests and examinations for easy and effective learning (Cristobal & Ventura, 2013). As smart education is the backbone of smart city development, big data analytics is nowadays being used to deploy digital technologies in the field of education in a better way. It includes the e-learning system and various teacher-centric tools for facilitating effective communication and innovative thinking of students (Elhoseny et al., 2018). It is also helpful for the teachers to better understand the interests and abilities of each individual student. Smart cities have acquired the smart classroom infrastructure, which provides digitized education to the students. This technology assists the teachers in preparing lectures and responding to students' queries efficiently and professionally.

5) ***Smart Security***

ICT is a vital component of a smart city's various sectors such as healthcare, transportation, security, and waste management. Big data analytics is an appropriate technique to deal with the massive data of these sectors, but with the rising volumes of data, public safety and security has become a major concern in smart city development. The government is taking necessary steps to administer these issues effectively and efficiently. Smart cities have implemented smart surveillance systems for the reinforcement of public welfare and democratic systems. These systems can alert citizens about irregular incidents and anomalous circumstances. Government officials and authorities are using online forums and web portals to access the suggestions and complaints people share on these platforms. Smart city citizens can obtain information about various government policies and schemes like GST and RTI and report on various issues such as corruption or women safety. Therefore, big data analytics is playing a major role in building a safe and secure city (Gahi et al., 2016).

11.3.2 Smart City-Based Big Data Analytics Platforms

Reliable and scalable platforms are required to perform data analytics on smart cities. These platforms are incorporated to execute efficiently the tasks of data assembling, data analysis, and data visualization. Selecting an appropriate analytics platform is an exigent task that depends on a variety of aspects like types of optimal infrastructure, storage facilities, and intensive applications required by the reliable smart city. The two commonly used scaling approaches for these platforms are horizontal scaling and vertical scaling (Tsai et al., 2016). Nowadays, most of the big data analytics frameworks are designed around horizontally scalable platforms (Osman, 2019). Hadoop (Diaconita et al., 2018) is a data analytics platform that can handle massive parallelism of data processing and fault tolerance due to data repetition. Another platform that can handle huge datasets is Cloudera (Vera-Baquero & Colomo-Palacios, 2018), but it can't handle privacy and security issues as efficiently as other platforms. Real-time data analysis can be efficiently done using

SAP-HANA (Hopkins & Hawking, 2018) and MongoDB (Santos et al., 2018), but they are not as powerful as Hadoop. Therefore, the data analytics platforms should be enhanced with new characteristics, functionalities, and services to improve the reliability and economic development of smart cities. Different big data analytics platforms for smart cities are listed in Table 11.3.

11.3.3 Related Work

A smart city uses an interconnected network of sensors for data collection and analysis. The main objective of a smart city is to enhance the social and economic development of its citizens (Pal et al., 2018). In this section, the work related to various proposed approaches has been discussed, and these approaches are presented in Table 11.4. Pramanik et al. (2017) proposed a smart healthcare framework, BSHSF, to emphasize the role of big data and smart systems in the growth of a sustainable healthcare environment. But this work has highlighted some challenges of this approach involves some challenges such as increased technological complexities, financial and cultural constraints, and reliability improvement. According to Gohar et al. (2018), new architectures capable of storing and analyzing big data generated by sensors are required. So, a four layered BDA architecture for an intelligent transportation system has been proposed, but its scope is limited to measuring the average speed of a vehicle. Elhoseny et al. (2018) explored the area of Learning Analytics by proposing a framework that integrates

TABLE 11.3
BDA Platforms used in Smart Cities

BDA platforms	Sources	Advantages	Disadvantages
Hadoop	Diaconita et al. (2018)	• Economical • Most scalable • Fast speed	• Not applicable to small datasets • Lack of encryption methods
Cloudera	Vera-Baquero and Colomo-Palacios (2018)	• Ability to handle data duplication • Mostly apt for smart cities	• Lack of independent hardware and software systems • Lack of security and privacy
SAP-HANA	Hopkins and Hawking (2018)	• Ability to handle real-time analytics for smart cities • Low latency	• Lack of flexibility • Lack of support to handle complex queries
MongoDB	Santos et al. (2018)	• Horizontally scalable • Ability to instantly respond to online queries	• Less powerful than Hadoop • Very costly
Map Reduce	Gohar et al. (2018)	• Less expensive • Better security and privacy	• Very complex

TABLE 11.4
Existing Smart Cities Approaches

S. no.	Sources	Purpose	Target area
1.	Strohbach et al. (2015)	To propose an initial version of BDA framework for smart city applications to deal with data velocity and volume issues.	Smart grid
2.	Uribe-Perez and Pous (2017)	To propose communication architecture stimulated in the human nervous system to fulfill the requirements of a real smart city.	Communication and scalability issues for smart cities
3.	Hashem et al. (2016)	To present a future business model to manage big data used by smart cities and discuss the technical challenges associated with the proposed model.	Big data management
4.	Silva et al. (2018)	To propose a smart city framework based on BDA that is operational on data generation, data processing, and application levels.	Real-time decision making
5.	Pramanik et al. (2017)	To propose a big data enabled framework titled BSHSF to conceptually represent the intra and inter organizational operations related to the healthcare business.	Smart healthcare
6.	Ahmed M. Shahat Osman (2019)	To propose a framework named Smart City Data Analytics Panel (SCDAP) to analyze the characteristics and design principles of BDA.	Model management and model aggregation
7.	Elhoseny et al. (2018)	To propose a framework using big data learning analytics platform to enhance the smart learning environment.	Smart education
8.	Pal et al. (2018)	To present a smart city paradigm based on big data comprising of four hubs architecture. Certain case studies of smart cities are also discussed.	Big data solutions for smart cities
9.	Silva et al. (2018)	To propose a generic four layered bottom-up architecture of a smart city.	Data security
10.	Gohar et al. (2018)	To propose a four layered BDA architecture for intelligent transportation system of a smart city.	Smart transportation

the traditional e-learning systems with the smart city environment. Garg (2016) recognized the growing complexities of an industrial system due to increasing number of components and developed a framework to analyze and predict the pattern of system failure. Using this framework, uncertainties in the applications can be minimized for realistic decision making and reliability enhancement. Niwas and Garg (2018) proposed a mathematical approach based on the Markov process to analyze various parameters like reliability, expected profit, availability, and rate of system failure. The future work highlights to work upon the analysis based on these parameters for multiple and complex industrial systems.

11.4 RELIABILITY-BASED BIG DATA ANALYTICS (BDA) FOR SMART CITIES

Big data analytics has huge potential to enhance the reliability of smart city services with the support of ICT. It is quite evident that data lies at the heart of smart cities. Large amount of data is being generated and made available from the devices used in everyday life to the environmental sensors. Big data analytics is employed to convert this hefty amount of unstructured and unorganized data into valuable and meaningful information for smart city citizens, visitors, and the business community (Osman, 2019). This section highlights the various analytical methods, requirements, and parameters for a reliable smart city, and the proposed framework is illustrated in Figure 11.8.

11.4.1 Big Data Analytical Techniques for Reliable Smart Cities

A brief description of various types of big data analytics methods (Sivarajah et al., 2017) is provided in this section, followed by a diagrammatic illustration in Figure 11.6. All these techniques help smart city residents in the decision-making process.

1) *Descriptive Analytics:* As the name implies, descriptive analytics is used to describe or summarize the past results. It emphasizes "what happened and why it happened". It enables smart city residents to learn from past trends and patterns, and to understand how they might affect the future possible results. This phase of analytics uses data aggregation and data mining tools to achieve results. It deals with comprehensive, accurate, and live data. The most common examples of descriptive analytics are related to financial, sales, and performance analysis.
2) *Diagnostic Analytics:* This type of analytics highlights "why is it happening". The main focus of this field is on processes and causes rather than results. It has the ability to drill down the root cause of the situation and isolate the confounding information. Due to the increase in data volume, velocity, and variety, diagnostic analytics has become feasible with the help of machine learning techniques. Diagnostic analytics can be best used in the healthcare industry to diagnose possibilities of patient's ailments, which in turn can reduce the risk of hospitalization.

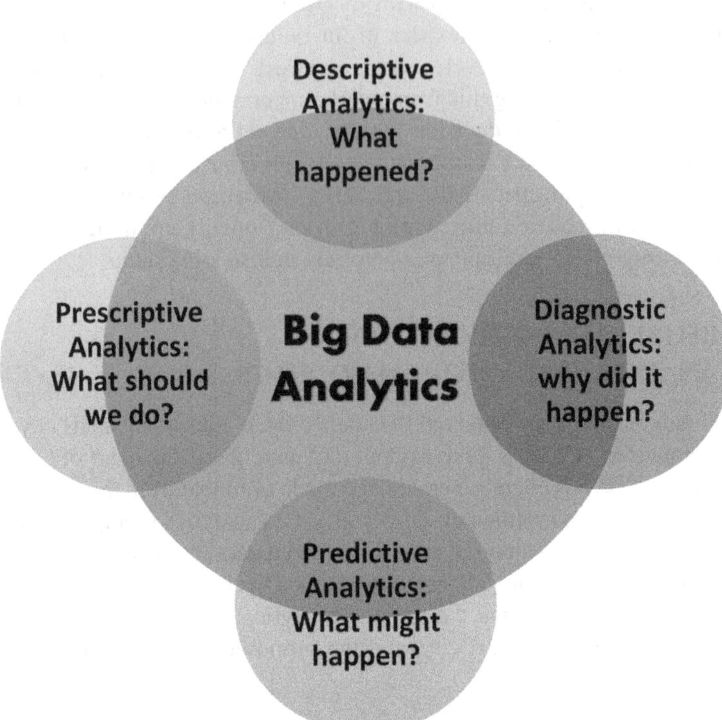

FIGURE 11.6 Analytics for smart cities

3) *Predictive Analytics*: It is used to make an educated guess to predict future outcomes based on past and present available data. It focuses on "what might happen". It is capable of using a stream of data generated from several sensors to predict the behavioral patterns of smart city citizens. It is based on statistical methods. Predictive analytical techniques can be regression techniques and machine learning techniques. Various factors like heterogeneity, noise accumulation, spurious correlations, and lack of computational efficiency are major concerns for developing new statistical methods (Gandomi & Haider, 2015). Sentiment analysis and credit score are the best examples of predictive analytics.
4) *Prescriptive Analytics*: Prescriptive analytics relates to both descriptive and predictive analytics. This field of analytics is used to apply advanced analytical techniques to make specific recommendations. It guides smart city citizens with feasible solutions to a problem. It focuses on "what should we do". This type of analytics is used for making important, time-sensitive, and complex decisions. Optimization and simulation-based models are used for prescriptive analytics. The main objective of prescriptive analytics is quality improvement and service enhancement.

11.4.2 Requirements

The accessibility of big data analytics provides imperative benefits and improves the overall process of decision making in smart cities. While considering big data-based smart city applications, numerous technological and social requirements need to be addressed (Al Nuaimi et al., 2015). This section summarizes several requirements to provide a suitable framework for a reliable smart city.

1) *Appropriate Data Processing Platforms*: Data is being generated from various smart city sectors such as energy, health, education, and traffic. All these data need to be accurately utilized to provide reliable urban services. Reliable and feasible data processing platforms are required for computational efficiencies, stream processing of data, and fault tolerance (Pal et al., 2018). Major advantages and disadvantages of these platforms are given in Table 11.3.
2) *Smart Networks*: Smart networks are required to transfer the data gathered from smart sensors and devices into the data processing module (Hashem et al., 2016). Reliable and high-speed connectivity of networks is the best approach to connect the basic infrastructures with well-equipped smart city services.
3) *Analytical Techniques*: To ensure effective decision making and enhanced economic development, appropriate analytical techniques should be applied to smart city data. The important analytical techniques used nowadays are descriptive, diagnostic, predictive, and prescriptive (Sivarajah et al., 2017), which have already been discussed above.
4) *Trustworthy Databases*: Smart mobiles and devices are accountable for enormous data generation that need to be assembled, processed, and interpreted for it to be valuable for users. Data need to be stored in upgraded and reliable databases for further use (Susmitha & Jayaprada, 2017).
5) *Citizen Awareness*: Smart city citizens must be conscious of the accurate usage of ICT for better security and privacy of data. The residents of urban suburbs should keenly participate in social activities and respond to various social issues. To promote this, the government must conduct social awareness programs and campaigns to develop a reliable city (Al Nuaimi et al., 2015).

11.4.3 Parameters

A reliable and sustainable smart city employs ICT to provide numerous services to its inhabitants. Some of these services are smart transportation, smart healthcare, smart energy, smart education, smart security, etc. Every smart city application is influenced by certain parameters for performance evaluation and future expansion. In a smart transportation system, the traffic surveillance system using signals and lights is an integral part of smart cities. Big data analytics is used to generate traffic patterns by interconnecting various traffic signals. It is helpful to make intelligent and timely decisions to avoid the occurrence of accidents. The smart education system includes smart classrooms, e-learning system, and 3D models to boost the

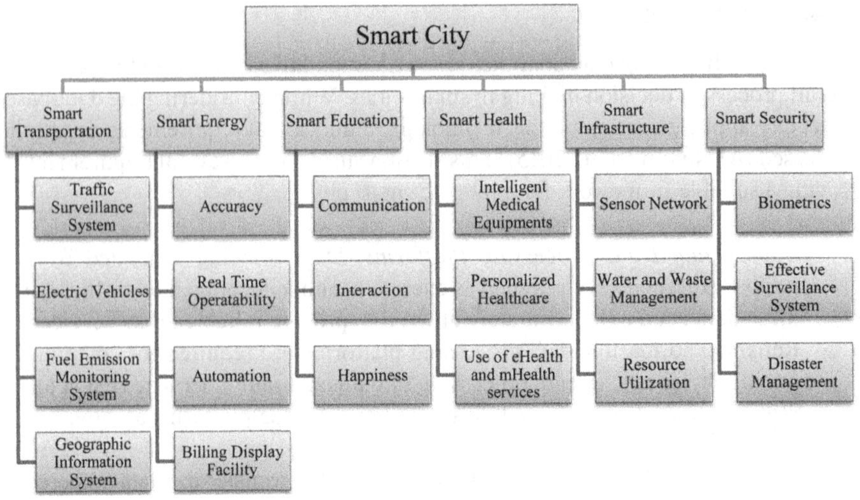

FIGURE 11.7 Taxonomy of smart city parameters.

overall development of students and help them to become responsible and active citizens. In the healthcare context, eHealth and mHealth services are playing a prominent role in effective communication between patients and medical professionals. Security and safety of smart city citizens can be efficiently provided by incorporating smart devices like biometrics and CCTVs with sensor networks. The smart city infrastructure depends on the proper utilization of resources, networks, and devices and on responding to citizens' needs in a real-time environment. Various parameters related to different smart city application areas are given in Figure 11.7.

11.4.4 Proposed Framework: Mode of Operation

A novel and ubiquitous framework has been proposed in this work by keeping in mind the abovementioned requirements. A detailed description of the proposed framework is provided below, followed by a brief diagrammatic representation in Figure 11.8.

1) The foremost step is to acquire smart city data from various sources. Presently, a variety of data sources (Hashem et al., 2016) such as radio-frequency identification (RFID), cameras, environmental sensors, global positioning systems (GPS), smart phones, and social media platforms are generating data at a rapid pace. All the data, including a large amount of redundant or unimportant data, are generated in this phase. Thus, data extraction is also a part of this phase. The challenges presented in data extraction are twofold: first, according to the nature and the initial context of the generated data, it has to be decided which data to keep and which one to discard. Second, the deficiency of a common platform presents its own set of challenges. Due to the broad range of existing data, a common platform has been established to standardize the data extraction process.

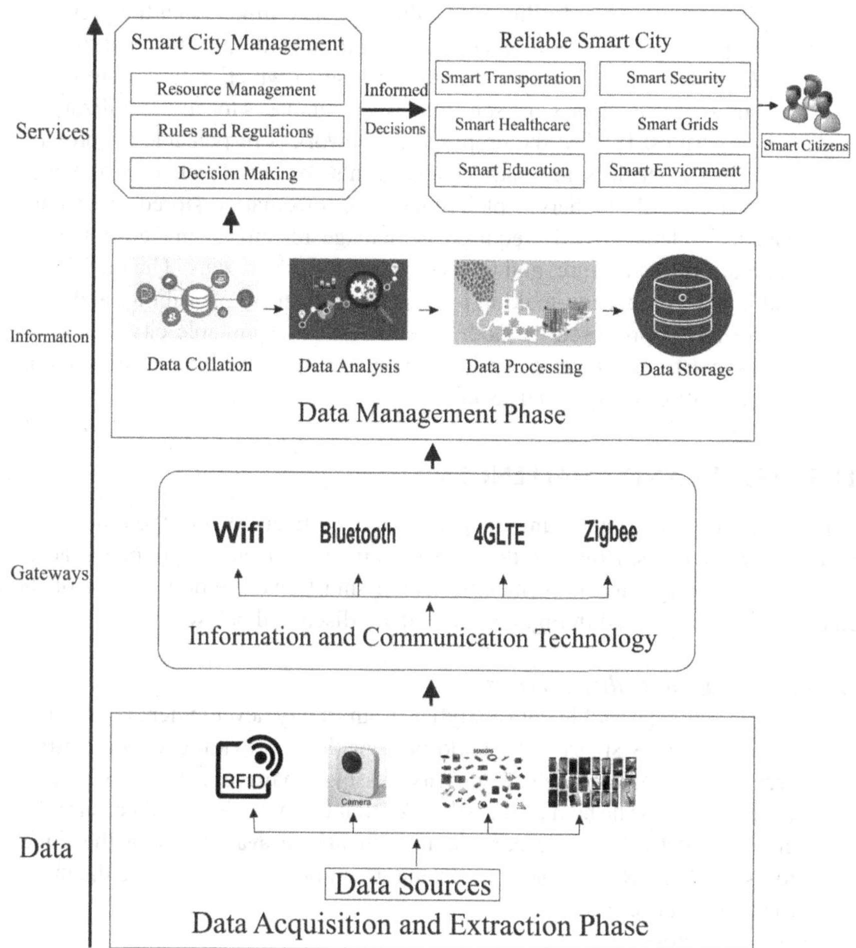

FIGURE 11.8 Reliability-based data analytics framework for smart cities.

2) Smart city applications need efficient ICT applications to connect accumulated data with the processing and storage phases. Smart gateways, such as Zigbee, Bluetooth, Wi-Fi (Yaqoob et al., 2017), and smart networks are used to transfer crucial information to data management level.
3) Data management is a transitional phase between data collection and reliable smart city development. It is the most intelligent phase of the proposed framework. It includes the following sublevels:
   I. *Data Collation*: It integrates multiple data sources of various sectors of a smart city and provides a better picture of the collected data to the next level for analysis.
   II. *Data Analysis*: It includes targeted areas of interest and provides results based on structured data.

III. *Data Processing*: In this phase, the results obtained from the previous step are processed and interpreted to gain meaningful information.
IV. *Data Storage*: It is the concluding phase in which data are stored in smart city databases and sensor network databases for future utilization.

4) The uppermost level of the proposed framework is responsible for providing reliable services to smart city residents. At this level, governments, authorities, and planners exploit smart city information stored in various databases. This stage is required to manage resources, make strategies, maintain infrastructure, and take decisions for the citizens. The intelligent and informed decision-making process strengthens the feasibility and reliability of the proposed method. A reliable and sustainable city suburb is competent to facilitate its various sectors such as healthcare, education, transportation, energy, and security.

11.5 ISSUES AND CHALLENGES

Smart cities are providing countless benefits to its citizens due to the rapid technological advancements. However, there are certain challenges that are being faced by developers to design and implement potential smart city solutions. Some of these challenges are illustrated in Figure 11.9, and are discussed below:

1) ***Reliability of Utility Services***
A key challenge for big data analytics in smart city development is ensuring its reliability. A smart city should be capable of providing various utility services, i.e. water supply, electricity, wastage management, and safety to its inhabitants (Hashem et al., 2016). A reliable smart city should ensure the effective and efficient resource management and availability of the same to its dwellers. Renewable sources and green buildings should be the main focus of smart cities.
2) ***Infrastructure***
Deploying smart city applications using big data analytics needs the support of a good ICT infrastructure (Silva et al., 2018). This infrastructure includes physical components, social aspects, and smart networks. Physical infrastructure needs proficient construction and maintenance of resources, whereas social infrastructure needs citizen awareness, rational investment, and so on. Smart networks are essential for the effective communication between these two to augment the quality of life of citizens.
3) ***Financing***
Project funding is a major issue in smart city development. Governments have to take the initiative to overcome this problem by implementing various financing strategies (Pal et al., 2018). It deals with the concerns about both the physical infrastructure and technological development. Lack of these financial resources can lead to the deficit of reliable and self-sustainable cities. Private financial sectors should also join hands with governments to overcome this barrier.

FIGURE 11.9 Challenges for smart cities.

4) ***Security and Privacy***
Security and privacy are among the major challenges being faced by smart cities (Bassoo et al., 2018). Although smart cities are providing numerous benefits to their inhabitants, a lot of threats are also associated with growing technologies. These days, the data accumulated by smart devices and sensors tend to be vulnerable to cyber attacks. Lack of trust and confidentiality between the government and citizens, unauthorized access to citizens' data, and fraud detection are other major serious issues that need to be addressed. Smart city developers and planners must include security and privacy guidelines and strategies to ensure the well-being of the society.

5) ***Planning***
Smart city planners come across numerous challenges while planning and developing comprehensive and feasible urban management strategies. Smart city solutions should be achievable in a short span of time for real-time applications (Hashem et al., 2016). The major obstructions in this context are lack of innovative ideas, funds, skilled manpower, and reliable databases. Therefore, authorities and developers must keep in mind these concerns for sustainability and reliability of smart cities.

6) ***Cost***
Future smart cities have to upgrade their technologies and infrastructure due to rapid urbanization. The major issue involved in acquiring, implementing, and integrating these components is the cost. Development of smart city includes designing, operational, and maintenance costs (Silva et al., 2018). Governments and authorities should be conscious of the cost optimization mechanisms to evade diverse issues like traffic congestion, increasing pollution, and scarcity of resources in the coming years.

7) ***Data Integration***
Nowadays, big data is generated from heterogeneous sources in structured as well as unstructured format. Therefore, it is essential to use reliable and realistic data sources. Effective smart city transformation can only be possible with the proliferation of big data. Data integration is an imperative challenge for smart cities which handle data quality management mechanisms (Gouveia et al., 2016). This concern needs to be addressed for well-organized urban cities.

11.6 DISCUSSION

The main scope of this work is to augment the reliability of smart cities by an efficient BDA framework. The traditional data processing techniques are not capable of handling the huge amount of unstructured data generated from smart city applications. Therefore, it is necessary to incorporate BDA into smart city environments. Various issues and challenges discussed in this chapter could be tested and employed in future studies for reliability management applications. The security and privacy of data need to be preserved to avoid the encumbrance to the growth of sustainable and reliable future smart cities. In a smart city, the recognition of urban reliability is based on citizen participation, technical adaptation, environmental aspects, and authorities' accountability. Therefore, the work done in this direction should focus on all these factors to improve the sustainability, resilience, and reliability of smart city services.

11.7 CONCLUSION AND FUTURE DIRECTIONS

Big data analytics has been progressively recognized as a new computing paradigm for smart city development. The smart city is exploiting the concept of BDA to enhance the livelihood of its inhabitants. The incorporation of BDA in the smart cities is treasuring the spectacular innovations and urban modernization. This chapter is presenting the fundamental concepts of Big Data Analytics and Smart Cities as well as the associated future challenges. A ubiquitous and reliability-based framework of BDA that is geared toward providing reliable services to smart city citizens has been proposed in this work. The authors have discussed the various data analytics techniques and platforms that can improve economic development and provide a sustainable smart city life. The smart city provides its numerous services to its residents, which have also been discussed.

Although extensive progress has been made in recent years, several challenges and issues associated with smart city development still need to be addressed. Therefore, resolving these concerns is a prominent research opportunity in the future to enhance the reliability of smart cities. Future research directions include the development of an efficient algorithm and user interfaces at the application level. Consequently, the proposed framework could be employed in real-world applications of a reliable smart city.

ACKNOWLEDGMENT

The authors would like to sincerely thank the editor and anonymous reviewers for their valuable and insightful comments and suggestions that improved this study.

REFERENCES

Al Nuaimi, E., H. Al Neyadi, N. Mohamed, and J. Al-Jaroodi. "Applications of big data to smart cities." *Journal of Internet Services and Applications* 6(1) (2015): 25.

Arroub, Ayoub, Bassma Zahi, Essaid Sabir, and Mohamed Sadik. "A literature review on smart cities: Paradigms, opportunities and open problems." In *2016 International Conference on Wireless Networks and Mobile Communications (WINCOM)*, pp. 180–186. IEEE, 2016.

Bassoo, V., V. Ramnarain-Seetohul, V. Hurbungs, T.P. Fowdur, and Y. Beeharry. "Big data analytics for smart cities." In Dey, N., Hassanien, A., Bhatt, C., Ashour, A., and Satapathy, S. (eds) *Internet of Things and Big Data Analytics Toward Next-Generation Intelligence*, pp. 359–379. Springer, Cham, 2018.

Belle, Ashwin, Raghuram Thiagarajan, S.M. Soroushmehr, Fatemeh Navidi, Daniel A. Beard, and Kayvan Najarian. "Big data analytics in healthcare." *BioMed Research International* 2015 (2015): 1–16.

Caragliu, Andrea, Chiara Del Bo, and Peter Nijkamp. "Smart cities in Europe." *Journal of Urban Technology* 18(2) (2011): 65–82.

Chen, C.L. Philip, and Chun-Yang Zhang. "Data-intensive applications, challenges, techniques and technologies: A survey on big data." *Information Sciences* 275 (2014): 314–347.

Corici, Andreea, Asma Elmangoush, Ronald Steinke, Thomas Magedanz, Joyce Mwangama, and Neco Ventura. "Utilizing M2M technologies for building reliable smart cities." In *2014 6th International Conference on New Technologies, Mobility and Security (NTMS)*, pp. 1–5. IEEE, 2014.

Cristobal, Romero, and Sebastian Ventura. "Data mining in education." *Wiley Interdisciplinary Reviews: Data Mining and Knowledge Discovery* 3(1) (2013): 12–27.

Dameri, Renata Paola. "Searching for smart city definition: A comprehensive proposal." *International Journal of Computers and Technology* 11(5) (2013): 2544–2551.

Debnath, Ashim Kumar, Hoong Chor Chin, Md. Mazharul Haque, and Belinda Yuen. "A methodological framework for benchmarking smart transport cities." *Cities* 37 (2014): 47–56.

Diaconita, Vlad, Ana-Ramona Bologa, and Razvan Bologa. "Hadoop oriented smart cities architecture." *Sensors* 18(4) (2018): 1181.

Elgendy, Nada, and Ahmed Elragal. "Big data analytics: A literature review paper." In *Industrial Conference on Data Mining*, pp. 214–227. Springer, Cham, 2014.

Elhoseny, Hisham, Mohamed Elhoseny, Alaa Mohamed Riad, and Aboul Ella Hassanien. "A framework for big data analysis in smart cities." In *International Conference on Advanced Machine Learning Technologies and Applications*, pp. 405–414. Springer, Cham, 2018.

Fan, Wei, and Albert Bifet. "Mining big data: Current status, and forecast to the future." *ACM SIGKDD Explorations Newsletter* 14(2) (2013): 1–5.

Gahi, Youssef, Mouhcine Guennoun, and Hussein T. Mouftah. "Big data analytics: Security and privacy challenges." In *2016 IEEE Symposium on Computers and Communication (ISCC)*, pp. 952–957. IEEE, 2016.

Gandomi, Amir, and Murtaza Haider. "Beyond the hype: Big data concepts, methods, and analytics." *International Journal of Information Management* 35(2) (2015): 137–144.

Garg, Harish. "Modeling and analyzing system failure behavior for reliability analysis using soft computing-based techniques." In Pham, H. (ed) *Quality and Reliability Management and Its Applications*, pp. 85–115. Springer, London, 2016.

Gartner Inc. *What is Big Data?* Gartner IT Glossary, 2013. [Online] Available at https://www.gartner.com/it-glossary/big-data.

Giffinger, Rudolf, Christian Fertner, Hans Kramar, and Evert Meijers. "City-ranking of European medium-sized cities." Centre of Regional Science, Vienna University of Technology (UT), Austria (2007): 1–12.

Gohar, Moneeb, Muhammad Muzammal, and Arif Ur Rahman. "SMART TSS: Defining transportation system behavior using big data analytics in smart cities." *Sustainable Cities and Society* 41 (2018): 114–119.

Gouveia, João Pedro, Júlia Seixas, and George Giannakidis. "Smart city energy planning: Integrating data and tools." In *Proceedings of the 25th International Conference Companion on World Wide Web*, pp. 345–350. International World Wide Web Conferences Steering Committee, 2016.

Harrison, Colin, and Ian Abbott Donnelly. "A theory of smart cities." In *Proceedings of the 55th Annual Meeting of the ISSS-2011, Hull, UK*, 55(1), 2011.

Hashem, Ibrahim Abaker Targio, Victor Chang, Nor Badrul Anuar, et al. "The role of big data in smart city." *International Journal of Information Management* 36(5) (2016): 748–758.

Hopkins, John, and Paul Hawking. "Big data analytics and IoT in logistics: A case study." *The International Journal of Logistics Management* 29(2) (2018): 575–591.

Kumar, A., and Anand Prakash. "The role of big data and analytics in smart cities." *International Journal of Scientific Research (IJSR)* 6(14) (2014): 12–23.

Laney, Doug. "3D data management: Controlling data volume, velocity and variety." *META Group Research Note* 6(70) (2001): 1. http://blogs.gartner.com/doug-laney/files/2012/01/ad949-3D-Data-Management-Controlling-Data-Volume-Velocity-and-Variety.pdf

Madden, Sam. "From databases to big data." *IEEE Internet Computing* 16(3) (2012): 4–6.

Mukherjee, Samiddha, and Ravi Shaw. "Big data–concepts, applications, challenges and future scope." *International Journal of Advanced Research in Computer and Communication Engineering* 5(2) (2016): 66–74.

Munshi, Amr A., and A.-R.I. Mohamed Yasser. "Big data framework for analytics in smart grids." *Electric Power Systems Research* 151 (2017): 369–380.

Nathali Silva, Bhagya, Murad Khan, and Kijun Han. "Big data analytics embedded smart city architecture for performance enhancement through real-time data processing and decision-making." *Wireless Communications and Mobile Computing* 2017 (2017): 1–12.

Niwas, Ram, and Harish Garg. "An approach for analyzing the reliability and profit of an industrial system based on the cost free warranty policy." *Journal of the Brazilian Society of Mechanical Sciences and Engineering* 40(5) (2018): 265.

O'Leary, Daniel E. "Ethics for big data and analytics." *IEEE Intelligent Systems* 31(4) (2016): 81–84.

Osman, Ahmed M. Shahat. "A novel big data analytics framework for smart cities." *Future Generation Computer Systems* 91 (2019): 620–633.

Osman, Ahmed M. Shahat, Ahmed Elragal, and Birgitta Bergvall-Kåreborn. "Big data analytics and smart cities: A loose or tight couple?" In *10th International Conference on Connected Smart Cities 2017 (CSC 2017), Lisbon, July 20–22, 2017*, pp. 157–168. IADIS, 2017.

Oussous, Ahmed, Fatima-Zahra Benjelloun, Ayoub Ait Lahcen, and Samir Belfkih. "Big data technologies: A survey." *Journal of King Saud University-Computer and Information Sciences* 30(4) (2018): 431–448.

Pal, Debajyoti, Tuul Triyason, and Praisan Padungweang. "Big data in smart-cities: Current research and challenges." *Indonesian Journal of Electrical Engineering and Informatics (IJEEI)* 6(4) (2018): 351–360.

Pramanik, Md Ileas, Raymond YK Lau, Haluk Demirkan, and Md Abul Kalam Azad. "Smart health: Big data enabled health paradigm within smart cities." *Expert Systems with Applications* 87 (2017): 370–383.

Ramaprasad, Arkalgud, Aurora Sánchez-Ortiz, and Thant Syn. "A unified definition of a smart city." In *International Conference on Electronic Government*, pp. 13–24. Springer, Cham, 2017.

Sager Weinstein, L. *Innovations in London's Transport: Big Data for a Better Customer Experience*. TfL, November 2015.

Santos, Jose, Thomas Vanhove, Merlijn Sebrechts, et al. "City of things: Enabling resource provisioning in smart cities." *IEEE Communications Magazine* 56(7) (2018): 177–183.

Sharma, Sunaina, and Veenu Mangat. "Technology and trends to handle big data: Survey." In *2015 Fifth International Conference on Advanced Computing & Communication Technologies*, pp. 266–271. IEEE, 2015.

Silva, Bhagya Nathali, Murad Khan, and Kijun Han. "Towards sustainable smart cities: A review of trends, architectures, components, and open challenges in smart cities." *Sustainable Cities and Society* 38 (2018): 697–713.

Sivarajah, Uthayasankar, Muhammad Mustafa Kamal, Zahir Irani, and Vishanth Weerakkody. "Critical analysis of big data challenges and analytical methods." *Journal of Business Research* 70 (2017): 263–286.

Strohbach, Martin, Holger Ziekow, Vangelis Gazis, and Navot Akiva. "Towards a big data analytics framework for IoT and smart city applications." In *Modeling and Processing for Next-Generation Big-Data Technologies*, pp. 257–282. Springer, Cham, 2015.

Susmitha, K., and S. Jayaprada. "Smart cities using big data analytics." *International Research Journal of Engineering and Technology (IRJET)* 4(8) (2017): 1617.

Tiwarkhede, Ankita S., and Vinit Kakde. "A review paper on big data analytics." *International Journal of Scientific and Engineering and Research* 4(4) (2015): 845–848.

Tragos, Elias Z., Vangelis Angelakis, Alexandros Fragkiadakis, et al. "Enabling reliable and secure IoT-based smart city applications." In *2014 IEEE International Conference on Pervasive Computing and Communication Workshops (PERCOM WORKSHOPS)*, pp. 111–116. IEEE, 2014.

Tsai, Chun-Wei, Chin-Feng Lai, Han-Chieh Chao, and Athanasios V. Vasilakos. "Big data analytics." In Furht, B. and Villanustra, F. (eds) *Big Data Technologies and Applications*, pp. 13–52. Springer, Cham, 2016.

Uribe-Pérez, Noelia, and Carles Pous. "A novel communication system approach for a smart city based on the human nervous system." *Future Generation Computer Systems* 76 (2017): 314–328.

Vera-Baquero, Alejandro, and Ricardo Colomo-Palacios. "Big-data analysis of process performance: A case study of smart cities." In Roy, S., Samui, P., Deo, R., and Ntalampiras, S. (eds) *Big Data in Engineering Applications*, pp. 41–63. Springer, Singapore, 2018.

Wang, Yichuan, LeeAnn Kung, and Terry Anthony Byrd. "Big data analytics: Understanding its capabilities and potential benefits for healthcare organizations." *Technological Forecasting and Social Change* 126 (2018): 3–13.

Yaqoob, Ibrar, Ibrahim Abaker Targio Hashem, Yasir Mehmood, Abdullah Gani, Salimah Mokhtar, and Sghaier Guizani. "Enabling communication technologies for smart cities." *IEEE Communications Magazine* 55(1) (2017): 112–120.

Index

Printed in the United States
by Baker & Taylor Publisher Services